P9-ECY-924

PRINCIPLES OF SYSTEMATIC ZOOLOGY

PRINCIPLES OF SYSTEMATIC ZOOLOGY

SECOND EDITION

Ernst Mayr

Alexander Agassiz Professor of Zoology, Emeritus
Museum of Comparative Zoology, Harvard University

Peter D. Ashlock

Late Professor of Entomology
University of Kansas, Lawrence

McGRAW-HILL, INC.

New York St. Louis San Francisco Auckland Bogotá
Caracas Hamburg Lisbon London Madrid Mexico Milan Montreal
New Delhi Paris San Juan São Paulo Singapore Sydney Tokyo Toronto

This book was set in Palatino by the College Composition Unit
in cooperation with Waldman Graphics, Inc.
The editors were Kathi Prancan and Holly Gordon.
The cover was designed by Rafael Hernandez.
R. R. Donnelley & Sons Company was printer and binder.

PRINCIPLES OF SYSTEMATIC ZOOLOGY

1 2 3 4 5 6 7 8 9 0 DOC DOC 9 5 4 3 2 1 0

ISBN 0-07-041144-1

Library of Congress Cataloging-in-Publication Data

Mayr, Ernst, (date).
Principles of systematic zoology / Ernst Mayr, Peter D. Ashlock.—2nd ed.
p. cm.
Includes bibliographical references and index.
ISBN 0-07-041144-1
1. Zoology—Classification. I. Ashlock, Peter D. II. Title
QL351.M29 1991
591'.012—dc20 90-42293

ABOUT THE AUTHORS

ERNST MAYR is Emeritus Professor of Zoology at Harvard University. After earning a Ph.D. from the University of Berlin, he led three expeditions (1928–1930) to New Guinea and the Solomon Islands. From 1932 to 1953 he was curator of the Whitney-Rothschild Collection at the American Museum of Natural History in New York. From 1953 to 1975 he was Alexander Agassiz Professor at Harvard, and he served as director of the Museum of Comparative Zoology from 1961 to 1970. He is a former member of the International Commission on Zoological Nomenclature. Mayr has written or edited 15 books and over 500 journal articles. He has received the National Medal of Science and the Balzan Prize (1983). He is also a foreign member of the Royal Society, the Académie des Sciences (Paris), and numerous other international academies and societies. He is the author of *Animal Species and Evolution* (1963), *The Growth of Biological Thought* (1982), and *Toward a New Philosophy of Biology* (1988).

PETER D. ASHLOCK, a widely recognized authority on taxonomy of the Lygaeidae, was also known for his work in theoretical systematics. He received a B.S. degree in entomology from the University of California (Berkeley), an M.S. degree at the University of Connecticut, and a Ph.D. from the University of California. He held appointments in the Department of Entomology and the Department of Systematics and Ecology in the University of Kansas and was a curator in the Snow Entomological Museum and curator of Hemiptera in the Bernice P. Bishop Museum in Honolulu, Hawaii. Dr. Ashlock, one of America's leading hemipterists, served as Editor of the *Journal of the Kansas Entomological Society* for three years. He was the author or co-author of numerous publications on the taxonomy and biology of lygaeid seed-bugs and other groups of Hemiptera.

CONTENTS

PREFACE

Systematics has had a remarkable renaissance in recent generations. The reasons for this are diverse. Taxonomists played a leading role in the new synthesis of evolutionary theory, and they have demonstrated that the study of organic diversity—the main concern of systematics—is a major integral branch of biology. Systematics has also been very important in initiating the entire field of population biology, including population genetics. The recognition of the importance of taxonomy has been helped by the realization that there are two great scientific methods: the experimental method and the comparative method (based on observations). Observational data are meaningless unless they are classified prior to comparison. An understanding of this methodological necessity has created a new interest in the methods and theory of classification in all comparative sciences. All this is superimposed on an increasing need for the applied aspects of taxonomy, such as the correct identification and classification of species in agriculture, public health, ecology, conservation, genetics, and behavioral biology.

The last 20 years has been a period of unprecedented activity in the field of taxonomy. This is documented by the founding of new journals and societies and by an almost exponential increase in the number of publications on the methods and principles of taxonomy. Hence, the publication of a revised edition of *Principles of Systematic Zoology* requires no justification. However, soon after the beginning of the revision it became evident that a mere updating of the 1969 text would not meet the need. What was required was the preparation of a virtually new book. The present volume is the result. It differs from the first edition in organization and emphasis.

Two major innovations characterize the new edition. The first is an entirely new arrangement of the subject matter. It is based on the insight, gained in the last several decades, that taxonomy at the species level is very different from taxonomy at the level of higher taxa. This has induced us to recognize two branches of taxonomy. One is *microtaxonomy*—taxonomy at the species level—to which Part A of this volume is devoted. The other is *macrotaxonomy*—the classification of higher taxa—to which Part B is devoted. The problems encountered by the taxonomist in the two areas are very different, and their separate treatment in this book permits a clearer focus on the specific problems at each level.

The second major departure in this edition is a much greater emphasis on concepts and theory. In the 1969 edition a special effort was made to acquaint inexperienced taxonomists with various methods of facilitating taxonomic practice. These practical suggestions were praised so widely that they have been retained in the current edition. However, we concluded that a modern text of systematics should be more than a recipe book showing "how to do it." Taxonomists can make intelligent use of the available methods, particularly the many new techniques (computers, molecular analysis, etc.), only if they fully understand the basic principles of biological classification. In many of the current controversies one has the impression that the opponents have lost sight of the forest in their preoccupation with trees. This inevitably leads to misunderstandings and dissension. To dispel some of the current confusion, our account concentrates on the basic underlying issues, some of which are hardly ever articulated in the recent literature. For this reason special attention is paid to the concepts of taxonomy; to the meaning of terms used in taxonomy, particularly the traditional terms; and to a precise statement of opposing views on controversial issues.

Taxonomy is a subject which is so operational that it cannot be learned merely by reading a book. Therefore, the main objective of this volume is to serve as a guide and companion for those who are learning the subject and perhaps even more for those who are teaching it. However, it cannot take the place of a laboratory in which the procedures of classification are actually demonstrated. The problems of taxonomy are different in every group of animals, and teachers will want to use material and illustrations from the zoological groups with which they are most familiar. This is why this volume does not cite more examples to illustrate basic principles and methods. The teacher will know best what examples are most instructive in a given case.

Current differences of opinion concern both the methodology of taxonomy and the most basic principles of the philosophy of classification. What should a classification reflect, genealogy only or the total aspect of the included taxa? What is best, and what is most natural? What components of the phenotype should one choose as the basis of one's classification, considering the frequency of mosaic evolution and homoplasy? How important is the facilitation of diagnosis in a classification? To what extent should different characters be given different weights in the construction of classifications? Should one give equal weight to the quasi-neutral replacement of base pairs in DNA and to evolutionarily significant changes?

Disagreements about the best methods for the construction and testing of trees are equally great. Each year new algorithms are proposed for finding the "best" tree. Are methods based on Wagner trees better than clustering methods? What is the best method for testing a tentative classification: parsimony, compatibility, or maximum likelihood? Or does this depend on the circumstances? Almost every taxonomic procedure has been the subject of recent controversies.

Even though there may never have been another period with as great a need for a volume dealing with these questions, this is a singularly difficult time to write one. Perhaps never before in the history of systematics have there been so many uncertainties and controversies. The representatives of each of the three major current schools of classification—phenetics, cladistics, and evolutionary classification—are trying to convince us that their philosophy and methodology are superior to those of the other two. All three schools have numerous followers, indicating that the methods or principles of each school have considerable appeal. It would appear that none of them could be all wrong.

It is not the mission of this text to decide these controversial issues. We are known to be supporters of the biological species concept and the so-called evolutionary method of classification. Inevitably, our account is somewhat biased in favor of those views. However, to permit comparison, we have made a sincere effort to give an objective account of opposing views. Fortunately, in recent years a certain amount of rapprochement has taken place between the various opposing schools. It is to be hoped that further analysis and a clarification of the underlying concepts will eventually permit a complete synthesis in the field of systematics. It would be gratifying if this volume, with its emphasis on basic principles and its critical analyses of most of the proposed methods, could make a contribution to the ultimate goal of a universal consensus.

The flood of literature in the field of taxonomy has become overwhelming in recent decades. It was not possible in this textbook to treat some topics in as much detail as would have been desirable. For readers who want to go deeper into some of these subjects, we have supplied extensive literature references. Furthermore, a real effort has been made to include all aspects of taxonomy. Quite deliberately, some of the literature cited in the 1969 edition has not been replaced by newer references to prevent excellent older papers from being forgotten.

In the preparation of the manuscript Ernst Mayr wrote a first draft of Parts A and C, and Peter Ashlock wrote a first draft of Part B. Each draft was thoroughly revised by the coauthor and was shown to other readers, whose suggestions were incorporated as far as possible. Chapters 12, 13, and 14 are revisions of Chapters 6, 11, 12, and 13 of the first edition.

It is with great sadness that I report the unexpected death of Professor Ashlock of a heart attack on January 26, 1989, while the manuscript was nearing completion.

ACKNOWLEDGMENTS

The authors owe a great debt of gratitude to numerous friends for critically reading parts of the manuscript and for providing constructive criticism and helpful comments. Special thanks are due to J. Carpenter, J. Felsenstein, and David Maddison, who carefully analyzed all the chapters in Part B even though in many respects they disagree with the views of the authors. Their numerous suggestions led to considerable improvement. W. Bock, G. W. Cottrell, B. Holmes, R. Honeycutt, Wayne Maddison, James A. Slater, Alex Slater, and Ward Wheeler read some of the chapters and made valuable comments. Important information was supplied by J. Lloyd, R. R. Sokal, G. C. Eickwort, G. Barrowclough, and R. Woollacott.

The provisionally completed manuscript of Chapter 11 was completely revised by David Maddison and brought up to current standards. In turn, this revision necessitated a revision of Chapters 6, 8, 9, and 10.

We are greatly indebted to Walter Borawski for the careful typing and word processing of numerous versions of the manuscript and for help in the literature search, the bibliography, and the index. Virginia Ashlock contributed in many ways, particularly by typing the first draft of Part B, helping with the glossary, and copyediting provisional versions of all the chapters. They all greatly lightened our burden and deserve our warmest thanks.

McGraw-Hill and I would like to thank the following reviewers for their many helpful comments and suggestions: Brooks Burr, Southern Illinois University; Thomas O. Duncan, University of California, Berkeley; Robert F. Inger, Field Museum of Natural History; Janus E. Lloyd, University of Florida; Brent D. Mishler, Duke University; and Arthur M. Shapiro, University of California, Davis.

Ernst Mayr

CHAPTER 1

THE SCIENCE OF TAXONOMY

The most impressive aspect of the world of life is its diversity and the uniqueness of its components. No two individuals in sexually reproducing populations are the same, nor are any two populations, species, or higher taxa. Wherever we look in nature, we find uniqueness, and uniqueness means diversity. To sort all this and determine its nature is the task of taxonomy.

More than 1 million species of animals and half a million species of plants and microorganisms have been described, and estimates on the number of undescribed living species range from 3 million to 10 million and even higher. An estimate of at least half a billion for extinct species is consistent with the known facts. Each species may exist in numerous different forms (sexes, age classes, seasonal forms, *morphs,*[1] and other *phena*). The diversity of faunas has numerous dimensions. There are rich faunas not only on land but also in freshwater and in the oceans down to the greatest depths. Every organism has its fauna of parasites, many of them species-specific. Evidently, the amount of diversity is immense.

The important biological discipline concerned with the scientific study of diversity is often indiscriminately referred to as systematics or taxonomy. There is a broad overlap in the application of these terms, but there is also a subtle difference. The term *taxonomy* is derived from the Greek

[1]Consult the glossary for unfamiliar terms.

words *taxis* (''arrangement'') and *nomos* (''law'') and was first proposed in its French form by de Candolle (1813) for the theory of plant classification. It agrees best with current thinking to define it as follows: *Taxonomy is the theory and practice of classifying organisms.*

The term *systematics* stems from the latinized Greek word *systema* as applied to the systems of classification developed by the early naturalists, notably Linnaeus (*Systema naturae,* 1st ed., 1735). We follow Simpson's (1961:7) modern redefinition of the term: ''*Systematics is the scientific study of the kinds and diversity of organisms and of any and all relationships among them.*'' More simply, *systematics is the science of the diversity of organisms.* The word *relationship* is not used here in a narrow phylogenetic sense but is broadly conceived to include all biological interactions among organisms. This explains why such a broad area of common interest has developed between systematics, evolutionary biology, ecology, and behavioral biology.

Systematics deals with populations, species, and higher taxa. No other branch of biology occupies itself in a similar manner with these levels of integration of the organic world. It not only supplies urgently needed information about these levels but, more important, cultivates a way of thinking, a way of approaching biological problems that is important for the balance and well-being of biology as a whole (Mayr 1968*a,* 1982*a*).

One of the major tasks of systematics is to determine by means of comparison what the unique properties of each species and higher taxon are. Another is to determine what properties certain taxa have in common and what the biological causes of the differences or shared characters are. Finally, systematics is concerned with variation within taxa. In all these concerns systematics holds a unique and indispensable position among the biological sciences. Classification makes organic diversity accessible to the other biological disciplines. Without it, most of them would be unable to give meaning to their findings.

THE CONTRIBUTION OF SYSTEMATICS TO BIOLOGY

A consideration of the contributions of systematics to other branches of biology and to humankind as a whole adds to an appreciation of its scope.

Before the rise of genetics, the study of evolution was carried out almost entirely by taxonomists. From Lamarck and Darwin on, nearly all the leading evolutionists were practicing systematists, and the names Chetverikov, Dobzhansky, Simpson, Mayr, Stebbins, and Grant prove that this is still true in modern times. Virtually all the major evolutionary problems were first pointed out, and often solved, by systematists. Even today, the study of organic diversity as practiced by systematists continues to reveal new evolutionary problems. No other branch of biology has made a greater contribution to our understanding of evolution.

Leaders in many other fields of biology have acknowledged their dependence on taxonomy. Elton (1947:166) made this statement regarding ecology:

> The extent to which progress in ecology depends upon accurate identification, and upon the existence of a sound systematic groundwork for all groups of animals, cannot be too much impressed upon the beginner in ecology. This is the essential basis of the whole thing; without it the ecologist is helpless, and the whole of his work may be rendered useless.

No thorough ecological survey can be conducted without the most painstaking identification of all species that are of ecological significance. A similar dependence on taxonomy is true in other areas of science. In the delimitation of geological strata, key fossil species have played a decisive role. Even the experimental biologist has learned to appreciate the need for sound taxonomy. There are many genera with two, three, or more very similar species. Such species often differ more conspicuously in their physiological traits or cytology than in their external morphological characters. Every biologist can recall examples in which two workers came to very different conclusions concerning the physiological properties of a certain "species" because, in fact, one specialist had been working with species *a* and the other had been working with species *b*.

Molecular biologists are vitally interested in sound classification. The evolution of molecules, an increasingly important area of research, can be understood only against the background of a sound classification. It is only in consultation with the taxonomist that the biochemist can determine what organisms may supply the key to important steps in the evolution of molecules. In turn, molecular biology continues to make valuable contributions to the classification of organisms and the discovery of *sibling species*. Let us single out some specific areas to which taxonomy has made noteworthy contributions.

Applied Biology

The contribution of taxonomy to the applied sciences—medicine, public health, agriculture, conservation, management of natural resources—has been both direct and indirect. Taxonomic breakthroughs have often supplied the key to the solution of previously perplexing problems in economic entomology. The famous case of the epidemiology of malaria is a good example. The supposed vector in Europe, the malaria mosquito, *Anopheles maculipennis* Meigen, was reported throughout the continent, yet malaria was restricted to local districts. Large amounts of money were wasted because no one understood the connection between the distribution of the mosquito and that of malaria. Careful taxonomic studies finally provided the key. The *maculipennis* complex was found to consist of several sibling species with different habitat preferences and breeding habits, only some of which are responsible for the transmission of malaria in a given area. This new information allowed control measures to be directed to the spots where they would be most effective.

With biological control of insect pests again receiving increased attention, the determination of the exact country of origin of insect pests and their total fauna of parasites and parasitoids has been restored to the great importance it had prior to the brief period in which at least some applied entomologists thought that they could completely control insects with pesticides.

Pemberton (1941) cites an outstanding instance of the value of insect collections assembled for taxonomic study in the solution of a problem involving biological control. Some 20 years earlier the fern weevil, *Syagrius fulvitarsis* Pascoe, had become very destructive to *Sadleria* ferns in a forest reserve on the island of Hawaii, and control measures became necessary. The entomological literature failed to reveal the occurrence of this weevil anywhere outside Hawaii except in greenhouses in Australia and Ireland. These records gave no clue to the country of origin. However, while engaged in other problems in Australia in 1921, Pemberton had an opportunity to examine an old private insect collection at Sydney; among the beetle specimens was a single *S. fulvitarsis* bearing the date of collection—1857—and the name of the locality in Australia from which it had been obtained. This provided the key to the solution, for a search of the forest areas indicated on the label revealed a small population of the beetles and, better still, a braconid parasite that was attacking the larvae. Collections were made immediately for shipment to Hawaii, and the establishment of the parasite was quickly followed by satisfactory control of the pest. The data borne on a label attached to a single insect specimen in 1857 in Australia thus contributed directly to the successful biological control of that pest in Hawaii 65 years later.

For a while it appeared that biological control had become obsolete owing to the success of chemicals. Recently, however, applied entomologists have had to revert increasingly to biological control because of the development of resistant strains among insect pests and the adverse effects of many of these chemicals on human health and the well-being of the entire ecosystem.

Theoretical Biology

The service functions of taxonomy are often stressed to such a degree that the important contributions of systematics to the conceptual structure of biology are overlooked. *Population thinking,* for instance, has come into biology through taxonomy (Mayr 1963), and indeed, one of the two roots of population genetics is taxonomy (Chapter 3). The problem of the multiplication of species was solved by taxonomists. They have made the greatest contributions to our understanding of the structure of species and the evolutionary role of peripheral populations and were important contributors to the evolutionary synthesis (Mayr and Provine 1980). It was taxonomists who continued to uphold the importance of natural selection when the early Mendelians thought that mutation had eliminated the role of natural selection as an evolutionary factor. Taxonomists such as H. W. Bates and F. Müller made significant contributions to the understanding of mimicry and related evolutionary phenomena and thus provided the first clear proof of natural selection. Taxonomists and naturalists in close contact with taxonomy were instrumental in the development of ethology and the study of the phylogeny of behavior. Taxonomists have consistently played an important role in counteracting the *reductionist* tendencies dominant in so much of functional biology. They have thus contributed to a healthy balance in biological science (Mayr 1974*a*, 1982*a*).

The Role of Taxonomy

The erroneous view is widespread among laboratory biologists that systematics consists merely of the pigeonholing of specimens. In this view, the taxonomist should be content with identifying material and devising keys. Beyond that, the taxonomist should keep collections in good order, describe new species, and have every specimen properly labeled. According to this view, systematics is a more or less clerical activity.

In actuality systematics is one of the major subdivisions of biology, as broad-based as genetics or molecular biology. It includes not only the service functions of identifying and classifying but also the comparative study of all aspects of organisms as well as an interpretation of the role of lower and higher taxa in the economy of nature and in evolutionary history. It is a synthesis of many kinds of knowledge, theory, and method applied to all aspects of classification. The ultimate task of the systematist is not only to describe the diversity of the living world but to contribute to its understanding.

Modern taxonomists are far more than the caretakers of a collection. They are well-trained field naturalists who study the ecology and behavior of species in their native environment. Most younger systematists have had thorough training in various branches of biology, including genetics and molecular biology. This experience in both field and laboratory gives them an excellent background for more fundamental studies.

The multiple role of taxonomy in biology can be summarized as follows:

1 It is the only science that provides a vivid picture of the existing organic diversity of the Earth.

2 It provides most of the information needed for a reconstruction of the phylogeny of life.

3 It reveals numerous interesting evolutionary phenomena and thus makes them available for causal study by other branches of biology.

4 It supplies, almost exclusively, the information needed for entire branches of biology (e.g., biogeography).

5 It supplies classifications which are of greatest heuristic and explanatory value in most branches of biology, e.g., evolutionary biochemistry, immunology, ecology, genetics, ethology, and historical geology. A sound classification is the indispensable basis of much biological research. It is a prerequisite for the application of the comparative method. All aspects of living organisms are of interest to systematists who adopt Simpson's (1961:7) definition of the field as "the scientific study of the kinds and diversity of organisms and of any and all relationships among them." Such studies are often meaningless without a sound classification. Studies of species formation, the factors of evolution, and biogeography are unthinkable without classification. Classifications are particularly important in applied biology, for instance, agriculture, public health, and environmental biology, because the correct identification of an important agricultural insect pest, disease vector, or major component of an ecosystem depends on the availability of a sound classification.

6 In the hands of its foremost exponents systematics makes important conceptual contributions (such as population thinking) that would not otherwise be easily accessible to experimental biologists. Thus it contributes significantly to a broadening of biology and to a better balance within biological science as a whole.

SYSTEMATICS AS THE SCIENCE OF ORGANIC DIVERSITY

The diversity of the living world is one of the most interesting and challenging aspects of nature (Mayr 1982*a*:133–146; Wilson 1988). This diversity, so strikingly in contrast with the ultimately uniform world of physics, is expressed in virtually unlimited forms of variation. First, there are the five major kingdoms with their myriad lower taxa and millions of species. There are organisms with nuclei (eukaryotes) and those without (prokaryotes). Some organisms are haploid during most of their life cycle, while others are diploid. Some reproduce sexually; others, by one of several forms of asexuality. There are primary producers; others are herbivores, carnivores, or parasites or have adopted other forms of specialization. No matter what aspect of the life history of organisms one looks at, one will find remarkable diversity.

The study of this diversity has on the whole been rather neglected in teaching. In discussions of this subject, two rather different aspects have usually been confused.

1 The teaching of taxonomy narrowly defined—that is, the teaching of proposed classifications of animals, plants, and microorganisms; the reconstruction of their phylogeny; and the methods of identifying specimens

2 The study of biological diversity in all its aspects, particularly in areas where systematics overlaps with evolutionary biology, ecology, and behavioral biology

In the teaching of systematics, emphasis has usually been placed on the description and survey of the major (and a few minor) animal types. Although at least a superficial acquaintance with the system should be part of the background of every biologist, the teaching of systematics should be much broader. It should include analyses of causations, theories of classification, discussions of the origin of diversity (speciation, adaptive radiation)—in sum, all the factors and processes that are causally responsible for organic diversity. It should also attempt to develop generalizations. Such an approach to systematics would be far more attractive to the student than are the largely descriptive surveys that are usually taught. It would permit an *organismic* treatment, a counterpart to the necessarily reductionist treatment of most of molecular biology. A modern course in the principles of systematics would provide an excellent foundation for more advanced courses in evolution, population genetics, morphology, and behavior.

The Study of Patterned Diversity

Organic diversity is not chaotic but patterned, revealing all sorts of regularities. These regularities have various causes, and it is one of the major tasks of systematics to discover the nature of the causation of these patterns. Rodents and lagomorphs (rabbits, etc.) have rootless gnawing incisors. Is this evidence of common descent or of adaptation to an equivalent adaptive zone? Under what conditions do descendants of a common ancestor diverge drastically, and under what conditions do unrelated taxa become convergently similar when invading the same adaptive zone? Similarity might have either form of causation, and it is the task of taxonomy to find out. There is perhaps not a single evolutionary process or phenomenon that can be studied reliably until a taxonomic foundation has been laid. The fact that the evolutionary answers are usually supplied almost automatically once the taxonomic analysis has been completed is especially rewarding. It is therefore not surprising that so many of the leading evolutionists of the past 100 years have been taxonomists by background and primary interest. It is still true that a taxonomist is singularly well qualified to point out evolutionary problems and solve them.

THE HISTORY OF TAXONOMY

Diversity has interested humans ever since the beginning of our species. No matter how ignorant a native tribe may be in other matters biological, invariably it has a considerable knowledge of local plants and animals as well as names for them and often even a rudimentary classification. However, the development of a scientific theory of classification is a remarkably recent phenomenon. Simpson (1961) gives a valuable survey of the history of taxonomy and the development of its concepts, and Mayr (1982*a*) describes in considerable detail the various periods in this development. There are also histories of the study of higher taxa, such as that of Smith, Mittler, and Smith (1973) for entomology.

Several early Greek scholars, notably Hippocrates (460–377 B.C.), enumerated types of animals, but there is no indication of a useful classification in the surviving fragments of their work. There is no doubt that Aristotle (384–322 B.C.) was the father of biological classification. He lived for some years on the island of Lesbos, where he seems to have devoted himself almost entirely to the study of zoology, especially the study of marine organisms. He not only studied morphology but also paid much attention to embryology, habits, and ecology. Emphasizing that all attributes must be taken into consideration, he said, "Animals may be characterized according to their way of living, their actions, their habits, and their bodily parts" (*Hist. Anim.* 1.1. 487*a*). He referred to such major groups of animals as birds, fishes, whales, and insects; in the insects, he made distinctions between mandibulate and haustellate types and winged and wingless conditions. He also used terms for lesser groups, such as Coleoptera and Diptera, which persist today. He established numerous collective categories, or genera, using as differentiating characters blooded versus bloodless, two-footed versus four-footed, hairy versus feathered, with or without an outer shell, and so forth. All this was a tremendous advance over anything that had previously existed, and Aristotle's thinking completely dominated animal classification for the next 2000 years. Nevertheless, he did not supply (or even attempt to supply) an orderly, fully consistent classification of animals (Mayr 1982*a*; Pellegrin 1986).

Interest in natural history and in the study of animals as things important in themselves steadily decreased after the death of Aristotle. Animals were written about not to provide knowledge about them but for the sake of moralizing; they became symbols of virtues (courage, diligence) or of objectionable behavior. Most animal books up to Gesner (1551) and Aldrovandi (ca. 1600) were encyclopedias. Only from about 1550 on did the knowledge of animals make more rapid progress, as documented by the writings of William Turner (1508–1568), Pierre Belon (1517–1564), and Guillaume Rondelet (1507–1566); however, the recognized taxa were on the whole those of folklore, such as birds, fishes (including all sorts of aquatic organisms), and shells.

Downward Classification

Plant classification experienced a great flowering in the period from Cesalpino (1519–1603) to Carolus Linnaeus (1707–1778), not only in the writings of these two great taxonomists but also in those of Magnol, Tournefort, Rivinus, Bauhin, Ray, and various lesser figures. Their method of *downward classification* was the principle of logical division, which consisted in dividing a larger (superordinated) group by *dichotomy* into two subordinated groups: animals—with or without blood, animals with blood—hairy or not hairy, and so forth. This principle dominated taxonomy up to the end of the eighteenth century. Animal taxonomy made little conceptual progress in the seventeenth and eighteenth centuries, although the work of Willughby (1635–1672) on birds and that of Reaumur (1683–1757) on insects revealed a remarkable advance in knowledge. Natural history in the eighteenth century was dominated by two great figures, Buffon (1707–1788) and Linnaeus.

Linnaeus, sometimes called the father of taxonomy, largely adhered to the principles of downward classification by logical division. His thinking was that of an *essentialist* for whom species reflect the existence of fixed, unchanging types (essences). However, in a period during which the number of new species and kinds of organisms grew at an exponential rate, he was a desperately needed methodological innovator. Speedy and correct identification was what the naturalist required most, and this was facilitated by Linnaeus's careful keys, his rigorous system of telegraphic-style diagnoses, his standardization of synonymies, and his invention of *binominal nomenclature*. Because of his authority Linnaeus was able to impose his methods, and this brought consensus and simplicity back into taxonomy and nomenclature, where there had been a threat of total chaos.

The actual classifications adopted by Linnaeus were of mixed value. For the groups with which he was most familiar, for instance, insects, he produced classifications that are still largely acceptable. By contrast, his classifications of other groups, such as birds, amphibians, and lower invertebrates ("Vermes"), were inferior to those of earlier authors.

Buffon was not a taxonomist and had little interest in classification and the higher categories. However, in some respects he had perhaps as great an impact on the ensuing history of systematics as Linnaeus did. First of all, by using the sterility barrier (instead of degree of morphological difference) as the species criterion, he prepared the way for the biological species concept. More important, by his attacks on scholasticism and his emphasis on the biological interpretation of characters (and on the utilization of as many characters as possible), he laid the foundation for a new approach to classification.

Upward Classification

By the middle of the eighteenth century the shortcomings of the method of downward classification by logical division were increasingly recognized. It was actually a method of identification, not of classification, and since the arrangement it produced depended entirely on the sequence in which the differentiating characters were used, it was blatantly artificial. This method was incapable of producing order in a large fauna. As a result, it was gradually replaced by the entirely different method of *upward classification*. This method consists of assembling species by inspection into groups of similar or related species and forming a hierarchy of higher taxa by again grouping similar taxa of the next lower rank. As stated by Buffon (1749): "It would seem to me that the only way to design an instructive and natural method is to group together things that resemble each other and to separate things that differ from each other." This thought was systematically applied by the botanist Adanson (1763) and was practiced by nearly all post-Linnaean zoologists, who delimited taxa by inspection and through an evaluation of numerous characters. Characters were weighted, usually not by a priori principles (such as physiological importance) but by an a posteriori determination of a covariance of characters. (For a discussion of the problems of weighting, see Chapter 7.)

Concurrent with the methodological shift from downward to upward classification was a major philosophical change. A strong belief in a linear, teleological aspect of the universe, as reflected in the *scala naturae* and in Lamarck's concept of evolution, was replaced by a belief in the existence of *archetypes* (idealistic morphology) (Desmond 1982). Cuvier recognized five phyla (embranchments), and von Baer, along with Owen and the comparative morphologists in the pre-Darwinian period, thought they were able to arrange all species of animals into a limited number of groups, each representing a distinct "type." As a descriptive device such "typology" was clearly legitimate (Schindewolf 1969), but it must not be confused with the typological thinking of essentialism.

Four other developments characterized the period between Linnaeus and Darwin. First, specialization became more pronounced. The days when authors such as Ray, Linnaeus, and Lamarck could successfully deal with the taxonomy of both animals and plants were over by 1800. Indeed, more and more authors became specialists in a single group, such as birds, beetles, or butterflies. Second, classifications became more hierarchical. Above the species Linnaeus recognized only genus, order, class, and kingdom, but soon the categories family and phylum were added, and numerous additional ones came later. Third, philosophical guidelines were expressly rejected, and classifying became an entirely empirical enterprise. Fourth, the search for a natural system was intensified, with the term *natural* serving as the antonym of *artificial*. That system was considered most natural which succeeded best in grouping together the species that had the most in common.

Impact of *The Origin of Species*

The one question that taxonomists were unable to answer before 1859 was why the members of a taxon are more similar to each other than they are to members of other taxa. Darwin supplied the explanation through his theory of evolution by common descent. "Natural" groups exist because the members of a natural taxon are descendants of a common ancestor and therefore have a much greater chance to be similar to each other than do unrelated species. Classifications proposed prior to 1859, based on the grouping of similar species, continued on the whole to be acceptable after 1859, since similar species are ordinarily descendants from a common ancestor. Darwin, however, did more than provide the theoretical basis for a natural system. He also provided in Chapter XIII of the *Origin* (1859) a set of clear, practical criteria to be applied during the construction of a classification. These criteria are discussed in Chapter 6.

A major preoccupation of taxonomists in the first 50 years after the publication of the *Origin* was to substantiate the theory of common descent. This was expressed in the search for missing links between seemingly unconnected taxa, in the reconstruction of "primitive ancestors," and more generally in the construction of phylogenetic trees. This endeavor led to a boom in the fields of comparative systematics, comparative morphology, and comparative embryology.

However, the period of the discovery of major new types of terrestrial animals was essentially over well before the end of the nineteenth century. By then the need to prove the reality of evolution had ceased to exist. Taxonomy no longer was an exciting bandwagon, and taxonomists were forced to concentrate on the necessary though tedious job of describing, diagnosing, and classifying the seemingly endless number of species. A minority of species describers were dilettantes who brought discredit to the field by the creation of numerous synonyms and an excessive splitting of families and genera. Others concentrated on the unearthing of long-forgotten synonyms, thus arousing the ire of general biologists, who complained rightly that this activity defeated the basic objective of nomenclature as the key to an information retrieval system. There is little question that taxonomy fell into some disrepute during the latter part of the nineteenth century and the early twentieth century. The situation was aggravated by the fact that the taxonomists sided against the immensely popular and powerful early Mendelians with their antiselectionist and saltationist interpretation of evolution. However, a turn for the better began to develop in the 1920s.

Population Systematics

For the sake of convenience, taxonomists had continued to treat species, in an essentialist manner, as invariant units until long after the invalidity of the typological dogma had become apparent. However, when population samples from different portions of the geographic range of a species were compared, smaller or greater differences were often found. This resulted eventually in the replacement in certain animal groups of the typologically defined species by the *polytypic species,* which is composed of different populations in the dimensions of space and time. The study and comparison of intraspecific populations became the objective of population systematics. The history of this development, beginning in the first half of the nineteenth century and reaching its climax in the 1930s and 1940s, has been described in detail by Mayr (1942, 1963).

Replacement of essentialist thinking by population thinking has had important consequences in many areas of taxonomy. The consideration of taxa as populations or aggregates of populations greatly facilitated the study of variation and the delimitation of lower taxa and categories. Labeled by J. S. Huxley (1940) as the *new systematics,* it led to a reevaluation of the species concept and to a more biological approach to taxonomy. The population systematist understands that all organisms occur in nature as members of populations and that specimens cannot be understood and properly classified unless they are treated as samples of natural populations.

The same period saw two additional aspects come to the fore. One is what might be called the biological approach to taxonomy. As taxonomists moved more and more from the museum into the field, they increasingly supplemented morphological characters with characteristics of living animals, such as behavior, voice, ecological requirements, physiology, and biochemistry. Taxonomy truly became biological taxonomy. The other development was the introduction of the experiment into taxonomy. Although this was far more widely practiced in botany than in zoology, the experimental analysis of isolating mechanisms—particularly in vertebrates, in *Drosophila,* and in protozoans—and the application of other experimental methods were very helpful.

Population systematics is not an alternative to classical taxonomy but an extension of it. Among groups in which the inventory taking of species is still in full progress and in which too few local populations have been sampled, one cannot easily apply the methods of population systematics. Being centered on the population level, the new systematics naturally had little impact on the theory of classification at the level of higher taxa. The population thinking of the new systematics was one of the major sources of population genetics, which in turn influenced the development of population systematics. Together they greatly helped to clarify our thinking about evolution at the species level and were instrumental in producing the great synthesis in evolutionary biology (Mayr and Provine 1980).

Current Trends

The new systematics dealt almost exclusively with the species level. *Macrotaxonomy* made little conceptual progress from the 1870s to the 1950s; indeed, it was neglected. This, as discussed in Chapter 6, changed dramatically with the rise of numerical taxonomy. Controversies concerning the merits of the newer theories of classification in comparison with the traditional approaches have monopolized journals of systematic biology in recent years.

Perhaps more important in the long run for the history of systematics is the growing interest among molecular biologists in problems of classification and the development of numerous molecular techniques for testing proximity of relationship. This topic also is treated in Chapter 6.

Taxonomy as Unfinished Business

Taxonomy is the oldest biological discipline, and one might well expect that its job would now be nearly finished. However, nothing could be further from the truth. Although the terrestrial and littoral faunas of the temperate zone of most groups of animals are reasonably well known, the fauna of the tropics is still largely uncollected and undescribed. The recently discovered richness of the fauna of the canopy in tropical forests was altogether unexpected. On the basis of these findings, Erwin (1983) estimated the total number of living species of insects in the tropics to be as high as 30 million, yet among all kinds of animals, only a little more than 1 million species have been described to date.

The inventory of species is not the only unfinished task. Even our knowledge of the higher taxa is still incomplete. New classes and even new phyla of animals continue to be discovered. When the deep oceanic "hot vents" were explored, whole new faunas were discovered, including giant tube worms (Pogonophora), giant clams, and living fossils (barnacles). The terrestrial soil fauna and the marine *meiofauna* (inhabitants of the interstitial spaces of sand and mud of the ocean floor) are still very imperfectly known. Strikingly new types have been discovered not only among living animals such as *Latimeria, Neopilina,* and the Gnathostomulida and Concentricycloidea but also among fossils. The late Precambrian Ediacara fauna (Glaessner 1984), the Burgess shale fauna (Gould 1989), and the Heliocoplacoidea (a new class of echinoderms) are examples.

Even more backward is our understanding of the relationship among the higher taxa of most groups of animals. Morphological clues to relationship are often insufficient, but the recently developed methods of molecular biology promise a solution of many old puzzles. Indeed, the classification of entire major groups (prokaryotes, birds) has been revolutionized with the help of molecular methods. However, a more intensive evaluation of morphological characters will continue to make major contributions. Electron microscopy has revealed that the aberrant "ciliate" *Stephanopogon* is a flagellate (Lipscomb and Corliss 1982) and that there are protists without mitochondria (Cavalier-Smith 1987).

SYSTEMATICS AS A PROFESSION

Opportunities and Difficulties

Positions for professional taxonomists exist in museums, at universities, and in various government agencies, particularly those dealing with crop and forest protection and public health. In the United States there are perhaps more taxonomists in teaching positions in various colleges and universities than there are in government or museum positions. Well-trained taxonomists are particularly qualified to teach a general course in zoology or biology, because they have a broad background in zoology, morphology, physiology, genetics, and ecology that specialists in cellular and molecular biology usually have not acquired. The situation in applied taxonomy, however, often changes rapidly. When the importance of Foraminifera for the stratigraphic determination of oil-bearing geological formations was discovered, the demand for micropaleontologists became so great that it took decades to fill it. With geophysical methods now dominating oil exploration, the demand for micropaleontologists has decreased sharply. During the height of chemical insect control, the demand for well-trained insect taxonomists lagged. With the renewed recognition of the importance of biological control, the need for qualified insect taxonomists has increased greatly. The recent stepping up of oceanographic research revealed an extraordinary shortage of marine zoologists, and new positions for specialists in marine invertebrates were created in several museums, universities, and marine laboratories. No one can predict where the next need will arise. It is conceivable, for instance, that institutes in molecular biology may begin to employ taxonomists for consultation on the numerous evolutionary and taxonomic problems that comparative biochemists encounter at every step. As in all fields of biology, there is always room for superior workers regardless of temporary fluctuations in the employment situation.

The number of pure research positions is as limited in taxonomy as it is in most branches of science.[2] Most taxonomists earn a living as teachers, curators, members of identification services, or workers in other branches of applied biology. Another group of taxonomists are "amateurs" who work as civil servants, businesspeople, lawyers, or doctors and conduct taxonomic research as a hobby. The role of the amateur in systematics has changed over the years. When collecting and "naming" were believed to be the essence of taxonomy, almost any untrained person could become a specialist in a group the knowledge of which was still at the level of *alpha taxonomy*. The stamp-collector type of amateur has been losing ground as the quality of taxonomic research has improved. However, when field systematics became an increasingly important branch of taxonomy, a new niche opened up for the amateur naturalist, who often supplies information on behavior and ecology which is very important taxonomically. The work of the better amateur is sometimes of the same high level of quality as that of the good professional. Considering the size of the taxonomic task still to be done, taxonomy would make slow progress without the dedicated amateur.

One of the most interesting current developments is the blurring of the borderline between taxonomists and other biologists. Population geneti-

[2]For further information, see *Careers in Biological Systematics* (1986, Society of Systematic Zoology, c/o National Museum of Natural History, Washington, DC 20560).

cists such as Sturtevant, Dobzhansky, Carson, and Wheeler have been most active in elucidating the classification of the genus *Drosophila* and in describing new species. Ethologists have discovered new sibling species of frogs by analyzing their vocalizations. Entomologists have done the same thing for cicadas and grasshoppers. Ecologists have sometimes taken up taxonomic studies in order to give a new dimension to their research. Virtually all taxonomists have a broad interest in biogeography, and most major contributions to this branch of biology have been made by taxonomists. Now that *taxonomic character* means not only aspects of morphology but every kind of manifestation of the genotype, there is continuous transgression of the borders between taxonomy, physiology, behavior, genetics, biochemistry, and so forth. This active interchange is evident in various symposia (Leone 1964; Handler 1964; Bryson and Vogel 1965). This means that there is room in taxonomy for biologists of the most diverse interests.

The Training of the Taxonomist

Regular courses in methods and principles of taxonomy are a comparatively new phenomenon. Formerly, the young taxonomist learned as an apprentice to a master. Perhaps this was in part responsible for the highly uneven quality of taxonomic work done in the past. A competence of such breadth is demanded from the modern professional taxonomist that the method of apprenticeship is no longer sufficient. One expects from the well-trained taxonomist a broad knowledge of zoology, a thorough knowledge of the comparative morphology of the group in which he or she specializes, and an understanding of genetics and evolutionary biology. The taxonomist must be versed in statistical methods and in the use of the computer. The discussion of taxonomic characters in Chapter 7 indicates some of the other areas of biology in which the taxonomist should be knowledgeable.

THE FUTURE OF SYSTEMATICS

The study of organic diversity will always remain an important and active branch of biology. There will always be a need for the ordering activities of the taxonomist no matter how ecology, neurobiology, and molecular biology flourish. Simpson (1945:1) put it very well when he said that taxonomy "is at the same time the most elementary and most inclusive part of zoology, most elementary because animals cannot be discussed or treated in a scientific way until some taxonomy has been achieved, and most inclusive because [systematics] in its various branches gathers together, utilizes, summarizes, and implements everything that is known about animals, whether morphological, physiological, psychological, or ecological."

The popularity of species describing is gone, even though there are more undescribed species in the world, particularly in the tropics, than there are described species. However, the interest in constructing reliable classifications is perhaps greater now than ever before. In addition, some of the eternal questions remain unsolved, such as the relation between classification and evolution and how suitable the method of binominal nomenclature is as the basis of an information retrieval system. This much is undisputed, however: Today's taxonomists are far more interested in a biological approach to their task than were most taxonomists of former generations.

STRATEGIES IN TAXONOMIC RESEARCH

Most taxonomists are overwhelmed with demands, and it is of the utmost importance for them to order their priorities. What proportion of time should a taxonomist devote to curating, the identification of specimens for others, research, and teaching? What curatorial practices could be modernized to save time? What types of research promise to be most productive? Mayr (1971) has discussed these questions in considerable detail.

The productivity of a taxonomist is vitally affected by the soundness of his or her decisions. As far as research is concerned, should the taxonomist concentrate on work in microtaxonomy, assign phena to species, and describe new species? Or is it better to concentrate on macrotaxonomy and prepare a basic revision or monograph on the basis of an examination of all included species and preferably of all type specimens, a study of ancestral and derived characters that culminates in a well-balanced, carefully reasoned reclassification? Finally, should the taxonomist do work in certain aspects of evolutionary taxonomy, such as making a detailed study of individual and geographic variation in a single species on the basis of numerous large population samples, study the behavioral or chemical characteristics of a set of species, or analyze trends in the evolutionary change of characters?

The answer to these questions is twofold. First, depending on the level of maturity reached in the group in which one is a specialist, one of these areas of research will permit the most productive work. Second, most specialists avoid mental fatigue by alternating between different approaches. When one gets tired of the tedious compilation of data necessary for the preparation of a generic revision, it is relaxing to do a study of geographic variation in a single species with rich material, a study of living material, or a study of certain distributional aspects. Field systematics offers countless challenges that are particularly stimulating when alternated with museum research.

In recent years nonsystematists have proposed abandoning most of taxonomy as "old-fashioned," "an exhausted mine," or "useless" and restricting their activity to a specialized area such as comparative protein chemistry, the taxonomy of behavioral traits, functional morphology, or the study of geographic variation. Advocates of such specialized approaches forget that (1) the various approaches are not mutually exclusive, (2) in many groups there is still an abundant need for the most basic descriptive taxonomy, and (3) *Homo* is one of the most polymorphic species. Some workers are interested in using computers, others in watching behavior; some are interested in insects, others in fish; some like to work with books, others with test tubes; some prefer to work with preserved specimens, others with living material. Human polymorphism cannot be controlled by means of regimentation. Systematics has room for students with the most diversified interests and talents.

A good biological education will reveal to the student where the problems are that are most exciting. The student's natural inclination will lead to the selection of a specific area. It will be beneficial for biological science if not too many students in a given year class jump on the same bandwagon. (For a useful bibliography of systematics, see Knutson and Murphy 1988.)

PART **A**

MICROTAXONOMY

The first and most basic task of the taxonomist is to sort the bewildering diversity of individuals found in nature into species. It is impossible to construct a classification until the several species that are to be ordered are correctly discriminated. The activities by which this is achieved constitute *microtaxonomy*.

From the beginning there has been controversy about how species should be discriminated; these arguments are usually referred to as the *species problem*. Evidently, the recognition of species is not an easy matter. There are difficulties at several levels. First, there are semantic confusions that involve the concepts underlying the terms *phenon, taxon,* and *category* (see below). A more profound problem is caused by the fact that the visible diversity of nature among whole organisms includes two levels of discontinuity that are of special importance to the taxonomist: individuals and reproductively isolated populations. Taxonomists must adopt criteria that permit them to distinguish between these two levels and must learn how to apply these criteria properly. Finally, insoluble problems are posed by incipient species, that is, populations that have some of the properties of species but lack others.

Since much of the confusion in the taxonomic literature is due to a misunderstanding and the consequent erroneous application of certain terms, we begin by offering definitions of a number of terms that are commonly used in this textbook.

PHENON

This is a convenient term for the different forms or phenotypes that may occur within a single population. It includes many of the "varieties" of the older literature, the sexes (when there is sexual dimorphism), age stages, seasonal varieties, and morphs (individual variants). The term *morphospecies* has sometimes been applied confusingly to what is designated here as a phenon. Recognition of a technical term for a phenotypically uniform sample greatly facilitates the description of the taxonomic procedure. The term *phenon* was introduced by Camp and Gilly (1943) to describe phenotypically homogeneous samples at the species level. It was later used in a very different sense by Sneath and Sokal (1973). Chapter 3 deals with the taxonomic treatment of phena.

TAXON

The words *bluebirds, thrushes, songbirds,* and *vertebrates* refer to groups of organisms. Such concrete objects of zoological classification are taxa. A taxon is defined by Simpson (1961:19) as "a group of real organisms recognized as a formal unit at any level of a hierarchic classification." The same thought can be expressed as follows: *A taxon is a named taxonomic group of any rank that is considered sufficiently distinct by taxonomists to be formally recognized and assigned to a definite category*. This definition calls attention to the fact that the delimitation of a taxon against other taxa of the same rank is usually subject to the judgment of the taxonomist.

Two aspects must be stressed. The term *taxon* always refers to concrete zoological objects. Thus the species is not a taxon but a category; for example, the robin (*Turdus migratorius*) is a taxon. Second, the taxon must be formally recognized by the taxonomist. Within any large genus, groupings of species can be recognized. They are taxa only if and when they are formally distinguished and named, for instance, by being recognized as separate subgenera. Similarly, demes and geographic isolates within a species become taxa only when they are formally recognized as subspecies.

Taxa are not classes but what philosophers call individuals or particulars. This includes all biopopulations (sensu stricto). They are characterized by internal cohesion and other aspects of the ontology of species discussed in Chapter 2. Since higher taxa lack the degree of cohesion shown by species, they are best referred to as *historical groups* (Wiley 1981), yet they are clearly individuallike and are definitely not classes.

We speak of higher taxa, such as thrushes, birds, and vertebrates, and lower taxa, such as bluebirds and robins. The taxonomist ordinarily classifies taxa of species rank, yet there is a great deal of variation within most taxa, as will be discussed in later chapters. The recognition of what belongs to a given taxon of species rank is often the most difficult step in classification owing to individual variation (existence of highly different phena) or extreme similarity of individuals in different species (sibling species).

CATEGORY

A *category* designates rank or level in a hierarchic classification. *It is a class whose members are all the taxa that are assigned a given rank.* For instance, the species category is a class whose members are the species taxa.

A full understanding of the meaning of *category* depends on an understanding of hierarchical classification, which is discussed in Chapter 6. Terms such as *species, genus, family,* and *order* designate categories. A category is thus an abstract term, a class name, while the taxa placed in categories are concrete zoological objects. Until the word *taxon* was introduced into the literature, the term *category* was often confusingly used both for group and for rank, just as the word *character* is often still confusingly used to mean both character in the original sense (a specific feature) and variable (a feature that varies from taxon to taxon) (Chapter 6).

The 20 or more categories that the taxonomist uses in classification are of unequal value and different significance. They fall naturally into three groups:

1 The species category (Chapter 2)

2 Categories for distinguishable populations within species (*infraspecific* categories) (Chapter 3)

3 Categories for taxa above the species level, that is, higher taxa (collective categories are higher categories) (Chapter 6)

In a number of different ways the species occupies a unique position in the taxonomic hierarchy.

SPECIES AND CLASSIFICATION

What is the relationship between classification and the study of species? Every phyletic line and every higher taxon originated through a speciation event. Speciation and macroevolution thus are part of a single continuum. However, in the everyday practice of the taxonomist, the study of species and the operation of classifying are remarkably independent. When one proposes a new classification, one does not have to ask whether the species one is arranging originated by peripatric, parapatric, stasipatric, or sympatric separation. Indeed, we are not aware of any case where a difference in the origin of a species would have affected the proposal of a classification (except possibly in cases of hybridization).

It is evident, then, that taxonomic research at the species level—*microtaxonomy*—is rather different from the process of classifying genera and higher taxa—macrotaxonomy. This conclusion is supported by the recent history of taxonomy. For instance, the new systematics of the 1930s and 1940s involved the species level almost exclusively. Geographic variation, the recognition of polytypic species, the definition of subspecies and species, the taxonomic status of incipient species, and the role of nonmorphological characters in the delimitation of species were the principal concerns of the new systematics. Authors such as Mayr who were active in the new systematics usually did not make a substantial contribution to the classification of higher taxa. By contrast, most authors who have been most active during the flowering of macrotaxonomy from the 1960s on have made few, if any, contributions to species-level taxonomy. This statement cannot be interpreted to mean that there is no connection between the two levels. Obviously, species are the vehicle of all macroevolution. As Mayr (1963:621) stated,

> The evolutionary significance of species is quite clear. Although the evolutionist may speak of broad phenomena, such as trends, adaptations, specializations, and regressions, they are really not separable from the progression of the entities that display these trends, the species. The species are the real units of evolution, as the temporary incarnation of harmonious, well-integrated gene complexes. . . . The species, then, is the keystone of evolution.

The delimitation and proper ranking of species populations and the sorting and evaluation of characters during the construction of classifications are two very different activities. This difference is acknowledged in the recognition of two remarkably independent domains of taxonomy: microtaxonomy and macrotaxonomy. Chapters 2 through 5 are devoted to various aspects of microtaxonomy; Chapters 6 through 11, to aspects of macrotaxonomy.

CHAPTER **2**

THE SPECIES CATEGORY

One of the most elementary urges of humankind is to identify things and name them. Even the most primitive peoples have names for kinds of birds, fishes, flowers, and trees. If only individual organisms existed and the diversity of nature were continuous, it would be difficult to sort them into groups and distinguish "kinds." Actually, at least in sexually reproducing higher organisms, the diversity of nature is discontinuous and consists in any local fauna of the more or less well-defined kinds of animals we call species. Around New York City, for instance, there are about 150 kinds of breeding birds. These are the species of the taxonomist. Natives in the mountains of New Guinea independently distinguish the same kinds of organisms as do specialists in the big museums of the western world (Diamond 1965). Clearly, biological species are not an arbitrary construct of the human mind.

The concept of species seems so simple that it always comes as something of a shock to a beginning taxonomist to learn how voluminous and seemingly endless the debate about the species problem has been. In zoology there is now fair agreement on the species concept, although heterodox views are still vigorously defended. For recent summaries, see Bocquet, Génermont, and Lamotte (1980), Grant (1971), Iwasuki, Raven, and Bock (1986), Mayr (1957, 1963, 1982*a*, 1987*a*, *b*), Osche (1984), Roger and Fischer (1987), Simpson (1961), Wiley (1981), and Willmann (1985).

The working taxonomist sorts specimens (individuals) into phena and decides which of them belong to a single taxon of the species category. To undertake the ranking of taxa, the taxonomist must have a clear conception of the species category.

If a taxon is defined (as a morphospecies) in such a way that it coincides with the phenon, the taxonomist may facilitate the task of sorting specimens, but this activity will result in "species" that are biologically, and hence scientifically, meaningless. The objective of a scientifically sound concept of the species category is to facilitate the assembling of phena into biologically meaningful taxa on the species level. (See pp. 20–21 for a discussion of the meaning of category and taxon.)

A short survey of the history of species concepts shows how different the taxa are if one adopts either one or the other species concept.

SPECIES CONCEPTS

The taxonomic literature reports innumerable species concepts, but they fall into four groups. The first two have mainly historical significance but are still upheld by a few contemporary authors.

Typological Species Concept

According to this concept, the observed diversity of the universe reflects the existence of a limited number of underlying "universals" or types (the *eidos* of Plato). Individuals do not stand in any special relation to each other, being merely expressions of the same type. Members of a species form a class. Variation is the result of imperfect manifestations of the idea implicit in each species. This concept, which goes back to the philosophy of Plato, was the species concept of Linnaeus and his followers (Cain 1958). Because this philosophical tradition is also referred to as *essentialism,* the typological definition is also sometimes called the essentialist species concept (Mayr 1982*a*:256–263).

Various attempts at a purely numerical or mathematical definition of species are the logical equivalents of this species concept. Degree of morphological difference is the criterion of species status for the adherent of the typological species concept, for whom a different species is that which is "different." Morphological evidence is used by all taxonomists, but there is an enormous difference between basing one's species concept entirely on morphology and using morphological evidence as an inference in the application of a biological species concept (Simpson 1961:68–69).

The essentialist species concept was accepted by taxonomists almost unanimously as late as the early post-Linnaean period. It included the acceptance of four postulates:

1 Species consist of similar individuals sharing the same essence.
2 Each species is separated from all others by a sharp discontinuity.
3 Each species is completely constant through time.
4 There are strict limits to the possible variation within any one species.

This typological species concept has been universally rejected for two practical reasons. First, individuals are frequently found in nature that are clearly conspecific with other individuals in spite of striking differences resulting from sexual dimorphism, age differences, polymorphism, and other forms of individual variation. Although often described originally as different species, they are deprived of their species status, regardless of the degree of morphological difference, as soon as they are found to be members of the same breeding population. Different phena that belong to a single population cannot be considered separate species no matter how different they are morphologically. Second, there are species in nature—*sibling species*—which differ hardly at all morphologically yet are good biological species. Degree of difference thus cannot be considered the decisive criterion in the ranking of taxa as species.

Its adherents abandon the typological species concept whenever they discover that they have named as a separate species something that is instead a conspecific phenon. The typological species concept is still defended by some writers who adhere to Thomistic philosophy. When there is a lack of biological information, a taxonomist may be forced to recognize species provisionally on the basis of strictly morphological evidence. Such species are subject to later reconsideration.

Nominalistic Species Concept

Nominalists (Occam and his followers) deny the existence of "real" universals. For them, only individuals exist while species are abstractions created by people. The nominalistic species concept was popular in France in the eighteenth century (Buffon and Lamarck in their early writings and Robinet) and has adherents to the present day (Mayr 1982*a*:264). Bessey (1908) expressed this point of view particularly well: "Nature produces individuals and nothing more . . . species have no actual existence in nature. They are mental concepts and nothing more . . . species have been invented in order that we may refer to great numbers of individuals collectively."

Any naturalist, whether a primitive native or a trained population geneticist, knows from practical experience that this is simply not true. Species of animals are not human constructs, nor are they types in the sense of Plato; rather, they are entities for which there is no equivalent in the realm of inanimate objects.

Biological Species Concept

In the late eighteenth century it began to be realized that neither of the two medieval species concepts discussed in the preceding sections is applicable to biological species. An entirely new species concept began to emerge after 1750. It is augured by statements made by Buffon in his later writings (Sloan 1987), Merrem, Voigt, Walsh (1864), and many other naturalists and taxonomists of the nineteenth century. K. Jordan (1905) was the first to clearly formulate the concept in all its consequences. It combines elements of the typological and the nominalistic concepts by stating that species have independent reality and are typified by the statistics of populations of individuals. It differs from both by stressing the populational nature and genetic cohesion of the species and by pointing out that the species receives its reality from the historically evolved shared information in its gene pool.

As a result, the members of a species form a *reproductive community*. The individuals of a species of animals recognize each other as potential mates and seek each other for the purpose of reproduction. A multitude of devices ensure intraspecific reproduction in all organisms. The species is also an *ecological unit* which, regardless of the individuals that constitute it, interacts as a unit with other species with which it shares its environment. The species finally is a *genetic unit* consisting of a large, intercommunicating gene pool, whereas the individual organism is merely a temporary vessel holding a small portion of the contents of the gene pool for a short period of time. These three properties show that species do not conform to the typological definition of a class of objects (Mayr 1963:21). Instead, species are biological populations and have all the properties that philosophers ascribe to individuals (Chapter 3). The biological species definition which results from this theoretical species concept is as follows: *A species is a group of interbreeding natural populations that is reproductively isolated from other such groups.*

The development of the biological concept of the species was one of the earliest manifestations of the emancipation of biology from an inappropriate philosophy based on the phenomena of inanimate nature. This species concept is called biological not because it deals with biological taxa but because the definition is biological. It utilizes criteria that are meaningless in the inanimate world.

When difficulties are encountered, it is important to focus on the basic biological meaning of the species: A species is a protected gene pool. It is a Mendelian population shielded by its own devices (*isolating mechanisms*) against unsettling gene flow from other gene pools. Genes of the same gene pool form harmonious combinations because they have become coadapted by natural selection. Mixing the genes of two different species usually leads to a high frequency of disharmonious gene combinations; mechanisms that prevent this are therefore favored by selection (Mayr 1988*b*).

This makes it quite clear that *the word* species *in biology is a relational term: A* is a species in relation to *B* and *C* because it is reproductively isolated from them. This concept has its primary significance with respect to sympatric and synchronic populations (*nondimensional species*), and these are precisely the situations where the application of the concept poses the fewest difficulties. The more distant two populations are in space and time, the more difficult it becomes to test their species status in relation to each other but the more biologically irrelevant this status becomes.

The biological species concept also solves the paradox caused by the conflict between the fixity of the species of the naturalist and the fluidity of the species of the evolutionist. It was this conflict that made Linnaeus deny evolution and Darwin deny the reality of species (Mayr 1957). The biological species combines the discreteness of the local species at a given time with an evolutionary potential for continuing change. The importance of the biological species concept lies in the fact that it is the concept employed in the largest number of biological disciplines, particularly ecology, physiology, and behavioral biology. In these disciplines one deals with the action of species in one locality at one time; one deals with coexisting gene pools and with the mechanisms that maintain their integrity and explain their capacity for coexistence. For all those who work with a species in a nondimensional situation, it is usually immaterial whether another population, well isolated in space or time, should be considered conspecific.

Paterson (1985) suggested that the biological species concept is faulty and should be replaced by a species recognition concept. However, Coyne, Orr, and Futuyma (1988) and Mayr (1988*b*) showed that Paterson's arguments are invalid, being largely based on misunderstandings. The unique position of species in the hierarchy of taxonomic categories has been pointed out by many authors. Taxa of the species category can be delimited against each other by operationally defined criteria, for example, interbreeding versus noninterbreeding of populations. It is the only taxonomic category for which the boundaries of the included taxa can be defined objectively.

Evolutionary Species Concept

Some authors, particularly paleontologists, are not satisfied with the biological species concept because of its strict applicability only to the nondimensional situation. They prefer a definition which, as Simpson has said, involves evolutionary criteria. Simpson (1961:153) therefore proposed the following definition: "An evolutionary species is a lineage (an ancestral-descendant sequence of populations) evolving separately from others and with its own unitary evolutionary role and tendencies."

As has been pointed out elsewhere (Mayr 1982*a*:294), this is the definition of a phyletic lineage, not of a species. It applies equally to almost any isolated population or incipient species; it also sidesteps the crucial question of what a "unitary role" is and why phyletic lines do not interbreed with each other. What apparently most concerned Simpson was the problem of delimitation of species taxa in the time dimension, but here also his definition is of little help. When we consider a sequence of morphotypes in a single phyletic lineage, how are we to know whether these morphotypes have different unitary evolutionary roles and should thus be considered different species or whether they all have the same unitary evolutionary role and should thus be treated as *chronospecies* (Mayr 1987*a*:310–313; Reif 1984)?

The principal weakness of the so-called evolutionary species definition is that it ignores the core of the species problem—the causation and maintenance of discontinuities between contemporary species—and concentrates instead on trying to delimit species taxa in the time dimension.

Wiley (1981) attempted to improve the evolutionary species definition ("maintenance of identity" is construed as reproductive isolation), but his definition is that of a species taxon, not of the species category. Neither Simpson, Wiley, nor Hennig has solved the problem of how to deal with the relationship of descendant populations in a single lineage. Simpson arbitrarily divides the lineage into species by inferring reproductive isolation from the degree of morphological difference. By contrast, Wiley (1981:34) believes that "no presumed separate, single, evolutionary lineage may be subdivided into a series of ancestral and descendant species," a proposal that led Simpson to comment that one "could start with man and run back to a protist still in the species *Homo sapiens*." Finally, Hennig (1966*a*) arbitrarily terminated every evolutionary species when a daughter species branched off the parental lineage, ignoring the fact that a peripatric speciation event usually leaves the parental species unchanged.

FROM PHENON TO TAXON TO CATEGORY

A failure to understand the meaning of these three terms and their theoretical foundation has led some taxonomists into great confusion. It has been the cause of most attacks on the biological species concept. When an author says, "As a paleontologist I cannot employ the biological species concept because I cannot test the reproductive isolation of fossils,'' he or she reveals a lack of understanding. What the taxonomist observes directly are individuals, which are then sorted into phena. On the basis of certain biological concepts and information, such as an awareness of the possibility of sexual dimorphism, growth, alternation of generations, and nongenetic modifications of the phenotype, a taxonomist assigns the phena to populations, which in turn are classified into taxa. The ranking by a taxonomist of a taxon in the appropriate category (subspecies, species, or genus) is based on inferences drawn from the available data.

This methodology of basing inferences on evidence and its justification was perceptively discussed by Simpson (1961:69):

> Here it is necessary again to emphasize the distinction between definition and the evidence that the definition is met. We propose to *define* taxonomic categories in evolutionary and to the largest extent phylogenetic terms, but to use evidence that is almost entirely non-phylogenetic when taken as individual observations. In spite of considerable confusion about this distinction, even among some taxonomists, it is really not particularly difficult or esoteric. The well-known example of monozygotic (''identical'') twins is explanatory and is something more than an analogy. We *define* such twins as two individuals developed from one zygote. No one has ever seen this occur in humans, but we recognize when the definition is met by *evidence* of similarities sufficient to sustain the inference. The individuals in question are not twins because they are similar but, quite the contrary, are similar because they are twins. Precisely so, individuals do not belong in the same taxon because they are similar, but they are similar because they belong to the same taxon. (Linnaeus was quite right when he said that the genus makes the characters, not vice versa, even though he did not know what makes the genus.) That statement is a central element in evolutionary taxonomy, and the alternative clearly distinguishes it from non-evolutionary taxonomy.

The reproductive isolation of a biological species—the protection of its collective gene pool against pollution by genes from other species—results in a discontinuity not only of the genotype of the species but also of its morphology and other aspects of the phenotype produced by the genotype. This is the fact on which taxonomic practice is based. Reproductive isolation cannot, of course, be observed directly in samples of preserved specimens. However, it can be inferred on the basis of various types of evidence, such as the presence of a discontinuity—a bridgeless gap—between two correlated character complexes. In living species, such inferences can be tested by means of observation and experiment.

The crucial difference between the reasoning of a typologist and that of an adherent of the biological species concept is as follows: The typologist says, ''There is a clear-cut morphological difference between samples *a* and *b*; therefore, they are by definition two morphospecies, that is, two species." Any list of synonymies will quickly reveal how often this philosophy has led to the description of phena as species. The biological taxonomist asks, "Is the morphological difference between samples *a* and *b* of the kind one would expect to find between two reproductively isolated populations, that is, between two biological species?" In other words, the biological taxonomist uses the amount and kind of morphological difference only as an indication of reproductive isolation, only as evidence from which to draw an inference. This is a legitimate and reliable technique. Where typologists would recognize phena as (morpho) species, biologists will draw the right inferences from largely morphological evidence, and their species are usually confirmed by subsequent research. When competent taxonomic work based on morphological evidence is reexamined in the light of findings in behavior or biochemistry, it is usually confirmed in its entirety.

It is not always realized that the classification of phena is based on evidence entirely different from that which leads to the classification of species. The classification of species is based on weighted similarity, which entails the evaluation of all sorts of comparative data, morphological, physiological, or behavioral. The classification of phena is based on their relation to the gene pool of the population to which they belong. Ultimately this can be established only by their breeding behavior, which in turn can be observed in nature or studied experimentally. It does not matter whether one deals with strikingly different sexes in birds, insects, or marine invertebrates, with larval forms, or with alternating generations of parasites; breeding (or the piecing together of growth stages) is the best way to establish what phena together form a population. Sometimes molecular data also provide important clues. The experienced taxonomist knows what variation to expect within a biological species. No computer method has been found that can empirically assign phena to species. The taxonomist does this rapidly and with a high degree of precision on the basis of accumulated knowledge of the biology of the species concerned. In this taxonomic operation the classical methods still reign supreme "because they are enormously faster than the numerical methods" (Michener 1963).

DIFFICULTIES IN THE APPLICATION OF THE BIOLOGICAL SPECIES CONCEPT

The fact that difficulties sometimes arise when the biological species concept is applied to natural taxa does not mean that the concept is invalid. This has been shown by Simpson (1961:150) and Mayr (1963:21–22). Many generally accepted concepts cause similar difficulties when they have to be applied in a particular situation or to a specific sample. The concept of a tree, for instance, is not invalidated by the existence of spreading junipers, dwarf willows, giant cacti, and strangler figs. One must make a clear distinction between a concept and its application to a particular case.

The more ordinary problems of taxonomic discrimination at the species level, in particular the criteria for ranking a taxon as a species rather than a subspecies, are dealt with in Chapter 5.

The most serious difficulties in the application of the biological species concept are those caused by insufficient information, uniparental reproduction, and evolutionary intermediacy.

Insufficient Information

Individual variation in all its forms often raises doubt about whether a certain *morphotype* is a separate species or only a phenon within a variable population. Sexual dimorphism, age differences, polymorphism, and other types of variation can be unmasked as individual variations through a study of life histories and through population analysis (Chapter 5). The neontologist who normally works with preserved material is confronted by the same problems that confront the paleontologist, who also must assign phena (morphotypes) to species.

Uniparental Reproduction

Systems of reproduction in many organisms are not based on the principle of an obligatory recombination of genetic material between parental individuals during the formation of a new individual. Self-fertilizing hermaphroditism and other forms of automixis, parthenogenesis, gynogenesis, and vegetative reproduction (budding or fission) are some of the forms of uniparental reproduction. They are not infrequent among lower invertebrates, with parthenogenesis occurring even among insects and lower vertebrates up to the reptiles.

A population as defined in evolutionary biology is an interbreeding group. By this definition, an asexual biological population is a contradiction, even though the word *population* has also other uses in which a combination with *asexual* would not be contradictory. Since interbreeding is the ultimate test of conspecificity in animals and since this criterion is available only in sexually reproducing organisms, determination of categorical rank is difficult in taxa of uniparentally reproducing organisms. How should the taxonomist treat clones, pure lines, biotypes, and so-called strains or stocks of such organisms?

Such uniparental lineages are sometimes designated as *agamospecies*, binoms (Grant 1957), or paraspecies (Mayr 1987*b*). Whatever designation one chooses, one must be aware that such entities are not subdivisions of biological species (Mayr 1963:27–29) but something quite different. Ghiselin (1987) quite rightly questioned the propriety of applying the word *species* to groups of asexual clones, and we are inclined to agree with him (Mayr 1988*a*:353–355).

In some groups of animals, particularly aphids, gall wasps (Cynipidae), *Daphnia* (Crustacea), rotifers, and digenetic trematodes, a regular alternation between sexual and parthenogenetic generations may occur. In such cases neither kind of generation qualifies for separate species status, as nomenclatural recognition is not given to temporary clones. However, the parthenogenetic generations sometimes seem to fail to return to sexuality, particularly in aphids, and become permanent. When they differ from the sexual races by host plant preference or even by color genes, the suggestion that these parthenogenetic taxa should be called species deserves serious consideration.

In the case of permanently uniparentally reproducing lines, it is customary to assign species status on the basis of degree of morphological difference. There are usually well-defined morphological discontinuities among kinds of uniparentally reproducing organisms. These discontinuities are apparently produced by natural selection among the various mutants that occur in the asexual clones. It is customary to utilize the existence of such discontinuities and the amount of morphological difference among them to delimit species among uniparentally reproducing types.

Species recognition among asexual organisms is based not merely on analogy but also on the fact that each morphological entity that is separated by a gap from other similar entities seems to occupy an ecological niche of its own; each one plays its own evolutionary role. In groups such as the bdelloid rotifers, all of which reproduce by obligatory parthenogenesis, there is evidence of a definite biological meaning to the recognized morphological species.

Examples are known in which a form that is as distinct as a good species reproduces strictly parthenogenetically and no biparental species is known from which it might have branched off. Nomenclatural recognition is justified in such cases. Whenever several reproductively isolated chromosome types occur within such a "species," as in various crustaceans (e.g., *Artemia salina* Linnaeus) (White 1978), it may be convenient to distinguish them nomenclaturally. Although they are conventionally referred to as races, it is more logical to designate reproductively isolated chromosomal populations as (micro)species.

There are approximately 1000 animal species known in which the male sex is absent or nonfunctional. Such all-female species reproduce by *thelytoky*, a special term for this type of parthenogenesis. Depending on the meiotic mechanisms, there is a trend toward either homozygosity or heterozygosity in such species. The homozygosity-generating type is rare, being restricted to a few insect groups, and there is sometimes a normally bisexually reproducing sibling species alongside the thelytokous species (White 1978). Many cases of heterozygosity-promoting thelytoky seem to be products of instantaneous speciation which results from a switch to parthenogenesis (thelytoky) in an individual that had originated as an interspecific hybrid (White 1978, Chapter 9). Existing species in this category seem to be products of relatively recent speciation events, not having been provided with sufficient time to accumulate enough individual variation to create taxonomic difficulties. This is the case with the all-female species of the lizard genus *Cnemidophorus,* which is found in the southwestern United States and Mexico. The known cases of thelytoky in salamanders and fishes are also the result of hybridization, as is one case in grasshoppers. In some special cases (e.g., *Rana esculenta, Poeciliopsis*) the male chromosomes are lost during meiosis, and fertilization by males of one of the parental species (gynogenesis, or *pseudogamy*) is needed to induce development of the egg even though the males do not contribute to the genotype of the developing zygote [kleptospecies (Dubois and Günther 1982)]. Hybridization between two distant species of animals apparently always results either in total sterility or in the abandonment of sexual reproduction (Chapter 4).

In groups with cyclical parthenogenesis, the sexual generation may be permanently lost in some species. Where this is correlated with a switch in a host species (as in the case of some aphids), it may raise doubts about species status. Hermaphrodites in most cases reproduce sexually; that is, fertilization of the eggs is effected by the spermatozoa of a different individual. However, an occasional species may practice complete self-fertilization (automixis). This results in increasing homozygosity, such as Foltz et al. (1982) found in several species (aggregates of clones) of slugs.

Evolutionary Intermediacy

The species, as manifested by a reproductive gap between populations, exists in full classical distinctness only in the nondimensional situation of a local fauna. As soon as one deals with species taxa extended in the dimensions of space (longitude and latitude) and time, the stage is set for incipient speciation. Populations may be found in these circumstances that are in the process of becoming separate species and have acquired some but not yet all of the attributes of distinct species. At what stage in this process of divergence should a diverging population be called a species? A decision is particularly difficult when the acquisition of morphological distinctness is not closely correlated with the acquisition of reproductive isolation. The various difficulties for the taxonomist that may result from evolutionary intermediacy can be summarized as follows.

1 *Acquisition of reproductive isolation without equivalent morphological change:* Reproductively isolated species without (or with very slight) morphological difference are called sibling species. Their taxonomic treatment is discussed in Chapter 5.

2 *Acquisition of strong morphological difference without reproductive isolation:* A number of genera of animals and plants are known in which morphologically very different populations interbreed at random wherever they come in contact. The typological solution of calling every morphologically distinct population a species is clearly inappropriate in such situations. Conversely, there are genera in which the isolating mechanisms between any two species break down occasionally. To consider such species conspecific would be to go to the opposite extreme. No generalized solution is possible in cases where morphological divergence and the acquisition of reproductive isolation do not coincide. The only recommendation to the specialist is to delimit species in such a way that they form biologically meaningful natural entities. The difficulty posed by the rapid morphological divergence of populations that lack reproductive isolation is well illustrated by the West Indian snail genus *Cerion* (Figure 2-1).

3 *The occasional breakdown of isolating mechanisms (hybridization):* Reproductive isolation may break down occasionally even between good species. Usually this leads only to the production of occasional hybrids, which, being either sterile or of lowered viability, do not cause any taxonomic difficulty. More rarely, there is a complete local breakdown of isolation that results in the production of an extensive hybrid swarm and more or less complete *introgression* (Mayr 1963:110–135).

Hybrid individuals are sometimes described as species before their hybrid nature is discovered. Such names lose their validity as soon as the hybridism is established (Chapter 14). Only populations are recognized as taxa, and hybrid individuals are not populations.

Situations where whole populations are formed as a result of hybridization are taxonomically more difficult. We recognize several types of natural populations that owe their origin to hybridization. The taxonomic treatment of secondary intergradation, which results from the fusion of previously isolated populations, is discussed in Chapter 5. Two other kinds of hybridism concern us here.

a *Hybrid swarms:* In certain species, the reproductive isolation that is

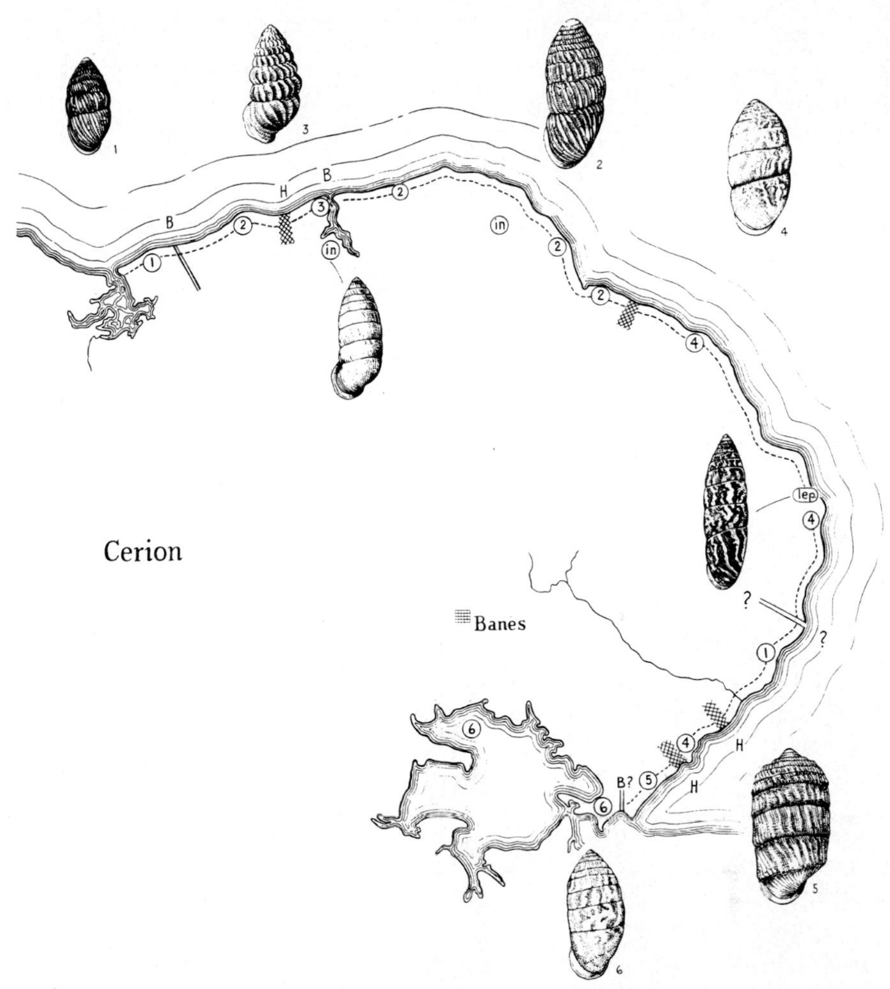

FIGURE 2-1
Irregular distribution of populations of the halophilous land snail *Cerion* in eastern Cuba. Numbers refer to different races or species. Where two populations come in contact (with the exception of *lepida*), they hybridize (H) regardless of difference. In other cases contact is prevented by a barrier (B). in = isolated populations. (*From Mayr 1963.*)

maintained over most of the area of sympatry may break down locally, resulting in the production of localized hybrid swarms. In such cases it is advisable to uphold the species status of the parental species. The example of the two Mexican towhees (*Pipilo erythrophthalmus* and *P. ocai*) provides an excellent illustration of this situation (Sibley 1954) (Figure 2-2). No taxonomic recognition is

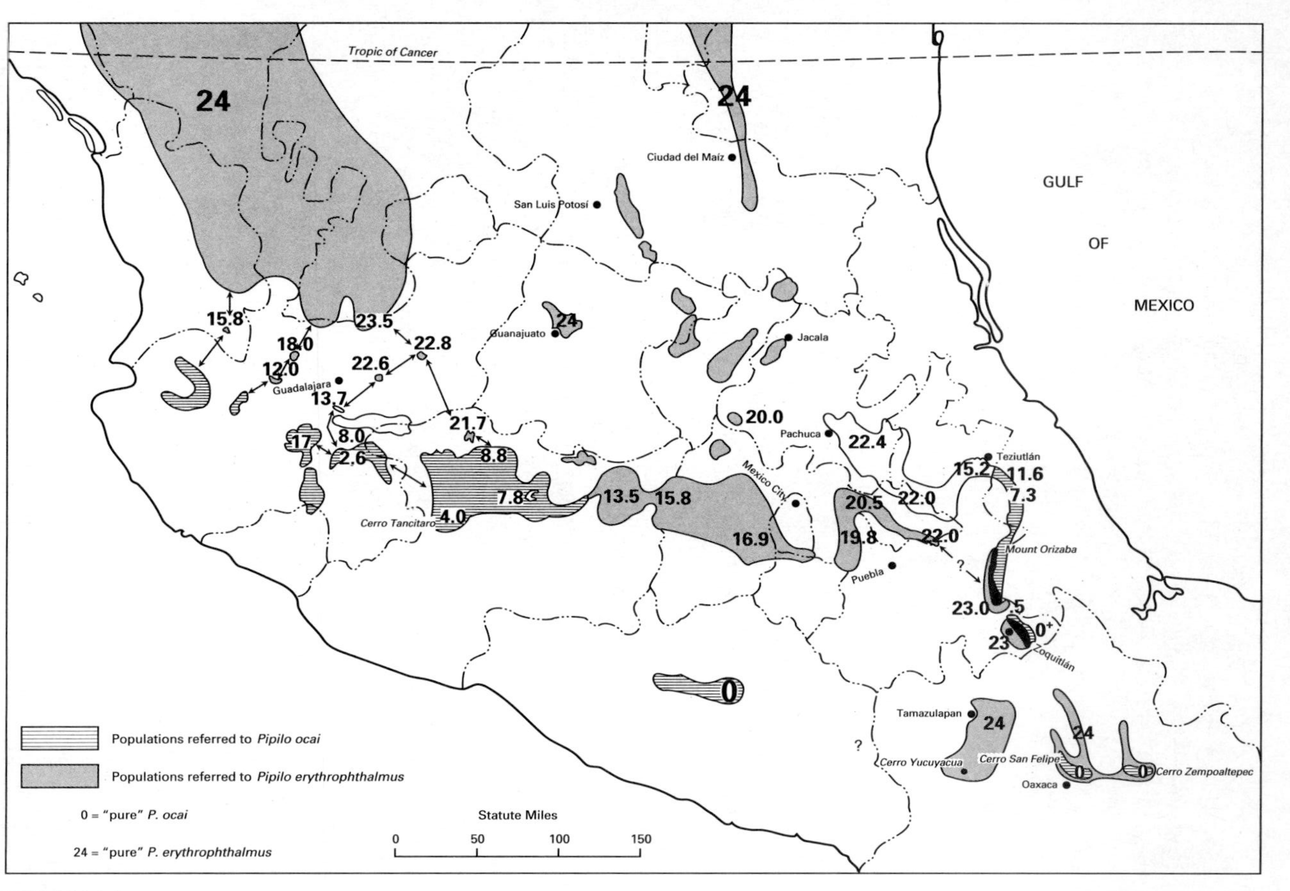

FIGURE 2-2
Sympatry and hybridization of two species of towhees (*Pipilo*) in Mexico: pure *erythrophthalmus* (24) in the north and southeast and pure *ocai* (0) in the south and southwest. The numbers (from 0 to 24) designate the mean character indexes of various hybrid populations. (*From Sibley 1954.*)

given to hybrid populations that result from such a local breakdown of reproductive isolation. The only possible exception would be a breakdown of isolation so complete that the two parental species fuse into a single new species. The taxonomic literature includes a number of instances that have been interpreted in this manner, but we are not aware of a single thorough analysis that would have established the fact unequivocally. For further discussion of the taxonomic aspects of hybridization, see Chapter 5.

b *Parthenogenetic species formed through hybridization:* Hybridization that results in the production of an individual with one chromosome set from parental species A and one from parental species B leads through doubling of the chromosome set to *allotetraploidy*. In plants, such individuals can immediately establish a self-fertilizing and eventually cross-fertilizing allotetraploid species. In animals, where cross-fertilization is almost always obligatory, a new allotetraploid can maintain itself only by switching to parthenogenesis (thelytoky). Such hybrid species are usually morphologically well characterized and are ordinarily recognized as valid species even after their mode of origin has been discovered.

More complex situations involve partial parthenogenesis, sometimes combined with triploidy or higher polyploidy, as is found in oligochaetes, planarians, weevils, moths (*Solenobia*) and other insects, mites, and crustaceans. Here one may find in a single "species" bisexual diploids, thelytokous diploids, and thelytokous polyploid "races" (White 1978). Even though these races may be morphologically indistinguishable, they are reproductively isolated. Most of these thelytokous races do not seem to be products of hybridization. In this case thelytoky seems to precede the origin of polyploidy.

4 *Semispecies and allospecies:* Geographic isolates occasionally have an intermediate status between subspecies and species. On the basis of some criteria, they would be considered species; on the basis of others, they would not. It is usually more convenient for the taxonomist (Chapter 5) to attach such doubtful populations as subspecies to the species with which they are most nearly allied. In other cases such isolates have evidently reached the species level but have remained members of a superspecies. Such populations may be designated as *allospecies*. Circular overlaps and other borderline cases (Mayr 1963:496–512) are other instances of evolutionary intermediacy. The method of ranking the resulting taxa has to be decided from case to case on the basis of convenience and degree of evolutionary intermediacy. Taxa intermediate between subspecies and species are often referred to as *semispecies*.

MEANING OF THE SPECIES CATEGORY

The basic role of the species definition is to function as a yardstick whenever the delimitation of a species taxon presents difficulties. Most species taxa, particularly monotypic taxa with limited variability and distribution, present no problems, but there are two sources of potential difficulties: the assembling of local phena into biological species (with the problems of polymorphism and sibling species) and the proper assignment of allopatric and allochronic populations. It is in these situations that the biological species concept facilitates the decision about which phena and populations should be considered full species and which should not.

In an analysis of the 607 species of North American birds, Mayr and Short (1970) found that the biological species concept was unable to clarify only a single case (*Pipilo*) (Figure 2-2). On the other hand, it permitted the settling of all controversies with respect to polymorphism and sibling species. In 45 cases of hybridization among species taxa, the criteria of the biological species permitted the classification of 27 cases as intraspecific and 18 cases as interspecific hybridization. Among the 607 North American species, 247 are polytypic, but only 46 of them contain populations (mostly peripheral isolates) that some authors consider full species and others (Mayr and Short included) consider subspecies. Since these populations are isolated, the biological species concept cannot be applied directly. This analysis illustrates how useful the biological species concept is in difficult situations.

In the case of populations that are isolated from each other in space or time, one can decide only by inference whether they would behave toward each other as members of the same species. Such inferences are often more robust (Chapter 5) than the outsider would suspect, but in most cases it is biologically not very important which decision is made (species or subspecies). It is the sympatric situation that is of particular importance for the ecologist and student of behavior. Furthermore, no other species definition can provide a foolproof system for the correct assignment of isolated populations or other cases of evolutionary intermediacy.

CHAPTER **3**

THE SPECIES TAXON

The word *species,* as conceived by the evolutionary taxonomist and defined in Chapter 2, describes the relationship of populations. When two populations coexist and do not interbreed, they belong to different species. Thus, in its modern conception, *species* is a relational term like the word *brother* (Mayr 1957). In a given locality, a species of animal is usually separated from other sympatric species by a complete reproduction gap. This is the species of the local naturalist, the species of Ray and Linnaeus. It may also be called the *nondimensional species* because it lacks the dimensions of space and time. Combining the properties of a species and of a single local population, a nondimensional species can usually be delimited unequivocally.

Every species taxon in nature consists, however, of numerous local populations, and this raises the problem of how to treat these populations taxonomically. Adding the dimensions of geography and time poses numerous problems. The totality of populations conforming to the species definition (Chapter 2) constitutes a species taxon. Since the beginning of the new systematics it has been one of the major tasks of taxonomy to delimit species taxa against each other by assigning populations to the correct species taxa with the help of the yardstick of the biological species definition.

THE ONTOLOGICAL STATUS OF THE SPECIES

Species are considered classes by all those who adopt either the essentialist or the nominalist species concept, in other words, by nearly all philosophers until recent times. These philosophers define a class, as the term is used in this context, as a set of discrete entities all of which possess the same unchanging defining property or properties. For naturalists, by contrast, a species is a population that is spatiotemporally localized, is reasonably discrete in space and time, and exhibits internal cohesiveness. Logicians call such a population an individual or particular; to them, the members of the species are actually parts of this species (individual) (Ghiselin 1966, 1974).

It seems incongruous to many biologists to refer to a species as "an" individual when it is composed of millions of individual organisms. Such an assemblage lacks the singularity and uniqueness one associates with the term *individual*. Therefore, species are also called *biopopulations*. This is largely a terminological difference, because such populations have the spatiotemporal properties and much of the internal cohesiveness of individuals. However, they lack altogether the defining properties of classes (Mayr 1987*b*). This is another difference between the typological and the biological species concepts.

POLYTYPIC SPECIES

Many species vary in space or time and consist of recognizably different populations. A population that the taxonomist considers to differ sufficiently from previously named populations of a species is described as a new *subspecies* (see below). Species that contain two or more subspecies are called *polytypic* species; species that are not subdivided into subspecies are called *monotypic* species. Recognition of the fact that many species taxa, particularly widely distributed species, are polytypic was one of the most important developments in taxonomy. For a full treatment of this development and of various aspects of polytypic species, see Mayr (1963, Chapter 12).

Importance of the Recognition of Polytypic Species Taxa

The shift from the morphological (typological) to the biological species concept necessitated a drastic change in taxonomic practice. While previously every population that could be diagnosed morphologically was called a species, it became necessary to infer the reproductive isolation of populations (Chapter 5). The discovery of the geographic variation of morphological and other characters led to the realization that many previously recognized morphospecies were not reproductively isolated from other allopatric populations and did not qualify as separate biological species. They were consequently ranked as subspecies in more or less widely distributed polytypic species.

A major benefit derived from the recognition of polytypic species taxa is the considerable simplification it has made possible in the classification of such well-studied groups of animals as birds, mammals, butterflies, and snails. The reclassification into polytypic species of geographically representative forms that originally were separately described as monotypic species led to a great simplification of the system. This reorganization of classification on the species level is virtually complete in birds, mammals, and some groups of insects and land mollusks but has hardly begun in most other groups of animals. The 19,000 monotypic species of birds listed in 1910, together with thousands of species described since then, have been reduced to approximately 9040 species. A similar simplification has been reported for many other groups of vertebrates and invertebrates. Of much greater significance than this practical benefit is the restoration of a definite biological meaning and homogeneity to the species category. Awarding species rank to every local population, no matter how slight its difference, destroys the biological significance of the species category.

The assembling of local populations into polytypic species, or, more broadly, the sorting of large numbers of "nominal species" and "varieties" into polytypic species, reveals many taxonomically and biologically interesting situations (Mayr 1963:343). It has provided the best available evidence for the process of allopatric speciation, the frequent origin of evolutionary novelties in peripherally isolated populations, and numerous intermediate stages in the evolutionary process, thus elucidating previously inexplicable discontinuities. Geographic variation, particularly in peripherally isolated populations, helps bridge the gap between microevolution and macroevolution. Some of the best proofs of the occurrence of evolution have emerged from the study of polytypic species taxa. To convert the nominal species of all groups of animals into well-delimited polytypic species taxa is therefore one of the major tasks of taxonomy.

Difficulties

In establishing polytypic species, the taxonomist encounters three sets of difficulties. Polytypic species are composed of allopatric or allochronic populations that differ from one another. However, all populations of sexually reproducing organisms differ slightly, and certain standards must be met before subspecies can be recognized. A second difficulty is that closely related species with similar ecological requirements occasionally replace each other geographically, and it is difficult to decide whether they are full species or subspecies. The method for choosing between these two alternatives is discussed in Chapter 5. Finally, many isolated populations are in the middle of the process of evolving into new species and are on the borderline between subspecies and species status. The appropriate choice between these alternatives is also discussed in Chapter 5.

THE OCCURRENCE OF POLYTYPIC SPECIES IN THE ANIMAL KINGDOM

The frequency of polytypic species differs from animal group to animal group. They occur most frequently where species are not contiguously distributed. In most well-studied groups of animals, between 40 and 80 percent of the species are polytypic, but some highly specialized groups, particularly certain host plant–specific insects, do not readily form polytypic species. Polytypic species are also scarce or absent in groups with slight species differences (e.g., groups of sibling species).

A number of other technical terms, such as *formenkreis* (Kleinschmidt) and *rassenkreis* (Rensch), were formerly applied to polytypic species but did not become established (Mayr 1963:339).

The polytypic species is in a sense the lowest of the higher categories. Being multidimensional, it lacks the simplicity and objectivity of the nondimensional species. Most of the difficulties involved in delimiting species of animals occur when it is doubtful whether two allopatric populations belong to the same polytypic species. Among birds, such borderline cases are few. The claim that such difficulties are more common in other groups of animals awaits verification.

Nomenclatural Problems

A polytypic species is often a compound of several "species" originally proposed as monotypic. It differs from the Linnaean species in that it is no longer the lowest category (which is now the subspecies) and in that it is a collective category. What scientific name should be given to this new collective taxon, and who should be the author? When Linnaeus named the white wagtail *Motacilla alba,* to give one example, he had in mind the Swedish population, which has the diagnostic characters he mentioned. The *M. alba* of Linnaeus is now called the *nominate* subspecies *M. alba alba* Linnaeus. When the *M. alba* of Linnaeus was combined with eight or more other subsequently named taxa from other geographic regions, all originally described as separate species (*lugubris* Temminck, *dukhunensis* Sykes, *baicalensis* Swinhoe, *leucopsis* Gould, *personata* Gould, *hodgsoni* Blyth, *ocularis* Swinhoe, *lugens* Kittlitz, etc.), the newly formed polytypic species was very different from the original *M. alba* of Linnaeus. If someone still associates the name Linnaeus with the new polytypic species *M. alba,* it is merely to indicate Linnaeus as the author of the name *alba,* not of the drastically reconstituted polytypic taxon to which we now attach the name.

INFRASPECIFIC CATEGORIES AND TERMS

The Variety

This term, as *varietas,* was the only subdivision of the species recognized by Linnaeus. It designated any deviation from the type of the species. As a consequence, the varieties of the early taxonomists were a heterogeneous potpourri of individual variants (Chapter 4) and various kinds of races. This confusion discredited the term *variety,* which is no longer used by animal taxonomists. For a further discussion, see Simpson (1961:177) and Mayr (1963:346).

The Subspecies

When the term *subspecies* came into general usage during the nineteenth century, it replaced the term *variety* in its meaning of "geographic race." It was considered a taxonomic unit like the morphological species, but at a lower taxonomic level. Many early authors used the term *subspecies* indiscriminately, almost like the term *variety,* for distinguishable entities that were less distinct than species. Ant specialists, for instance, employed the term not only for geographic races but also for sibling species and individual variants. When an author reports several subspecies of one species from the same locality, it strongly indicates an incorrect use of the term. Subspecies are normally allopatric and/or allochronic, with exceptions occurring, however, in migratory species and parasites with sympatric host subspecies. A purely morphological definition of the subspecies, as attempted by typologists, often results in a sympatry of the entities thus defined. In view of the many misuses of the term, it must be emphasized that the subspecies is a category quite different from the species. No nonarbitrary criterion is available to define the category subspecies, nor is the subspecies a unit of evolution except where it happens to coincide with a geographic or other genetic isolate.

The subspecies may be defined as follows: *A subspecies is an aggregate of phenotypically similar populations of a species inhabiting a geographic subdivision of the range of that species and differing taxonomically from other populations of that species.*

The reasons for the wording of this definition are as follows.

1 A single subspecies may consist of many local populations all of which, though very similar, are slightly different from each other genetically and phenotypically. A subspecies is therefore a collective category.

2 Every local population is slightly different from every other local population, and the presence of these differences can be established through sufficiently sensitive measurements, statistics, or molecular analysis. It would be absurd and would lead to nomenclatural chaos if each population of this type were given the formal trinominal name that is customary for subspecies. Therefore, subspecies are to be named only if they differ "taxonomically," that is, by sufficient diagnostic morphological characters.

3 Even when it is possible to assign populations to subspecies, an assignment on the basis of phenotype alone is not necessarily possible for every individual because of an overlap of the ranges of variation of neighboring populations.

The term *overlap* is often misused. The breeding ranges of two species may overlap geographically, but the breeding ranges of two subspecies of the same species do not. If two discrete breeding populations coexist at the same locality, they are full species, except in the rare case of *circular overlap*. Where two subspecies meet, intermediate or hybrid populations may combine the characters of both subspecies. To say that the two subspecies overlap in this area would be misleading, for the species is represented in this area only by a single population, no matter how variable.

Difficulties in the Application of the Subspecies Category Recognition of the polytypic species requires the use of the subspecies category, with all the concomitant benefits described above. However, various aspects of geographic variation cause difficulties for the taxonomist. Indeed, the subspecies has been misused in many ways. Some authors have applied the term to individual variants or to sibling species, many authors have named insignificantly different local populations as subspecies, and some authors have considered every subspecies a unit of evolution rather than realizing that the recognition of some subspecies is merely an arbitrary device to facilitate intraspecific classification. As a result, the practice of describing subspecies was criticized by numerous authors, most cogently by Wilson and Brown (1953) and Inger (1961). These authors pointed out four aspects of the subspecies that reduce its usefulness:

1 The tendency of different characters to show independent trends of geographic variation

2 The independent occurrence of similar or phenotypically indistinguishable populations in geographically separated areas (*polytopic subspecies*)

3 The occurrence of microgeographic races within formally recognized subspecies

4 The arbitrariness of the degree of distinction considered by different specialists as justifying subspecific separation of slightly differentiated local populations

Numerous articles on the pros and cons of the subspecies question can be found in Volumes 3 to 5 of *Systematic Zoology* (1954–1956). For general reviews of the subspecies question, see also Simpson (1961:171–176) and Mayr (1963:347–350). Recent arguments have led to a more critical attitude toward subspecies. It has also been shown that sensible use of the category of the subspecies is still a convenient device for classifying population samples in geographically variable species (Wiens 1982).

Practical problems that the population taxonomist frequently faces are discussed in Chapter 5. They concern in particular the following questions: How different must a population be in order to justify its recognition as a subspecies? How should one treat intermediate populations? How does one delimit subspecies against adjacent subspecies? Should one recognize polytopic subspecies for indistinguishable but geographically separated populations? When should geographical isolates be called species, and when should they be called subspecies?

Temporal Subspecies In paleontology slightly different populations separated in time are increasingly often assigned to the subspecies category. It does not seem advisable to make a terminological distinction between geographic and temporal subspecies because when different subspecies of a fossil species are found at different localities, it is usually impossible to determine whether they are precisely contemporary. Even when there is a sequence of subspecies at a single locality, the sequence need not be purely temporal. Subspecies found in succeeding strata may actually be geographic races that replaced each other owing to climatic changes (Figure 5-4).

When applying the subspecies category, the paleontologist faces certain difficulties not encountered by the neontologist. There may be a differential deposition of various age classes and sexes in different horizons as well as the occurrence of nongenetic habitat forms. However, treating fossils as the remains of formerly existing populations rather than as morphotypes usually leads to a deeper analysis and a better understanding of relationships and the meaning of evolutionary trends. As with living species, it must be kept in mind at all times that the subspecies is merely a classificatory device. For further discussions on the subspecies in paleontology, see Newell (1947, 1956), Sylvester-Bradley (1951, 1956), and Simpson (1961:175–176).

The Race

A race that is not formally designated as a subspecies is not recognized in the taxonomic hierarchy. However, the terms *subspecies* and *geographic race* are frequently used interchangeably by taxonomists working with mammals, birds, and insects. Other taxonomists apply the word *race* to local populations within subspecies.

The nature of ecological races among animals is still controversial (Mayr 1963:355). Since the environments of two localities are never identical, every subspecies is at least theoretically also an ecological race. However, some populations differ in their ecological requirements without acquiring taxonomically significant differences. More important from the taxonomic and evolutionary points of view are host races among parasites and species-specific plant feeders. If gene flow between populations on different hosts is drastically reduced, such host races are the equivalent of geographic races in free-living animals. Also, such host races often develop subspecific characters.

The Cline

This term was coined by Huxley (1939) to describe a character gradient. It is not a taxonomic category: A single population may belong to as many different clines as it has characters. A *cline* is formed by a series of contiguous populations in which a given character changes gradually. At right angles to the cline are the lines of equal expression of the character (points of identical phenotype); this type of line is called an *isophene*. For instance, if in the range of a species of butterfly the percentage of white specimens varies from north to south, the corresponding isophenes may be indicated on a map (Figure 3-1).

Any character, whether a morphological, physiological, or other genetically determined character, may vary clinally. Clines may be smooth, or they may be "step clines" with rather sudden changes of values. Clines do not receive nomenclatural recognition. Indeed, when the geographic variation of a species is clinal, it is usually advisable not to recognize subspecies, except possibly for the two opposite ends of the cline when they are very different or are separated by a pronounced step. For a further discussion of clines, see Simpson (1961:178) and Mayr (1963: 361–366).

Infrasubspecific Categories

The subspecies is the lowest taxonomic category recognized in the *International Code of Zoological Nomenclature* (Article 45e) (Chapter 14).

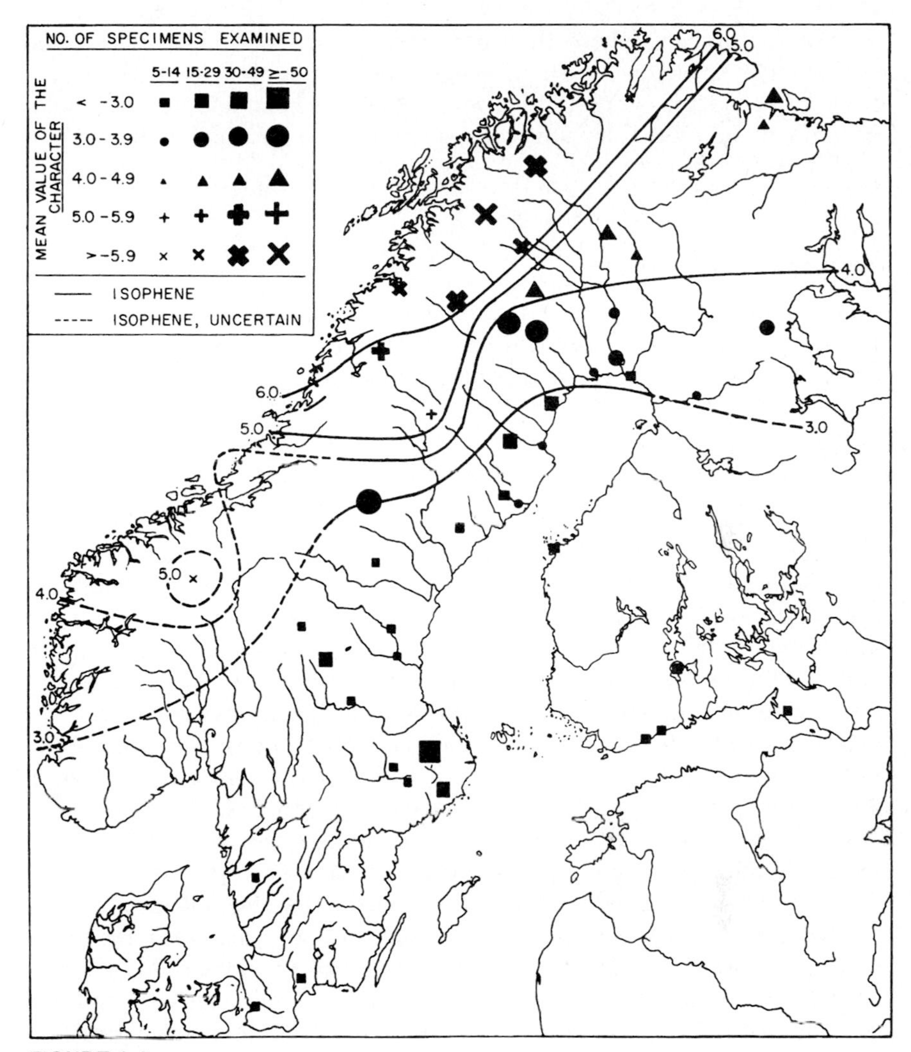

FIGURE 3-1
Cline in the darkness of the upper side at different Fennoscandian localities of *Pieris napi* females of the first generation. Isophenes of various darkness values are indicated on the map. (*From Petersen 1949.*)

In the days when the subspecies was still defined typologically, many proposals were made to subdivide heterogeneous subspecies into still smaller, hopefully uniform taxa, and terms were proposed for such taxa, e.g., *natio*. Now that it is being realized that every local population is genetically different from every other even if they live only a few miles apart or less and that these populations are not sharply separated from each other (except where they are separated by barriers), there is no longer any excuse for a formal recognition of innumerable local subdivisions of subspecies. The term *deme,* which has been adopted by zoologists for the evolutionary unit corresponding to a local population (Mayr 1963:137), is not the name of a taxonomic category.

Intrapopulation Variants

Taxa are populations, and populations are the material of classification. Phena, which are composed of intrapopulation variants, have no taxonomic status and deserve no formal recognition in nomenclature (Chapter 4).

Neutral Terms

It is very convenient in taxonomic work to have some terms that can be used informally for phena or populations, particularly in incompletely analyzed cases. These are the so-called neutral terms. The ones that are most frequently used in taxonomy are *form* for a single unit and *group* or *complex* for a number of units. We may speak of a form when we do not know whether the phenon in question is a full species or a subspecies or whether it is a subspecies or an individual variant. Seasonal variants and morphs are often referred to as forms. The term is also used in the plural to refer to two unequal units. For instance, when describing attributes common to a species and a subspecies of another species, one may refer jointly to the species and the subspecies as "these two forms."

The term *group* is more commonly applied to an assemblage of closely related taxa that one does not want to place in a separate category. In the large genus *Drosophila,* for example, numerous species groups are recognized, such as the *melanogaster* group, the *virilis* group, and the *obscura* group. A *species group* is a group of closely related and presumably recently evolved species. The use of the species group in formal taxonomy has been spreading in recent years because it reduces the need to recognize subgenera. In large polytypic species the term *group* is also applied to subspecies groups. The common Palearctic jay, *Garrulus glandarius,* has a total of 28 subspecies, which can be arranged in seven subspecies groups: the *garrulus* group, the *bispecularis* group, and others. The term is used more rarely for aggregates of genera and other higher categories. The word *complex* is sometimes used synonymously with the word *group*.

Terms such as *section, series,* and *division* are generally used for the higher categories. Their use is not standardized, however, and in different branches of systematic zoology these terms may be used above or below family, order, or class.

POPULATION TAXONOMY

The increasing preoccupation with the description of new subspecies and the establishment of polytypic species that began in the second half of the nineteenth century and has continued into the present has resulted in a subtle change of emphasis and outlook. The emerging understanding of the species taxon as a geographically variable aggregate of populations accelerated the replacement of the typological species concept and its taxonomic equivalent, the morphospecies, by the biological species concept. Taxonomists were no longer satisfied to separate collections into types and duplicates. They began to sample species at many localities and tried to assemble large series from every locality. This type of study was initiated in the second half of the nineteenth century almost simultaneously by ornithologists, entomologists, and malacologists.

Although the study of populations reached its dominant position in systematics only in recent generations, its roots go back to the pre-Darwinian period. Short histories are found in Mayr (1963, Chapters 11 and 12, 1982*a*:251–279).

Populations are variable; consequently, the description, measurement, and evaluation of variation has become one of the principal activities of the taxonomist who studies taxa in the lower categories. Typologists needed only one or two "typical" specimens of a species; when there were more, they could be disposed of as "duplicates." Modern taxonomists attempt to collect large series at many localities throughout the range of a variable species. Subsequently they evaluate this material by using the methods of population analysis and statistics (Chapter 4).

The work of population taxonomy not only led to a simplification of taxonomy through the introduction of polytypic species, it also led to a new approach to the study of evolution. Systematics has made many important conceptual contributions, and one of its greatest achievements is to have assisted in the introduction of the population concept into biology (Mayr and Provine 1980).

POPULATION STRUCTURE

Study of the population structure of species shows that the conventional division of species into subspecies is an inadequate and sometimes misleading representation of the actual situation. A species does not consist of a number of little species called subspecies. Rather, a species consists of innumerable local populations or demes that stand in a certain relationship to each other. When species are studied strictly from the standpoint of population structure, it is found that they can best be described in terms of three major population phenomena (Mayr 1963, Chapter 13).

The Population Continuum

A large part of the range of many species, particularly the central part, is occupied by a series of essentially contiguous populations. Even when there are minor breaks in distribution caused by the unsuitability of the habitat, such breaks are bridged by steady dispersal, resulting in copious gene exchange among populations. Variation in such a population continuum is essentially clinal. Terminal populations at the opposite ends of a continuum may be rather different phenotypically and may deserve recognition as subspecies.

The Geographic Isolate

This term designates all geographically isolated populations, or groups of populations that have only limited or no gene exchange with other populations of their species. Any insular population is normally such an isolate, and isolates are therefore particularly common near the periphery of a species range. Isolates are frequently of sufficient difference to be ranked as subspecies. The biological importance of the geographic isolate is that every isolate, regardless of its taxonomic rank, is an incipient species; it is an important unit of evolution (Figures 3-2 and 3-3).

The Zone of Secondary Intergradation

Whenever a geographic isolate reestablishes contact with the main body of the species or with another isolate, the two will interbreed if the isolate has not yet acquired an effective set of isolating mechanisms. Depending on the degree of genetic and phenotypic difference achieved by the previously isolated populations, a more or less well-defined hybrid belt or zone of secondary intergradation will develop. Fusion lines between ex-isolates can be found in many species (Short 1969).

An alternative explanation for zones of secondary intergradation has been advanced by authors who believe that the morphological change and increased variability in these zones are primary and are caused by selection along an environmental threshold (escarpment). No case is known to us, however, in which this explanation would fit the evidence better than does the secondary intergradation theory; indeed, it can sometimes be definitively refuted (Mayr and O'Hara 1986; Barton and Hewitt 1989).

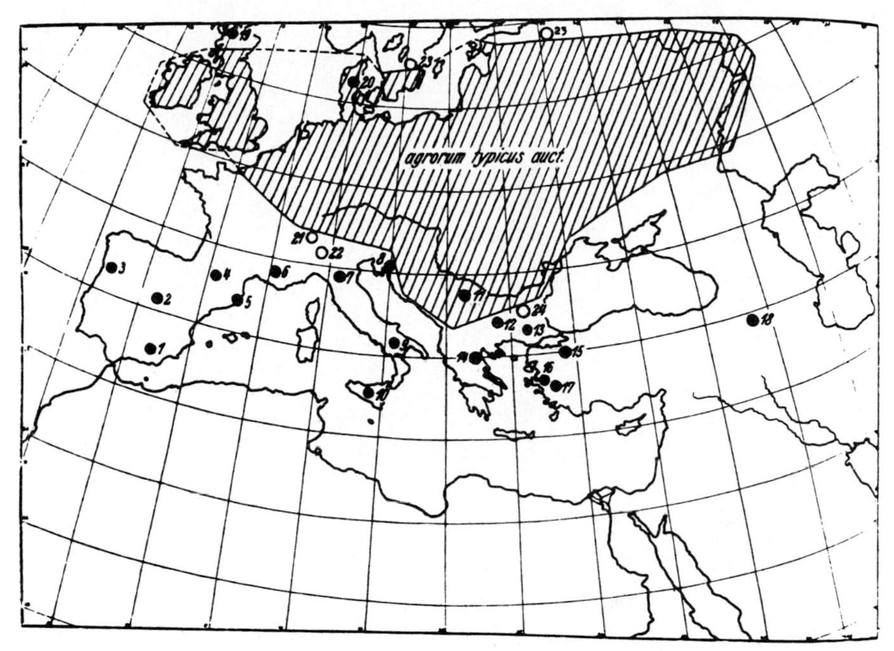

FIGURE 3-2
Pattern of geographic variation in the bumblebee, *Bombus agrorum.* There is little geographic variation in the continuous range of the nominate subspecies *agrorum,* while each peripherally isolated population (numbers 1 through 24) is distinct and is generally recognized as a separate subspecies. (*After Reinig.*)

There are only a few groups of animals in which the population structure of species is sufficiently well known to permit their analysis in terms of the three major population phenomena. This was attempted by Keast (1961) for all the species in a number of families of Australian birds. Such a study of the population structure of species does not replace classical taxonomy but is a superimposed refinement of classical methods. It is possible only in groups in which taxonomic analysis and population sampling have reached a high degree of maturity.

THE NEW SYSTEMATICS

The approach of the population taxonomist differs rather drastically from the simple pigeonholing of classical Linnaean taxonomy. To emphasize the difference, Huxley (1940) introduced the term *new systematics* for the newer approach; actually, its roots go back to the first half of the nineteenth century, and some traces of the new systematics can be found in the writings of taxonomists who worked as long as 150 years ago. Every generation has its own new systematics, and what might have been

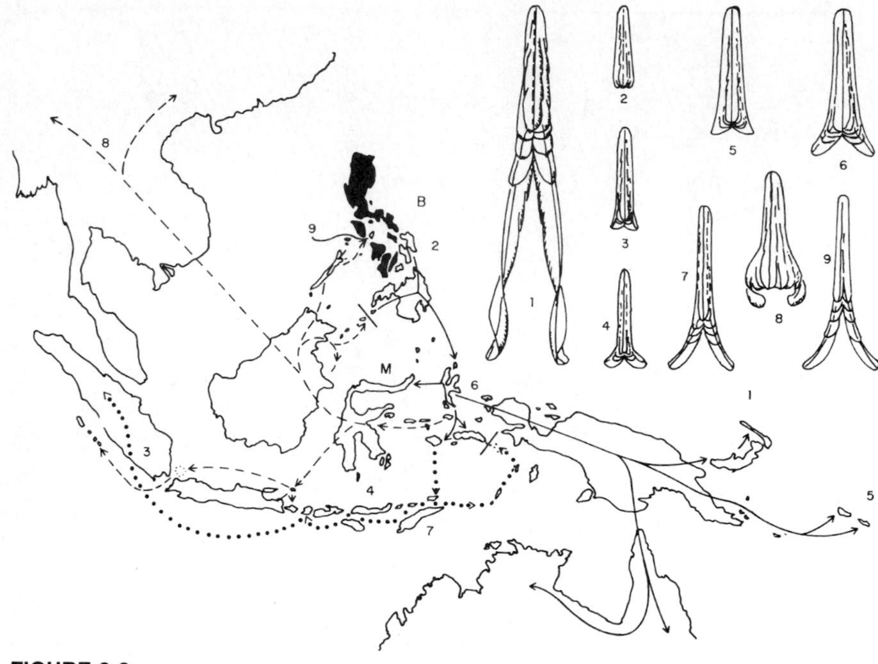

FIGURE 3-3
Peripheral isolates at the ends of various lines of expansion in the polytypic bird species *Dicrurus hottentottus.* The figures indicate the ranges of the nine forms, the tails of which are shown in the insert. The tails of 4 and 6 are typical for most populations of the species; the tails of the peripheral forms 1–3, 5, and 7–9 are aberrant and are specialized in various directions. (*From Mayr and Vaurie 1948.*)

considered new in 1940 may indeed be very old systematics now. To dispel misconceptions about the meaning of the new systematics, Mayr (1964:14–15) wrote: "What then is the new systematics? Perhaps it is best described as a viewpoint, an attitude, a general philosophy. It started primarily as a rebellion against the nominalistic-typological and thoroughly non-biological approach of certain, alas all too many, taxonomists of the preceding period."

Workers in the new systematics consider themselves biologists rather than filing clerks. This has a number of well-defined consequences in their attitude toward their material and toward various techniques.

1 These workers are conscious at all times that they are classifying organisms, not the remains of organisms or mere names.

2 As a consequence, they place considerable emphasis on so-called biological characteristics, that is, on nonmorphological information derived from behavior, physiology, biochemistry, ecology, and so forth.

3 They appreciate the fact that all organisms occur in nature as members of populations and that specimens cannot be understood and properly classified unless they are treated as samples of natural populations.

4 As a consequence, they attempt to collect statistically adequate samples, which in the case of variable species often amount to hundreds or thousands of specimens, in order to be able to undertake a study of individual and geographic variation with the help of the best biometric and statistical tools.

If we attempted to describe the current model of the new systematics, we would see at once that every single item is merely the continuation of a trend which in most cases started more than 100 years ago. Some of these points follow.

1 The utilization of an ever-increasing number of kinds of characters and a continued depreciation of key characters and single-character classifications, in contrast to the typological approach
2 A ready acceptance of new tools and techniques, including the following:
  a The visual analysis (by sonagrams and other means) of sounds in insects (cicadas, orthopterans), frogs, and birds
  b The analysis of courtship displays and other behavior
  c The utilization of biochemical characters, particularly those yielded by various methods of protein analysis
  d The utilization of computers to reduce the danger of subjectivity in character evaluation
3 A further clarification of concepts, for instance:
  a A clearer separation of taxon from category
  b The recognition of the subspecies as a category, not an evolutionary unit
  c A clearer understanding of the causes of similarities and differences between taxa

It is evident that the new systematics is neither a special technique nor a special method but a viewpoint or attitude which can be applied at every taxonomic level. However, its major impact has been in microtaxonomy.

For a somewhat different analysis of the characteristic trends in taxonomy, see Simpson (1961:63–66).

THE SUPERSPECIES

Allopatric populations are often so distinct from each other that there is little doubt about their having reached the species level. Rensch (1929) proposed the German word *Artenkreis* for groups of allopatric species. Since the literal translation "circle of species" was frequently misunderstood, Mayr (1931) introduced the term *superspecies* as a convenient international equivalent. *A superspecies is a monophyletic group of closely related and largely or entirely allopatric species.*

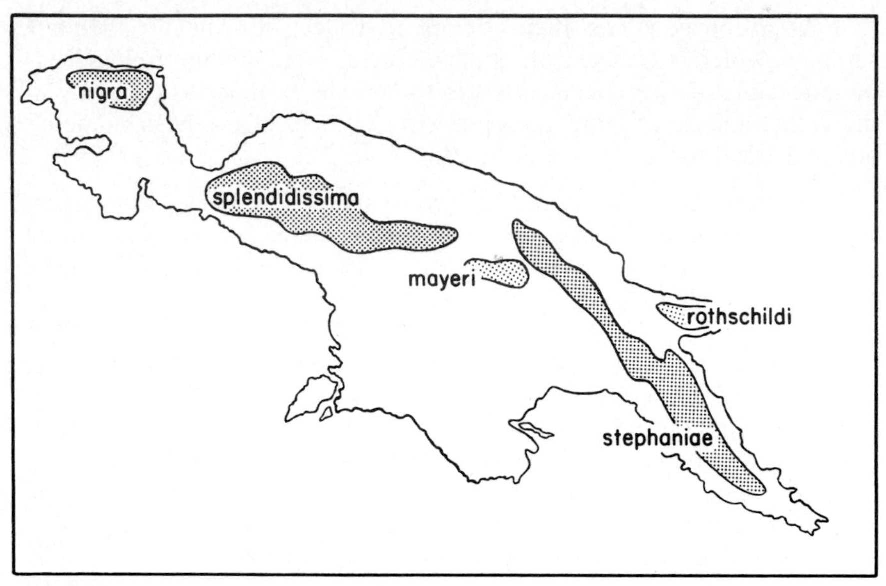

FIGURE 3-4
A superspecies of paradise magpies (*Astrapia*) in the mountains of New Guinea. Some hybridization has been recorded in the zone of contact between *mayeri* and *stephaniae*. (*From Mayr 1963.*)

When the ranges of the component species are plotted on a map, the superspecies usually presents the same picture that a polytypic species presents, yet there are three kinds of evidence to indicate that the component species have attained reproductive isolation. These species, although completely isolated from each other, may be morphologically as different as are sympatric species in the respective genus, they may be in geographic contact in some areas (*parapatry*) without interbreeding, or there may actually be a slight distributional overlap.

Superspecies are not given nomenclatural recognition but are listed as such in monographs and catalogues by an appropriate use of headings or symbols (Figure 3-4). They are important chiefly in zoogeographical and speciation studies. See Mayr (1963:499–501) for examples of superspecies and further discussion (see also Mayr and Short 1970).

The component species of a superspecies were originally designated *semispecies*. However, various authors have suggested a broadening of this term to include not only members of superspecies but all borderline cases in speciation (Mayr 1963:501). To maintain the original meaning, Amadon (1967) introduced the term *allospecies* for the component species of a superspecies.

CHAPTER **4**

INTRAPOPULATIONAL VARIATION AND THE COMPARISON OF POPULATION SAMPLES

The traditional method of classical taxonomy was to group individuals found at a given locality by similarity and call each similarity class a different species. This procedure was legitimate during the period when species status was determined solely by degree of morphological difference. As a consequence, Linnaeus described the brightly colored male mallard as *Anas boschas* and the drab female as *An. platyrhynchos,* and the barred adult goshawk as *Accipiter palumbarius* and the striped immature goshawk as *A. gentilis*. However, as soon as their biological status was recognized, each pair of morphological species was combined into a single biological species.

The male and the female mallard are two different phena, as are the adult and the immature goshawk. Indeed, the populations of most species of animals contain several different phena as a result of sexual dimorphism, age variation, seasonal variation, polymorphism, and other causes. One of the primary tasks of the taxonomist is to unmask such variants and assign them to the species to which they belong. Owing to the great variability of most species, this is by no means an easy task, as proved by the long list of synonyms in many groups of animals, names that were given to phena when they were believed to be different species.

Students of inanimate objects establish classes merely on the basis of similarity. Asked to sort them, they would not hesitate to place objects as different as the caterpillar, chrysalis, and imago stages of a butterfly in three different classes while also placing the adults of two similar species of butterflies in the same class. Biologists, however, know that they must take into consideration factors other than mere similarity. The continuity of the genotype from the fertilized egg through all the juvenile stages to the adult is one of these factors. *Sexual reproduction*—which results in genetic cohesion among all the individuals belonging to the same local population—is another. All members of a local population are products of the same gene pool and thus belong to a single taxonomic entity.

Knowledge of these biological phenomena permits the taxonomist to assign phena correctly to species. In a species with *sexual dimorphism,* for instance, males and females belong to two different phena. The fact that one phenon consists entirely of males while the other consists entirely of females suggests that these phena are not different species. Additional information should be brought to bear on the situation. For instance, if the two phena are the only ones in a collection made at a given locality that represent a certain genus and were collected simultaneously in the same habitat, the probability is high that they represent males and females of the same species. Breeding tests, the raising of young, and the study of courtship and copulation in nature furnish additional sets of biological information that permit the correct assignment of phena. Powerful evidence is often provided by proteins and other molecules when living material is available. The number of possible inferences that can be drawn from the available information is usually large. Knowledge of the nature and amount of sexual dimorphism found in living (recent) species often permits the correct assignment of fossil phena to species, for instance, among fossil ostracods and vertebrates.

To avoid premature decisions about which forms encountered in nature are genuine species rather than phena and to permit the inclusion of taxa of different rank in a single analysis, numerical pheneticists introduced the term *operational taxonomic unit* (*OTU*) (Chapter 8). It has turned out, however, that a clear-cut distinction between phena and species is absolutely necessary before a numerical analysis can be undertaken and that after the elimination of phena, the term *taxon* is fully sufficient when one is referring to taxa of different rank. For these reasons the term *OTU* is usually an unnecessary synonym of the term *taxon*.

The greater the amount of information available about the phena in question, the easier their classification. This information includes the correct locality, habitat (and other relevant ecological information), and season of capture. The reasons why precise information is needed are twofold:

1 Many aspects of ecology and life history are species-specific.

2 The phenotype of animal populations of the same species often varies according to locality, season, or habitat (see below).

Differences between phena thus may reflect either a species difference or intraspecific variation. A full understanding of intraspecific variation is therefore necessary before we can make the probabilistic statement that phenon *B* belongs to a species different from that of phenon *A*. This is the reason for the immense importance of a thorough understanding of individual and geographic variation.

The taxonomic literature still contains numerous named phena that have not been correctly combined into biological species. These phena include males and females in sexually dimorphic groups of insects, workers and sexual castes in social insects, stages in the life cycle of parasites, and juvenile stages as well as morphs. It is one of the continuing activities of the taxonomist to unmask nominal species that are not genuine biological species.

A distinction must be made between the analysis of phena from a single locality (sympatric phena) and that of phena from different localities (allopatric phena). Only sympatric phena are considered in this chapter.

When one is sorting specimens from a single locality, it must be remembered that one is potentially dealing with four possibilities (Table 4-1). Two of these (classes 1 and 4) pose no problems. However, difficulties arise when individuals are morphologically different but belong to the same species (class 2) or are morphologically identical (or exceedingly similar) yet belong to different biological species (class 3). Many errors in the taxonomic literature (synonyms) are due to the fact that individuals belonging to class 2 were considered to belong to class 4 or that individuals belonging to class 3 were considered to belong to class 1. Discrimination between classes 1 and 3 is discussed in Chapter 5. The information that must be obtained to permit assignment either to class 2 or to class 4 is discussed in this chapter. Since direct evidence of *reproductive isolation* is usually not available, it must be inferred from the pattern of variation. It is therefore necessary to undertake a detailed analysis of individual variation. Such variation is far greater than the beginner realizes, and it sometimes deceives even the experienced taxonomist.

TABLE 4-1
DISCRIMINATION GRID FOR SYMPATRIC SAMPLES

Morphology	Not reproductively isolated	Reproductively isolated
Identical	**1** Same phenon of a single species	**3** Sibling species
Different	**2** Different phena of the same species	**4** Different species

SYMPATRIC SAMPLES

In a study of several phena from the same locality, only two alternatives are possible. Either the phena are individual variants of a single species or they belong to different species. There are three reasons why it may be difficult to make a ready choice between these two possibilities: (1) an extreme difference in phena belonging to a single species (e.g., caterpillar and butterfly); (2) an extreme similarity of good biological species (sibling species); and (3) extreme variability and wide phenotypic overlap of two species. Clues as to whether only a single species is involved in spite of extreme phenotypic difference, or, rather, indications that more than one species is involved in spite of extreme similarity, are usually provided by behavioral, ecological, or distributional data. Molecular or chromosomal data, breeding tests, and other evidence must then be obtained to confirm or refute the earlier supposition.

PHENA (INDIVIDUAL VARIANTS) OR DIFFERENT SPECIES?

Most species contain several phena if not dozens. To add to the complexity of the situation, several other species with a similar assortment of phena may be sympatric. Often a phenon of one species resembles a corresponding phenon of another species much more closely than it does any other phenon of the same species. For instance, the female in some species of many bird and insect genera is more similar to females of closely related species than it is to males of her own species. Nothing in the phenotype of caterpillars permits correct association with imagoes. Only observation of breeding or the careful evaluation of other biological information can do this.

The correct assignment of phena often can be achieved by a correct interpretation of morphological information. If a large sample of a population is available, forms intermediate between the more extreme variants are usually present. Also, certain characters in every group are less subject to individual variation than are others. The genitalic armature in most insects, the palpus in spiders, the radula in snails, and the structure of the hinge in bivalves are examples of such stable characters. When several sympatric phena agree in their genitalic armatures or in one of the other characters mentioned above, it adds to the probability that they are conspecific. However, even here one has to apply balanced judgment. Although in most genera of Diptera there are characteristic differences between the genitalia of related species, there are cases in which forms have identical genitalia even though they are different species by every other criterion (e.g., *Drosophila pseudoobscura* and *D. persimilis*). Similarly, in many other orders of insects closely related species may have indistinguishable genitalia. Parasitic animals present some rather special problems (Manwell et al. 1957).

The establishment of correlations is often very helpful. If two forms that differ in character *a* can be shown to differ also in the less conspicuous and functionally unrelated characters *b, c,* and *d* (principle of covariation), it is very probable that they are different species. Some years ago Mayr found that among birds identified as the southeast Asiatic minivet (*Pericrocotus brevirostris* Vigors), some had innermost secondaries that were all black while others had a narrow red margin on those feathers. A detailed study revealed that birds with red on the innermost secondaries had seven additional minor characters: a more yellowish red of the underparts, a different distribution of black and red on the second innermost tail feather, a narrow whitish margin along the outer web of the first primary, and four other minor characters. Slight though they all were, these characters were well correlated with each other and with geographic and altitudinal distribution. The conclusion that two full species are involved has since been confirmed by several authors.

As a general rule, one finds that the decision of an experienced taxonomist, when based on a careful evaluation of the morphological evidence, is confirmed when a species recognized by that taxonomist is subjected to genetic tests or to an evaluation of molecular and other nonmorphological characters. Various forms of individual variation are listed in Table 4-2. The two most important types are nongenetic and genetic variation.

NONGENETIC VARIATION

It is of course impossible in a preserved museum specimen to determine directly whether a given variant has a genetic basis. Nevertheless, it is important for the taxonomist to understand that many types of variation exist and that in better-known groups it is usually possible to make a valid inference about the status of a given variant on the basis of field observations and available experimental evidence. For a discussion of the evolutionary aspects of individual variation, see Mayr (1963:138–158).

Animals as a whole are developmentally more strongly canalized than are plants and thus are less subject to nongenetic modification. In addition, through their power of locomotion and sensory abilities, they have the capacity for habitat selection. As a result, some well-known exceptions notwithstanding, nongenetic changes of the phenotype are far less of a problem for animal taxonomists than for plant taxonomists. However, every zoologist must be familiar with the types of nongenetic variability that may be encountered in a particular group.

TABLE 4-2
MAJOR TYPES OF VARIATION WITHIN A SINGLE POPULATION

1. Nongenetic variation
   a. Individual variation in time
      (1) Age variation
      (2) Seasonal variation in an individual
      (3) Seasonal variation in consecutive generations
   b. Social variation (insect castes)
   c. Ecological variation
      (1) Habitat variation (ecophenotypic)
      (2) Variation induced by temporary climatic conditions
      (3) Host-determined variation
      (4) Density-dependent variation
      (5) Allometric variation
      (6) Neurogenic or neurohumoral variation
   d. Traumatic variation
      (1) Parasite-induced variation
      (2) Accidental and teratological variation
2. Genetic variation
   a. Sexual dimorphism
      (1) Primary sex differences
      (2) Secondary sex differences
      (3) Gynandromorphs and intersexes
   b. Reproductively different generations (sexual and uniparental strains)
   c. Ordinary genetic variation
      (1) Discontinuous variation (genetic polymorphism)
      (2) Continuous variation

Individual Variation in Time

Age Variation Whether they are born more or less developed or hatch from an egg, animals in general pass through a series of juvenile or larval stages in which they may be quite different from their adult forms. The catalogues of any group of animals list numerous synonyms that resulted from the failure of taxonomists to recognize the relationship between various age classes of the same species.

In reptiles, birds, and mammals there are no larval stages, but immature individuals may be rather different from adults, particularly in birds. Several hundred bird synonyms are based on juvenile plumage. Finding specimens that molt from the immature into the adult plumage usually clears up this difficulty.

In many fishes the immature forms are so different that they have been described in different genera or even different families. The immature stages of the eel (*Anguilla*) were originally described as *Leptocephalus brevirostris* Kaup. The unmasking may be especially difficult in *neotenic* animals, that is, animals that become sexually mature in a larval stage.

The difficulties for the taxonomist are even greater in groups with larval stages so different that they have not the faintest resemblance to the adult (e.g., caterpillar and butterfly). The floating or free-swimming larvae of sessile coelenterates, echinoderms, mollusks, and crustaceans are often extremely different from the adults. The taxonomic status of such larval stages can be settled either by assembling a complete sequence of intermediate stages or by rearing an organism through a complete life cycle.

The taxonomic identification of larval stages of parasites is particularly difficult in groups in which the different stages occur on different hosts. It is customary in helminthology to assign formal taxonomic names to the larval (*Cercaria*) stage of flukes (trematodes) in order to facilitate their identification (Figure 4-1). Such dual nomenclature is dropped as soon as it becomes known to what trematode species a given cercaria belongs. Such a relationship can usually be established only through rearing.

Age variation is not restricted to differences between larval stages and adults but occurs also between "young" and "old" adults. For example, in various species of deer (*Cervus*, etc.) older stags often have antlers with more points than those of younger stags. The shape of the antlers may also change. This age variation must be considered when the antlers of different species or subspecies are compared. There is probably no further addition of points (or only an irregular one) after a certain age has been reached. It would be as futile to try to determine the exact age of a stag by the number of points of its antlers as it would be to attempt to determine the age of a rattlesnake (*Crotalus*) by the number of rings in its rattle.

The taxonomist aims to work with samples that are as homogeneous as possible. It is much easier to achieve this in animals that have a definite adult stage (after the larval ones) than it is in those which show continuous growth, such as snakes and fishes, which may reach maturity after having attained only half their potential size or less. In such forms it is advisable to work with regressions rather than with absolute measurements. In many species, *meristic* characters (e.g., number of scales or fin rays) do not increase after they are formed in spite of the enormous subsequent growth. These characters are therefore especially useful in herpetology and ichthyology.

Seasonal Variation in an Individual In animals that live as adults through several breeding seasons, the same individual sometimes has a very different appearance in different parts of the year. Many birds have a bright nuptial dress which they exchange for a dull ("eclipse") plumage at the end of the breeding season. Among North American examples are many ducks, shore birds, warblers, and tanagers. In many species a change of plumage is restricted to the males.

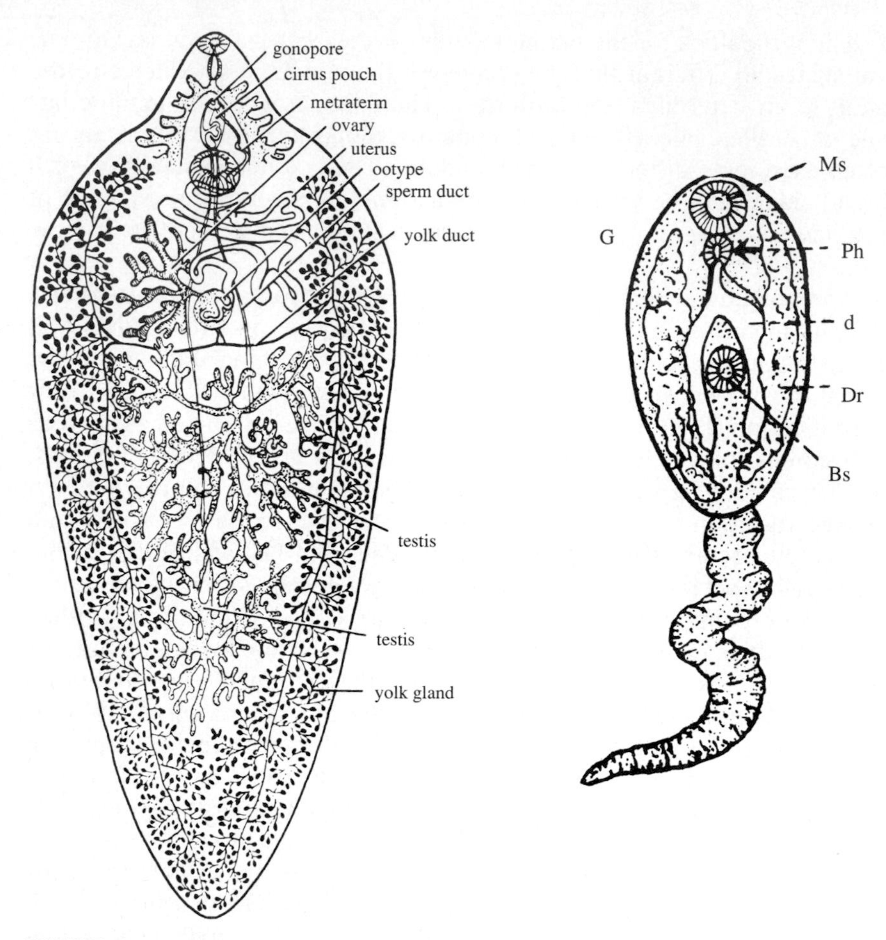

FIGURE 4-1
Difference between the adult liver fluke (*Fasciola hepatica*) and its larval stage (*Cercaria*). (*After Chandler and Read.*)

In arctic and subarctic birds and mammals such as ptarmigans (*Lagopus*) and weasels (*Mustela*), there may be a change from a cryptic white winter dress to a "normally" colored summer dress. In other birds the colors of the soft parts change with the seasons. In the common egret (*Ardea alba* Linnaeus) and the European starling (*Sturnus vulgaris* Linnaeus), the bill may change from yellow to black. Plumage changes in birds are usually effected by molting, but wear alone may produce striking changes. In the European starling (*Sturnus vulgaris*), for example, the freshly molted bird of October is covered with white spots and all the feathers show whitish or buffy margins. During the winter the pale edges of the feathers wear off, and in the spring, at the beginning of the breeding season, the whole bird is a beautiful glossy black without its having molted a single feather. A similar process of wear brings out the full colors of the nuptial plumage in the males of many other birds. In arid regions, particularly in real deserts, the sun bleaches the pigments. A bird before the molt looks much paler than does one in freshly molted plumage.

In all these cases it is the same individual that looks very different in different parts of the year. Such seasonal variation is particularly common among vertebrates, with their elaborate endocrine systems. Many seasonal variants were described as distinct species before their true nature was recognized.

Seasonal Variation in Consecutive Generations Many species of short-lived invertebrates, particularly insects, produce several generations in the course of a year. In such species it is not uncommon for the individuals that hatch in the cool spring to be quite different from those produced in the summer or for the dry-season individuals to be different (e.g., paler) from the wet-season population. Some tropical butterflies may have very different dry-season and wet-season phena.

Such seasonal forms can usually be recognized not only by the occurrence of intermediates in the intervening season but also through identity of wing venation, genitalia, etc.

Cyclomorphosis A special kind of seasonal variation is found in certain freshwater organisms, particularly rotifers and cladocerans. The populations of a species undergo regular morphological changes through the seasons in connection with changes in the temperature, turbulence, and other properties of the water. Many "species" have been named, particularly in the genus *Daphnia,* that are nothing but seasonal variants. In rotifers, different morphs may also result from different kinds of food. For a recent review, see Kerfoot (1980).

Social Variation (Insect Castes)

In the social insects, such as some bees and wasps, but particularly among ants and termites, castes have developed. These are definite groups of individuals within a colony in addition to the reproductive castes (queens and males or drones); there are workers (sometimes of different types) and soldiers (also sometimes of different types). In the Hymenoptera these castes are most commonly modified females and are genetically identical, but in the Isoptera (termites) both sexes may be involved. The structural types observed may result from different larval food or may be due to hormonal or other controls. Obviously, taxonomic names should not be applied to these intracolonial variants, but invalid species have sometimes been described because it was not realized that there were different types of soldiers or workers in the same colony.

Ecological Variation

Habitat Variation (Ecophenotypic) Populations of a single species that occur in different habitats in the same region are often visibly different. The taxonomic treatment of such local variants has fluctuated between two extremes: Some authors have described them as different species, while others have considered them all to be nongenetic variants. Actually they may be (1) microsubspecies (or ecological races) or (2) nongenetic ecophenotypes. The latter are particularly common in plastic species, such as some mollusks.

Dall (1898) gave a very instructive account of all the variations he observed in a study of the oyster (*Crassostrea virginica* Gmelin). Such habitat forms are particularly common in freshwater snails and mussels. The upper parts of rivers, with cooler temperatures and a more rapid flow of water, have different forms than do the lower reaches, which have warmer and more stagnant waters. In limestone districts the shells are heavy and of a different shape from those which grow in waters low in lime. This dependence of certain taxonomic characters of mussels on environmental factors was, curiously enough, entirely overlooked by some earlier workers, a fact which resulted in absurd systematics. Schnitter (1922:4–5), who largely cleared up the situation, described these absurdities as follows:

> The last step in the splitting of the freshwater mussels of Europe was done by the malacozoologists Bouguignat and Locard. According to the shape and the outline of the shell, they split up the few well-known species into countless new ones. Locard lists from France alone no less than 251 species of *Anodonta*. On the other hand, two mussels were given the same name, if they had the same outline of the shell, even though one may have come from Spain and the other from Brittany. It seems incredible to us that it never occurred to these authors to collect a large series at one locality, to examine the specimens, to compare all the individuals and to record the intermediates between all these forms. It is equally incomprehensible that these people did not see the correlation between environment and shape of shell, even though they spent their entire lives in collecting mussels.

All these "species" of *Anodonta* are now considered to be habitat forms of two species, and the other names have sunk into the synonymy of the two valid species.

Whether a given habitat form is an *ecophenotype* or a *microgeographic race* is not always evident. It is sometimes necessary to transplant the organism or raise it in the laboratory to answer this question. Much work of this sort remains to be done.

Variation Induced by Temporary Climatic Conditions Some animals with a highly plastic phenotype may produce year classes that differ visibly from the norm owing to unusual conditions (drought, cold, food supply, etc.) in a given year. Fish of a given year class may be stunted or, on the contrary, may have proportions indicative of particularly rapid growth. Samples of susceptible species must be collected in a way that will compensate for distortions caused by this factor (Harrison 1959; Mayr 1963:145).

Host-Determined Variation Host-determined variation in the parasites of plants and animals provides a potential source of taxonomic error and permits confusion with microgeographic races or sympatric species. This phenomenon is most commonly expressed in size differences but may involve other morphological or physiological characters.

Host-induced variation is not as common as was formerly believed. Many so-called host races in the literature have been found in recent decades to be valid sibling species (Mayr 1963, Chapter 3).

Density-Dependent Variation The effects of crowding are sometimes reflected in morphological variation, especially when crowding leads to a shortage of food materials. However, density-dependent variation need not be related to food supply. Uvarov, Kennedy, and others have shown that gregarious species of locusts exist in various unstable biological phases (Kennedy 1961; Albrecht 1962). These phases differ in anatomy, color, and behavioral characteristics and have often been described as distinct species. When newly hatched nymphs are reared under crowded conditions, they develop into the transitional phase; when they are isolated and reared separately, they develop into the solitary phase. Similar phases have been reported for two species of armyworms (Lepidoptera), *Laphygma exigua* (Hübner) and *L. exempta* (Walker).

Allometric Variation Allometric growth may result in the disproportionate size of a structure in relation to the size of the rest of the body. If individuals in a population show allometric growth, animals of different size will show allometric (heterogonic) variability. This phenomenon is particularly marked among insects. It involves such features as the heads of ants (Figure 4-2), the mandibles of stag beetles (Lucanidae), the frontal horns and thoraxes of scarabs, and the antennal segments of thrips. Failure to recognize the nature of such variations has resulted in much synonymy.

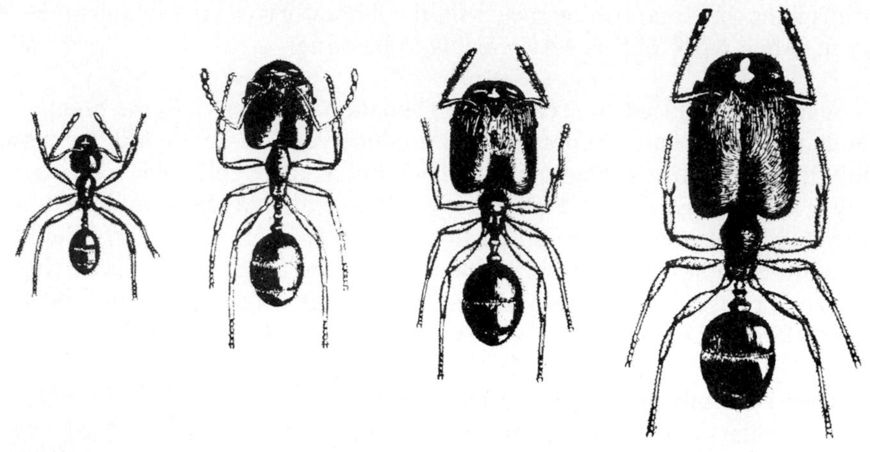

FIGURE 4-2
Allometric variability. Neuters of *Pheidole instabilis,* showing an increase in the relative size of the head with absolute size of the body. (*After Wheeler 1910.*)

Neurogenic or Neurohumoral Variation Neurogenic or neurohumoral variation refers to color change in individual animals in response to the environment. These changes are accomplished through the concentration or dispersal of color-bearing bodies known as *chromatophores*. This type of variation was first thoroughly studied in the chameleon. It occurs sporadically in the lower animals but is best developed among crustaceans, cephalopods, and cold-blooded vertebrates (cyclostomes, elasmobranchs, teleost fishes, amphibians, and reptiles). Space does not permit a discussion of this specialized type of variation. The reader is referred to Fingerman (1963), Gersch (1964), and Waring (1963) for details.

Traumatic Variation

Traumatic variation occurs with varying frequency in different groups of animals. The abnormal nature of this type of variation is usually obvious, but in some cases it is subtle and may be misleading.

Parasite-Induced Variation Aside from such familiar effects of parasitism as swelling, distortion, and mechanical injury, parasites may produce conspicuous structural modifications. In the bee genus *Andrena,* for instance, parasitism by *Stylops* frequently results in a reduction in the size of the head, enlargement of the abdomen, and changes in puncturation, pubescence, and wing venation. It also commonly results in intersexes. Since *Andrena* is markedly sexually dimorphic, these intersexes have been a source of taxonomic confusion and synonymy. However, in one case (Linsley 1937), a stylopized intersex proved to be of value in associating the sexes of a bee that had been described as two different species.

Salt (1927) has conducted the most comprehensive study of the morphological effects of stylopization in *Andrena*. In females he found reduction of the pollen-collecting organs, loss of anal fimbriae, changes in the relative length of antennal segments, reduction of facial foveae, reduction of the sting and accessory organs, paling of ventral abdominal pubescence, acquisition of angular cheeks, and yellow on the normally dark clypeus. In males he reported the development of long hairs resembling those of the female, flocculi, broadening of the posterior basitarsus, changes in proportions of antennal segments, loss of cheek angles and some yellow from the clypeus, indications of facial foveae, and reduction in the size of genitalia.

Some strikingly different termite soldiers from the orient were assigned to a new genus and species, *Gnathotermes aurivillii*. Later it was shown that these modified soldiers are nothing but parasitized individuals from colonies of *Macrotermes malaccensis* (Haviland).

Postmortem Changes

The taxonomist must be aware of one further type of individual variation. In many groups of animals it is impossible to prevent postmortem changes of preserved specimens. Some extreme cases are known in birds. The deep orange-yellow plumes of the twelve-wire bird of paradise (*Seleucidis ignotus* Forster) fade to white in collections. The plumage of the Chinese jay (*Kitta chinensis* Boddaert), which is green in life, turns blue in collections because of the loss of the volatile yellow component in the pigment. Many birds that are clear gray or olive-gray when freshly collected become more and more rufous through oxidation of the black pigment ("foxing"). Numerous synonyms have been created in ornithology through the comparison of freshly collected material with old museum specimens.

Other postmortem changes result from the chemical action of preservatives or killing agents. A common color change of this nature takes place when certain yellow insects, especially wasps, are overexposed to cyanide. The specimens turn bright red, and no method has been found for reversing this reaction. When one is preserving specimens with evanescent colors (corals, marine slugs, etc.), it is essential to take full notes and preferably color photographs or make color sketches. These records will allow an accurate description of the living animal.

GENETIC VARIATION

In the cases of variation discussed in the preceding sections, the same individual is actually or potentially subject to a change in appearance. Other kinds of intrapopulation variation are due to differences in genetic constitution. This genetically induced individual variation can somewhat arbitrarily be divided into three classes.

Sexual Dimorphism

Among genetically determined variants within a population, many of the variants are sex-associated. They may be sex-limited (express themselves in one sex only) or be otherwise associated with one sex or the other. Some of these are as follows.

Primary Sex Differences These are differences involving the primary organs utilized in reproduction (gonads, genitalia, etc.). When the two sexes are otherwise quite similar, primary sex differences are rarely a source of taxonomic confusion.

Secondary Sex Differences There is more or less pronounced sexual dimorphism in most groups of animals. The differences between males and females are often very striking, as in birds of paradise, hummingbirds, and ducks. In many cases the different sexes were originally described as different species and retained this status until painstaking work by naturalists established their true relationship. A celebrated case is that of the king parrot, *Eclectus roratus* (Muller), of the Papuan region, in which the male is green with an orange bill and the female is red and blue with a black bill. Described in 1776, the two sexes were considered different species for nearly 100 years, until naturalists proved conclusively in 1873 that they belonged together.

Striking sexual dimorphism is particularly common in the Hymenoptera. The males of the African ant *Dorylus* are so unlike other ants that they were not recognized as ants and were for a long time considered to belong to a different family. In certain families of wasps the small wingless female and the large winged male are so different that some taxonomists use a different nomenclature for the two sexes. Whole "genera" consist entirely of males; others, of females. The best way of determining with which female of genus *B* a given male in genus *A* belongs is to find a pair *in copula* or to watch a female in the field and catch the males as they are attracted to her. Once it has been established that *B* is the female of *A*, it is sometimes possible to associate several other "species pairs" in the same genus by utilizing additional information on distribution, frequency, color characters, and so on.

Gynandromorphs and Intersexes *Gynandromorphs* are individuals that show male characters in one part of the body and female characters in another part (Figure 4-3). Thus the two halves of the body may be of opposite sexes (this is the most conspicuous example of gynandromorphism), the division may be transverse, or the sex characters may be scattered in a mosaic. In the latter case symmetrical variants may be produced. Usually gynandromorphs are easily recognized as such and rarely provide a source of taxonomic confusion. Gynandromorphism is produced by an unequal somatic distribution of chromosomes, particularly sex chromosomes.

Unlike gynandromorphs, *intersexes* are likely to exhibit a blending of male and female characters. Intersexes are generally thought to result from an upset in the balance between male-tendency and female-tendency genes. This upset may result from irregularities in fertilization or mitosis or from physiological disturbances associated with parasitism. Intersexes are particularly likely to appear in populations of interspecific or intersubspecific hybrids. They have been studied in greatest detail in *Lymantria* (Goldschmidt 1933) but are well known in many other animals.

FIGURE 4-3
Gynandromorph of *Papilio dardanus*; left wing female, right wing male. (*From Wells and Huxley.*)

Reproductively Different Generations

In many insects there is an alternation of generations that is very confusing to the taxonomist. In the genus *Cynips* (gall wasps), the agamic generation is so different from the bisexual one that it has been customary to apply different scientific names to the two (Kinsey 1930). In the aphids (plant lice) the parthenogenetic wingless females are usually different from the winged females of the sexual generations (Figure 4-4).

Sexual and Uniparental Strains In some species of slugs (*Arion*) several strains are found which differ in their sexuality. The sexually reproducing strains are highly polymorphic in their enzymes, while self-fertilizing hermaphroditic strains are highly homozygous (Foltz et al. 1982).

Ordinary Genetic Variation

Much intrapopulation genetic variation is not sex-limited and does not primarily involve sex characters.

Discontinuous Variation (Polymorphism) The differences between individuals of a population are in general slight and intergrading. In certain species, however, the members of a population can be grouped into very definite classes that are determined by the presence of certain conspicuous characters. Such discontinuous individual variation is called *polymorphism*. Such polymorphism is frequently controlled by a single gene that is subject to simple Mendelian inheritance.

Polymorphism is more pronounced in some groups of animals than it is in others. In many species of Hemiptera and Coleoptera the same population may contain flying and flightless (wingless or short-winged) individuals. It is evidently advantageous to have both residents and dispersers in such populations. The spotting in lady beetles (Coccinellidae) is a well-known example of genetic polymorphism, as is industrial melanism in moths. Polymorphism has great biological importance because it proves the existence of selective differences between apparently neutral characters. For a more detailed discussion of this topic, see Ford (1945, 1965) and Mayr (1963). The practical importance of polymorphism to the taxonomist lies in the fact that it has led to the description of many so-called species that are actually polymorphic variants (morphs). In ornithology alone, about 100 species names were given to morphs. The recognition of their true nature has led to a considerable simplification of taxonomy.

Perhaps the most spectacular cases of polymorphism are to be seen in the Lepidoptera, particularly in certain species of butterflies. The com-

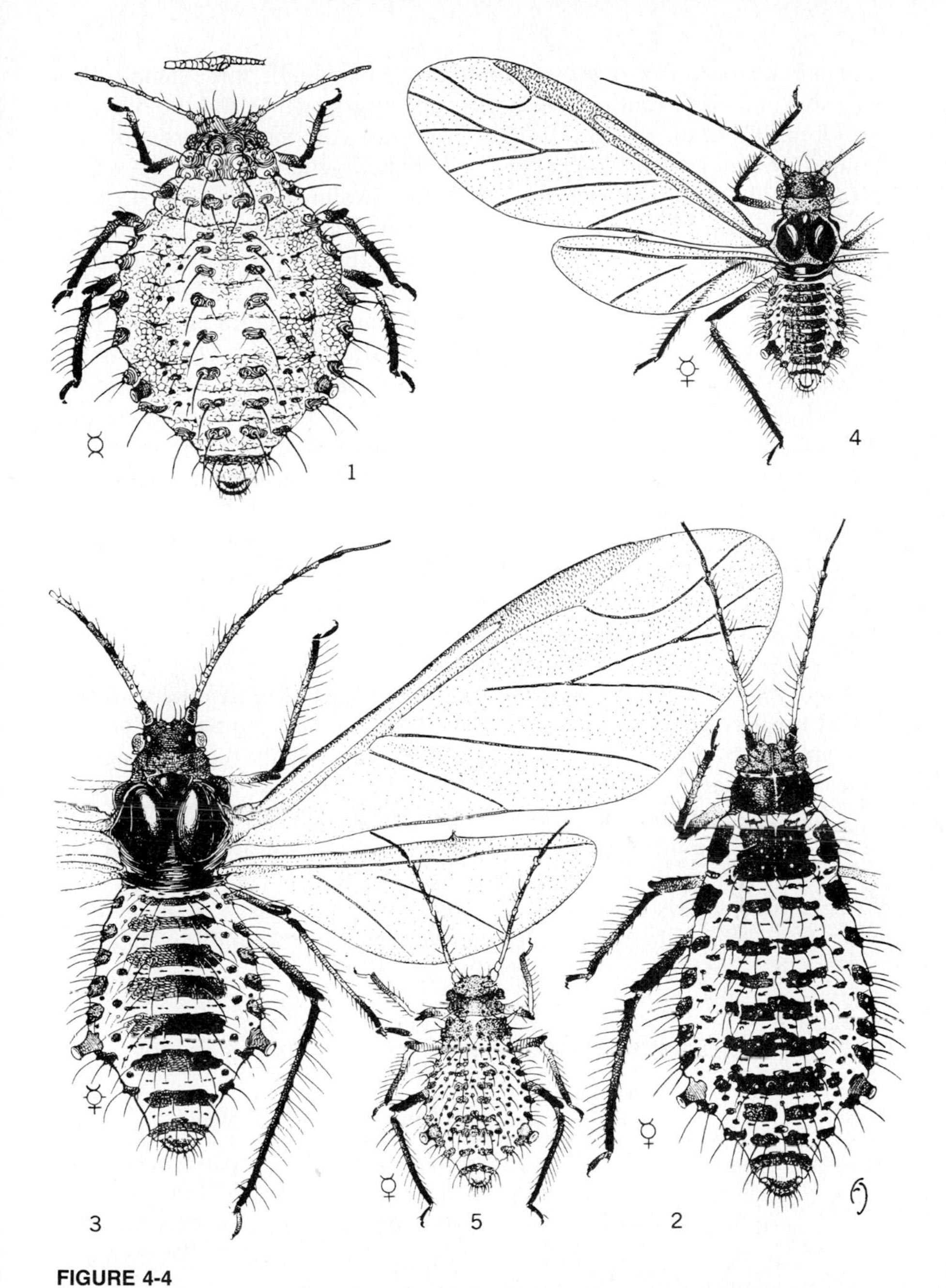

FIGURE 4-4
Periphyllus californiensis (Shinji). 1 = fundatrix or stemmother; 2 = normal apterous parthenogenetic viviparous female; 3 = alate of same; 4 = smallest spring alate viviparous female. (*From Essig and Abernathy 1952.*)

mon alfalfa butterfly, *Colias eurytheme* (Boisduval), for example, has two strikingly different female forms, one largely white and the other resembling the orange-colored male. The most complicated cases of sex-limited polymorphism that have been studied genetically are the examples of mimetic polymorphism in African swallowtail butterflies of the genus *Papilio*. Quite apart from the fact that allopatric populations throughout Africa show distinct subspecific differences that are correlated with differences in the species of the butterflies that they mimic, several distinct female forms may exist within a single population. Thus in west Africa one finds, in the same population of *P. dardanus* Brown, one male form and five female forms, with three of the female forms mimicking different models that belong to the families Danaidae and Nymphalidae (Table 4-3 and Figure 4-5). The most remarkable feature of this polymorphism is that although the various forms are sufficiently distinct to resemble representatives of three different families of Lepidoptera, breeding experiments have shown that the differences are caused by a few Mendelian genes. Other celebrated cases of mimicry are those of the butterfly *Pseudacraea eurytus* Linnaeus (Carpenter 1949) and of the species of *Heliconius* (Turner 1981).

Continuous Variation The most common type of individual variation is due to slight genetic differences among individuals. No two individuals (except monozygotic twins) in a population of sexually reproducing animals are exactly alike genetically or morphologically. One of the outstanding contributions of population genetics has been the establishment of this fact. The differences are in general slight and are often not discovered unless special techniques are employed.

The study of this variation is one of the foremost tasks of the taxonomist. It is now evident that no single individual is "typical" of the char-

TABLE 4-3
MIMETIC POLYMORPHISM IN WEST AFRICAN *PAPILIO DARDANUS* BROWN

Male	Nonmimetic females	Mimetic females	Models
Typical *dardanus*		*hippocoon* Fabricius	*Amauris niavius* Linnaeus
	dionysus Doubleday and Hewitson	*trophonissa* Aurivillius	*Danaus chrysippus* Linnaeus
		niobe Aurivillius	*Bematistes tellus* Aurivillius

Source: From Goldschmidt 1945.

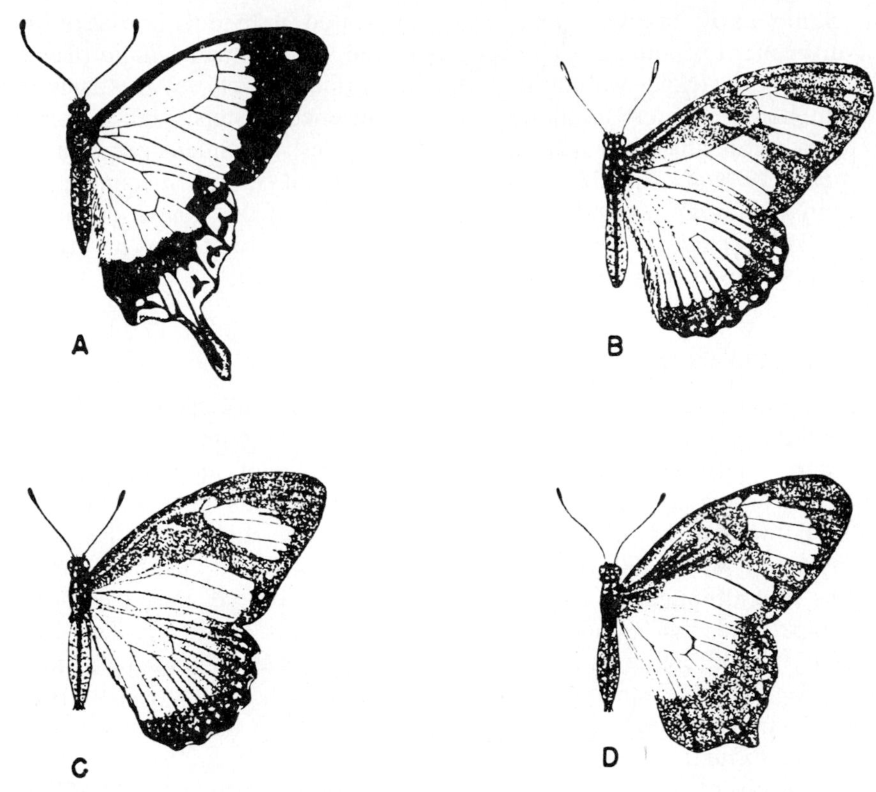

FIGURE 4-5
Mimetic polymorphism in the *Papilio dardanus* complex. (A) Male of *cenea*—also basic type of nonmimetic female, ground color yellow; (B) *dionysus,* nonmimetic female, ground color of forewings white, hind wings yellowish; (C) *trophonissa,* mimetic female, ground color for forewings white, hind wings brownish; (D) *hippocoon,* mimetic female, ground color white. (*Redrawn from Eltringham 1910 by Goldschmidt 1945.*)

acters of a population. The continuous variation among the members of a population manifests itself most conspicuously in linear measurements and proportions. Mean values, variances, and coefficients of variability for each trait are characteristic of each population and species. A model study of variability based on 2877 skins of the house sparrow (*Passer domesticus*) has been presented by Selander and Johnston (1967).

Each character may show a different degree of variability within a single population. Similarly, there are different degrees of variability among related species. Just why one species should be highly variable while another one is not is not always clear. A taxonomist who has adequate material for one species should not hastily assume that this makes it possible to be certain of the variability of related species.

Some early taxonomists vastly underrated individual variation in many genera of animals. The species of the snail genus *Melania* (freshwater and brackish water) have been described largely on the basis of such shell characters as the presence or absence of spines and diagonal versus spiral ribs. However, spined and spineless specimens occur in *M. scabra, M. rudis,* and *M. costata,* sculptured and smooth specimens occur in *M. granifera,* and so forth. In a revision of this genus, no fewer than 114 "species" were found to be individual variants and had to be added to the synonymy of other species (Riech 1937).

THE COMPARISON OF POPULATION SAMPLES

Taxonomic research at the species level consists to a large extent in the comparison of populations, or, more correctly, the comparison of samples from different populations. It is on the basis of such comparisons that important taxonomic decisions are made, such as whether the samples belong to the same species or whether they belong to the same subspecies or two different ones. Very often it is found in such comparisons that although the samples are somewhat different, they cannot be distinguished by a single all-or-none diagnostic character. In other words, the difference is only quantitative. Sample B differs from sample A by being lighter or more greenish, by being more distinctly patterned with somewhat larger extremities, or by any of the many thousands of possible differences among populations.

A purely qualitative statement such as "more hairy" or "larger" is unsatisfactory and indeed useless to another taxonomist who does not have the same material available. This is why taxonomists attempt to express all gradual or relative differences among taxa (and phena) as much as possible in quantitative terms. To do so, they make measurements of dimensions as well as counts of items showing numerical differences (numbers of teeth, scales, segments, etc.).

The replacement of typological thinking by population thinking has had a far-reaching impact not only on the concepts of taxonomy but also on its methods. Considering taxa as populations and aggregates of populations led automatically to a statistical approach. The taxonomist deals with samples from natural populations and can make estimates on the characteristics of these populations only with the help of statistics.

A quantitative analysis adds greatly to the precision of a description. Actual measurements of a series of specimens are required to give meaning to the statement "of medium size." "Sex comb with seven teeth" is more precise than is "sex comb present." Such precision is all the more important when related species differ not by the presence or absence of a character but rather by its size, proportions, or number. No concrete statement on the degree of difference between two populations or two other taxa can be made until the amount of variation within the compared entities has been determined.

Statistical methods should be used whenever they can contribute to taxonomic analysis. They are not to be used as window dressing. Taxonomic papers occasionally include highly elaborate statistics that make no contribution to the taxonomic analysis; such calculations are a waste of time. Statistics cannot improve heterogeneous original data or unreliable measurements. The simplest method yielding the desired information is always the best.

A detailed presentation of the principles of statistics and the application of the various statistical methods is beyond the scope of this manual of taxonomy. None of the available textbooks is entirely satisfactory, particularly for multivariate methods. The best for the everyday needs of the animal taxonomist is still *Quantitative Zoology* by Simpson, Roe, and Lewontin (1960), referred to in this text as SRL. A far more extensive treatment is provided by Sokal and Rohlf (1980), but no truly comprehensive treatment of multivariate methods is available.

Statistical analysis may provide important information on the nature of data. Experience has demonstrated that morphological measurements usually show a normal distribution. A strong deviation from normality requires that we examine the material for bias and the possibility of heterogeneity (see statistics texts). Statistical analysis may also provide important information about the weight we should assign to certain characters. Highly variable characters as well as some that are closely correlated with other characters are given low weight in classification (for weighting, see Chapter 7).

STATISTICAL ANALYSIS

While this book cannot present what is covered in a formal statistics text, we must discuss the recommendation that the samples used by taxonomists should be as homogeneous, adequate, and unbiased as possible.

The sample should not be *biased*; that is, the method of collecting the sample should ensure that variations of the pertinent characters in the sample occur at the same frequency with which they occur in the population. Fossils, for instance, are sometimes deposited according to size, and a sample from one of these size classes is not an unbiased representation of the population from which it is drawn. Collectors sometimes concentrate on unusual specimens and thus introduce bias. Instead, they should adopt the same safeguards of randomization used by experimentalists and pay due respect to the variation introduced by locality, season, and time of day. One should never discard part of a collection and keep only those specimens which seem either typical or particularly interesting because they are atypical. In polymorphic populations special efforts should be made to collect the different morphs in their true population frequency. To reduce collecting bias, it is often advisable to employ several different collecting techniques at the same locality.

A heterogeneous sample can often be segregated into smaller homogeneous samples by separating the specimens according to age, sex, locality, or other factors that may have introduced heterogeneity. To avoid bias, great care must be taken in segregating a sample (see below). Homogeneity is particularly important in comparative studies since samples that differ in their components because of heterogeneity cannot be legitimately compared.

What constitutes an adequate sample depends on the nature of the studied taxon and the objective of the investigation. In the case of the living coelacanth (*Latimeria*), a single specimen was adequate to prove that this was a new species and genus of a class of vertebrates believed to have been extinct for more than 70 million years. In some groups of animals diagnostic species characters are well defined in only one sex; the description of a new species may not be feasible if the collection contains only representatives of the other sex.

The study of polytypic species requires a comparison of samples from different populations that may differ only in quantitative characters. Large samples are a necessity in such studies. Large samples are also needed for the study of variation in polymorphic species. As a general rule one can say that an adequate sample is one that allows a reasonable estimate of the total variability of a species. When knowledge of a group has reached the degree of maturity where taxonomic analysis can concentrate on a study of individual and geographic variation of species, the availability of large samples for study is a necessity. Knowledge of the variability of species is valuable not only for the taxonomist but for anyone dealing with the biology of species. Evolutionists in particular, but also ecologists and population biologists, are interested in the nature and extent of variation within and between populations of species.

Measurements and Counts

Only quantitative data can be subjected to statistical analysis. In this fact lies the importance of characters that can be counted or measured. Meristic (countable) characters, such as number of spines, scales, or fin rays, permit greater accuracy than do measurements. They are therefore favored wherever possible by students of echinoderms, fishes, and reptiles. SRL (pp. 20–30) lists various requirements of good measurements. It is most important that measurements be standardized (applying to a specified distance) and accurate. For instance, the length of the bill in birds may be measured in several ways: (1) from the nostril to the tip, (2) from the beginning of the feathering to the tip, or (3) from the beginning of the bony forehead to the tip. Observations have shown that the first item can be measured very accurately but does not give the full length of the bill, the third can be measured fairly accurately in all birds with a steep forehead, and the second can rarely be measured with any degree of accuracy. Consequently, in some genera of birds the third item is the preferred measurement, whereas in other genera it is the first. In measuring bills and in all similar cases, the record should show which of several possible measurements was actually taken.

In measuring an important lot of specimens or measuring specimens before the method of doing so has been completely standardized, one should measure each variate repeatedly. The duplicate sets of measurements should be taken on different days and entered on new record sheets. When completed, the various sets of measurements should be compared and averaged. Measurements that deviate strongly should be checked for possible errors in the measuring technique or in recording.

Different measurements are used for nearly every category of animal. In mammals, for instance, body and tail length are measured, along with the length of the hind foot and ear and the various dimensions of the skull. In birds, wing, tail, bill, and tarsus are the most commonly measured variates. In most groups of insects not only length but also width and antennal and tarsal formulas should be given. These data should be recorded routinely regardless of their immediate diagnostic value. Special measurements are traditionally given in particular taxonomic groups, for example, the length of the rostrum in Hemiptera and the length of the wings in some Diptera. It is important for comparative purposes to give measurements that conform to the system that is customary in the group under study and to present them in a standardized sequence.

It is important to carry measurements only to the degree of refinement that is appropriate, not to waste effort in striving for meaningless accuracy. It would be useless, for example, to give the height of a person as 176.583 cm. How, then, should the proper degree of refinement be decided? The recommended unit of measurement is one-twentieth of the difference between the largest specimen and the smallest if an adequate series is available. Thus if the measurements range from 10 to 12 mm, one should measure to 0.1 mm; if they range from 40 to 50 mm, one should measure to 0.5 mm. If they range from 70 to 90 mm, no decimal places need to be recorded. If fractions are rounded off, they should consistently be rounded to the nearest full number; halves, to the nearest even number. When fractions are measured, a bias in favor of integral numbers should be avoided.

STATISTICS OF A POPULATION SAMPLE

In the physical sciences and in much experimental research the emphasis is on the mean value and the error of the mean. Although measures of centrality (mean, median, mode) are also important in taxonomic research, particularly in the comparison of populations, variation is usually the chief interest of the taxonomist. The two most commonly used statistics of variation are the *standard deviation* (*SD*) and the *coefficient of variation* (*CV*). Statistics textbooks describe what these statistics represent and how they are calculated.

The numerical value of the CV depends on the measured character and the particular taxonomic group. There are different CVs for meristic quantities, linear measurements, and ratios. The number of eyes (a meristic quantity) in the human species has a CV that is virtually 0; height in people, even in a homogeneous sample, has a CV that exceeds 4. The CV is often a sensitive indicator of the homogeneity of samples. If, for instance, the CV of a certain statistic fluctuates around 4.5 in a series of samples but is 9.2 in one sample, that sample should be reinvestigated because it may include an additional sibling species, wrongly sexed specimens, or another alien component. Zones of secondary intergradation between subspecies are often characterized by a greatly increased CV. The calculation of CV is particularly useful when comparable samples of the same species from different localities are investigated or when the variability of different variates of the same sample is compared.

Even though it is widely used by taxonomists, the CV has the great disadvantage that there are no statistical tests for comparing different CVs (Sokal and Braumann 1980). Lewontin (1966), following earlier authors, therefore proposed that the *variance* (the standard deviation) of the logarithms of measurements be used instead. This measure of intrinsic variability is invariant under a multiplicative change. Thus, it does not matter what units of measurement (metric or nonmetric) are used. A simplified approach is possible for CVs less than 30, which includes most that are used in taxonomy. Here the squared CV equals the variance of the natural logarithms (to the base of e) and can be used in statistical tests.

Measurements, Ratios, and Frequencies

Linear measurements, which reflect size, are among the most frequently used characters in animal taxonomy. Valuable information on questions such as which structures should be measured and how measurements should be taken and recorded can be found in SRL and Mayr (1969). Ratios between the dimensions of two parts of the body, for instance, wing and tail in birds and trunk and extremities in mammals, are often more informative than are linear measurements. Ratios are most easily compared when they are presented visually in the form of scatter diagrams (Figure 4-6), in which one value is plotted on the abscissa and the other is plotted on the ordinate.

When a sample is rather heterogeneous with respect to age and size or when the various compared body parts display allometric growth, it is better to undertake a *regression analysis* than to calculate simple ratios. In the case of allometric growth it is best to plot variation on logarithmic paper. When one wishes to deal simultaneously with more than two variables, some method of *multivariate analysis* must be employed (Sokal and Rohlf 1980). Numerous methods of multivariate analysis, particularly for computer use, are now available. They include *principal component analysis* and *discriminant functions analysis*. Statistics texts explain the use of such methods (see also Hand 1981; Dunn and Everitt 1982:112–121).

FIGURE 4-6
Separation of two subspecies of *Parus carolensis* on the basis of wing and tail length. Triangles = *agilis*; circles = *atricapilloides*; AB = line of best separation. (*From Lunk 1952.*)

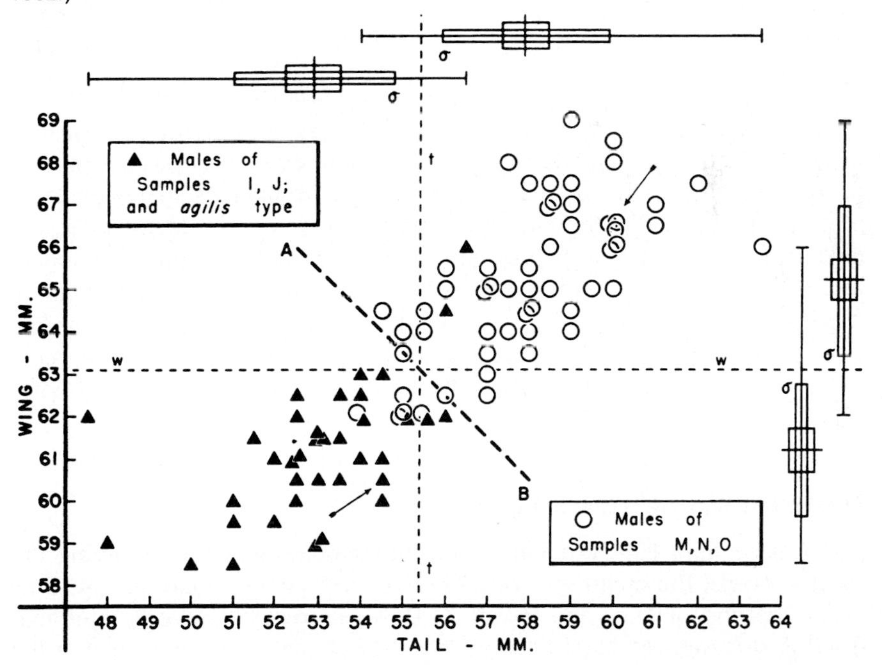

Comparison of Frequencies

The taxonomist is often confronted with the problem of having to determine whether two morphs or other variants occur at the same frequency in two or more populations. If the taxonomist had infinitely large samples, it would be possible simply to express frequencies in percentages. With samples of limited size, a test is employed that indicates the probability that the sampled populations are really different or the probability that the observed differences are merely due to accidents of sampling. This test is called the *chi-square* (χ^2) *test* (SRL, pp. 306–338).

The degree of significance of a given chi-square value is given in *P* tables, which can be found in all standard statistical textbooks. Chi-square tests are highly sensitive to sample size.

GRAPHIC PRESENTATION OF QUANTITATIVE DATA

It is often desirable to present numerical data visually. Such a graphic presentation not only permits a rapid visual survey of all the data but actually often brings out fine points that are not apparent in the raw data. Chapter 14 of SRL gives an excellent survey of such graphic methods. In this text only the few simple methods that are most frequently used in taxonomic publications will be mentioned.

Histograms

Unreduced samples are best shown as histograms. A *histogram* consists of a set of rectangles in which the midpoints of class intervals are plotted on the abscissa and the frequencies (usually number of specimens) are plotted on the ordinate. This presentation has several advantages, the principal one being that it presents the original data in minimum space. Whatever form of statistical analysis a subsequent author may want to apply, that author will find the actual number of specimens given for each size class. A quick comparison of different populations is made possible by the arrangement of a series of histograms above one another (Figure 4-7).

Population Statistics Diagrams

Even more data can be compressed into minimum space by giving the sample range, the mean, one or more SDs, and two standard errors. This is the method of Dice and Leraas (1936) (Figure 4-8). Several modifications of this method have been proposed. For instance, one can give the size of the sample with each bar and replace the standard error with the

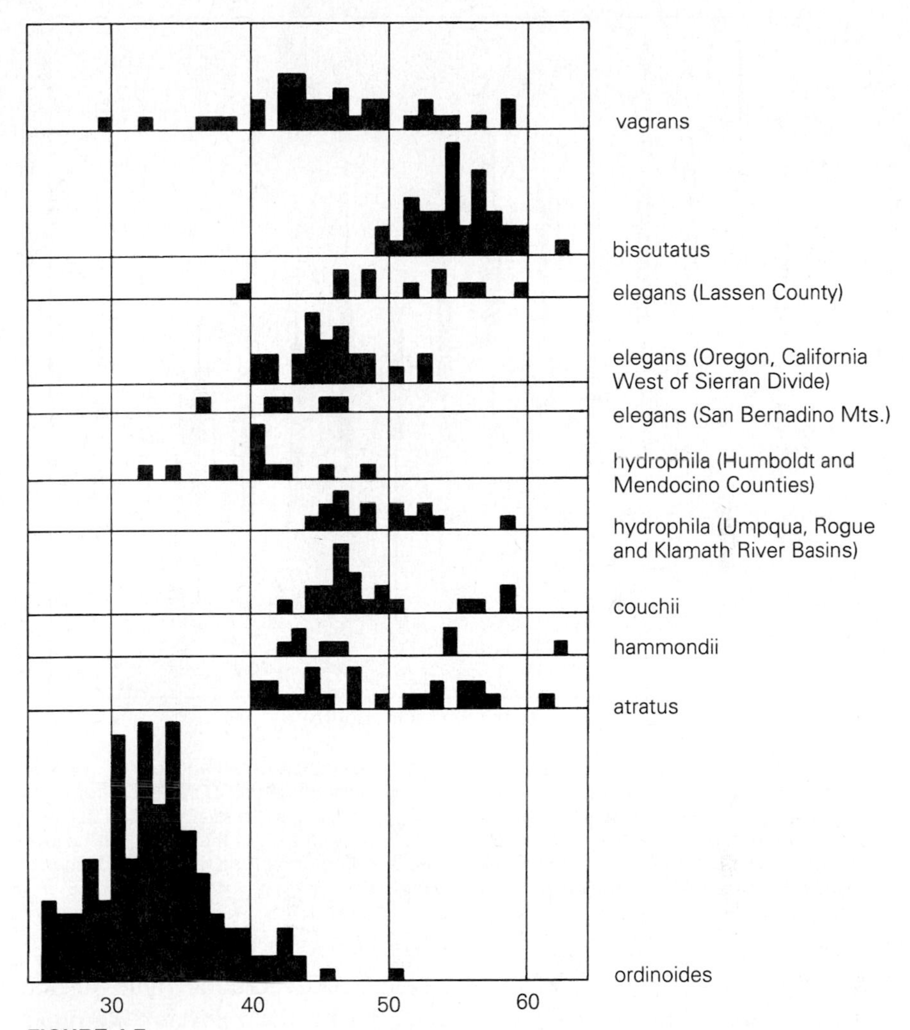

FIGURE 4-7
Histograms showing head and body length in centimeters of adult males of *Thamnophis ordinoides.* Each square represents a specimen. (*From Fitch 1940.*)

95 percent confidence limits of the mean. Nonoverlap of 1.5 SD (of each compared sample) indicates a degree of difference that is usually considered sufficient for subspecific separation, as discussed later (Chapter 5).

Scatter Diagrams

The difference between two or more populations in respect to two characters is best illustrated by a *scatter diagram.* Each individual is indi-

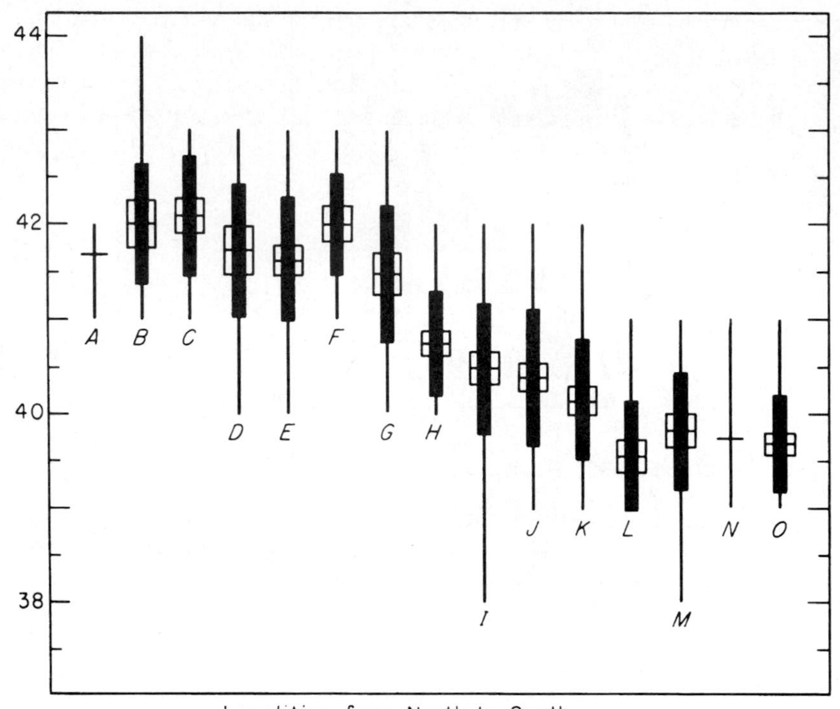

FIGURE 4-8
Population-range diagram. Variation in the number of vertebrae of the anchovy, *Anchoviella mitchilli.* The letters *A* to *O* refer to 15 population samples, arranged from north (*A*) to south (*O*). In each sample the vertical line indicates the total variation of the sample, the broad portion of the line indicates 1 SD on each side of the mean, the hollow rectangle indicates twice the standard error on each side of the mean, and the crossbar indicates the mean. (*From Hubbs and Perlmutter 1942.*)

cated by a spot or other symbol which is placed where the value for one character (read off the ordinate) intersects with the value for the other character (read off the abscissa); each population is indicated by a different symbol (circles, squares, triangles, etc.). Scatter diagrams have many advantages. They help the reader visualize allometric relationships and facilitate the plotting of regression lines. They also sometimes disclose errors of measurement or sexing that might otherwise go undiscovered.

If three characters are involved, triangular charts can be employed. In this case the actual values are not plotted; rather, one plots their percentage of contributions to the sum of the characters. For example, if character $a = 80$ mm, $b = 32$ mm, and $c = 48$ mm,

$$a + b + c = 160 \text{ mm} = 100\%$$

Then a = 50 percent, b = 20 percent, and c = 30 percent of the whole. These percentages are plotted on the graph, which thus shows proportions rather than absolute sizes. In each individual case the triangular graph is scaled so as to produce a maximum spread of the points. As an illustration, a triangular chart from a paper by Burma (1948) is reproduced in Figure 4-9.

Mapping of Quantitative Data

It is often desirable to illustrate the geographic relationships of populations that differ in quantitative characters. In the case of continuous characters such as size, the simplest method is to record the means of the various populations on a base map and, if there is regularity, to draw in

FIGURE 4-9
Triangular graph of the length (*L*), height (*H*), and distance to maximum down-bulge (*D*) of four species of *Anthracomys*. (*From Burma 1948.*)

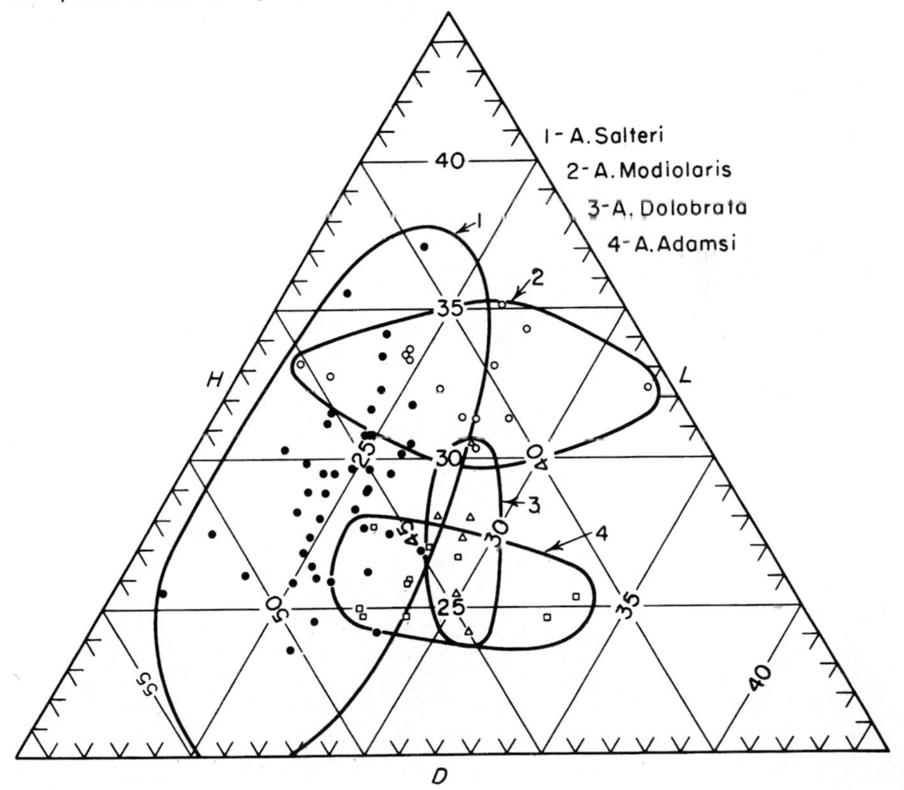

the *isophenes* (lines connecting points of equal expression of a character). For instance, if the means of a series of populations in a species vary from 142 to 187, it is helpful to draw in the isophenes of 140, 150, 160, 170, 180, and 190 (Figure 3-4). Subspecies borders frequently fall where the gradients are particularly steep. A method for determining such contour lines is described by Lidicker (1962).

If qualitative or semiqualitative characters are to be plotted, it is sometimes helpful to choose a different symbol for each class of characters. The relative size of the symbol can be used to indicate sample size.

The *pie graph* is the most convenient method of presenting frequencies of polymorphic characters on a map. The percentage occurrence within the population is indicated by the size of the sectors (Figure 4-10).

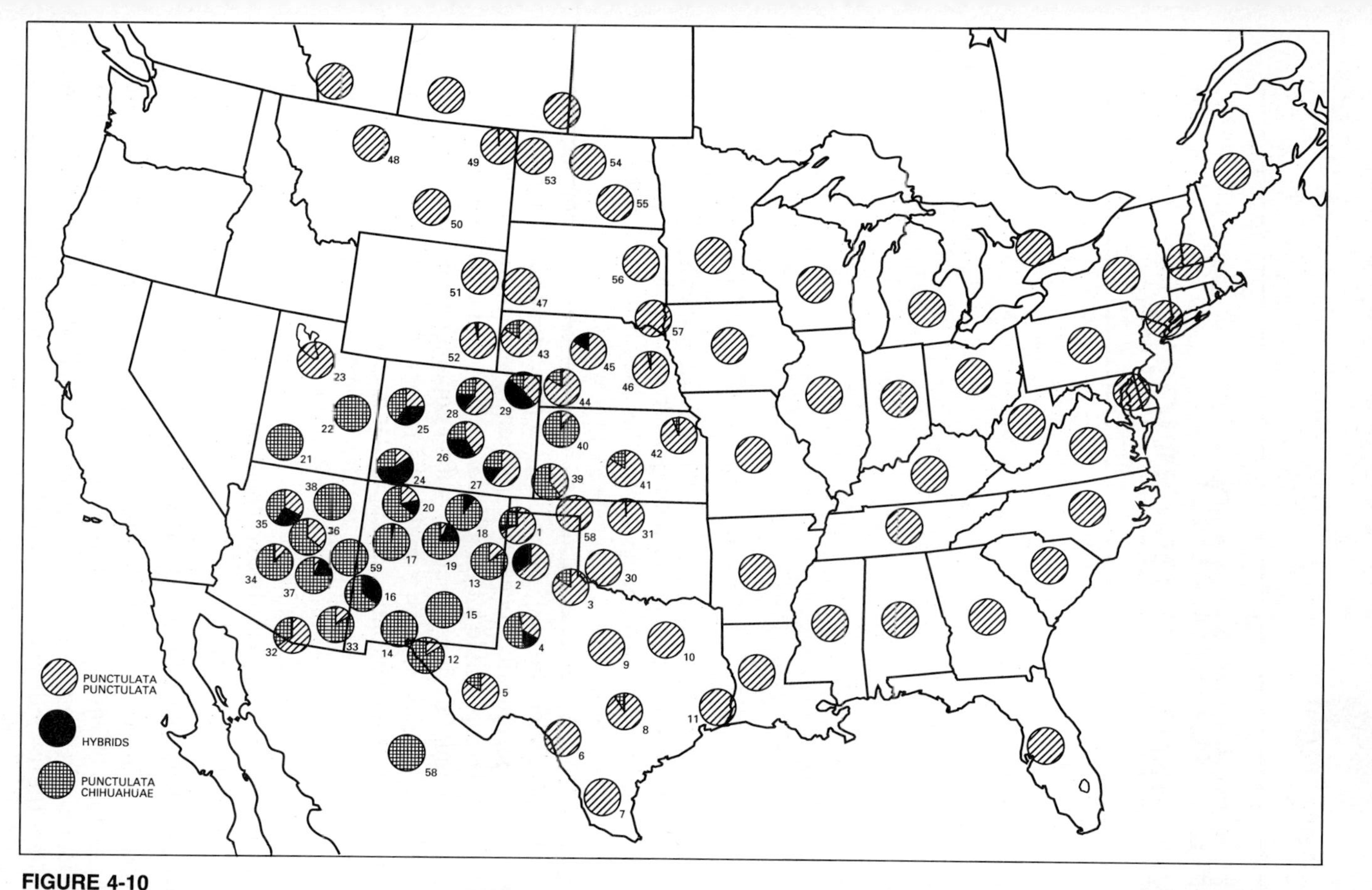

FIGURE 4-10

Map of the distribution of the tiger beetle *Cicindela punctulata,* illustrating the pie graph method. The size of the sectors of each circle indicates the relative frequency in the collected samples of three variants: *punctulata, chihuahuae,* and intermediates. Many populations are pure *punctulata*; others are pure *chihuahuae.* (*Unpublished, courtesy of Dr. M. A. Cazier.*)

CHAPTER **5**

SPECIATION AND TAXONOMIC DECISIONS

What taxonomists encounter in nature are concrete things: individuals and populations. The taxonomist knows that they belong to species, and the first task is to assemble populations in a way that leads to the best delimitation of species taxa. If species had been created according to a plan, this task might be easy. However, it is not, for species are the product of the opportunistic process of evolution. Two aspects of evolution in particular are responsible for the difficulties the taxonomist encounters in trying to delimit species: variability within populations and the existence of incipient species, that is, populations that have evolved only part of the way toward species status. Infrapopulational variation is discussed in Chapter 4. The emphasis in this chapter is on the consequences of the fact that species are the product of evolution. In order to be able to interpret correctly the difficulties created by opportunistic evolution, it is necessary to analyze the process by which new species taxa originate.

THE ORIGIN OF NEW SPECIES TAXA

Although it is now universally believed that geographic or *allopatric* speciation is the most common process by which new species of animals originate, other modes have been proposed. It is still controversial how frequent these modes are and indeed whether some of them occur at all.

Many of the controversies concerning speciation are due to the fact that the opponents did not see clearly that speciation has two aspects, genetic and populational. Contrary to the beliefs of some authors, these two aspects are not mutually exclusive alternatives; both are always involved simultaneously.

As far as the genetics of speciation is concerned, the evolutionist is compelled to confess that more than 80 years after the rediscovery of Mendel's laws, there is still almost complete uncertainty regarding its mechanisms (Barigozzi 1982). Chromosome structure is sometimes involved in speciation, but in other cases it is not. The best understood of all categories of genes, the enzyme genes, seem to be minimally involved in episodes of speciation. What other kinds of DNA are causally connected with the acquisition of isolating mechanisms is completely unknown. Regulatory genes are sometimes suspected, but there is no solid evidence for this. It is only in polyploidy that the genetic situation is reasonably well understood. In this case the doubling of the chromosomes—or in the case of triploidy the addition of a third set of chromosomes—may induce instantaneous *speciation,* that is, the production of a reproductively isolated individual. As deplorable as this ignorance of the genetic mechanisms of speciation is in regard to evolutionary studies, it seems to raise no major difficulties for the taxonomist.

Speciation, however, is not only a genetic but also a populational phenomenon, allowing for two very different processes.

1 New species may originate by the instantaneous production of a reproductively isolated individual within a population, as in *polyploidy*. This process may raise two problems. First, the new species individual, owing to polyploidy or another form of chromosomal restructuring, may have acquired total cross-sterility with the parent species but no morphological difference. Its recognition as a biological species will thus be very difficult except when artificial breeding or cultivation is possible. Sexually reproducing polyploid species occur in turbellarians, oligochaetes, leeches, pulmonate mollusks, some groups of insects, fishes, and anurans, but except in oligochaetes, polyploidy is a relatively rare phenomenon in animals. A second difficulty exists when the new karyotype does not produce complete sterility in crosses with individuals of the parent karyotype. White (1978) postulated that new species can originate under these conditions within the parental species population by means of a process he called *stasipatric speciation,* but this possibility has been questioned by Key (1981), Mayr (1982*b*), and others. There is no sound evidence for such a process. (Figure 5-1*b*).

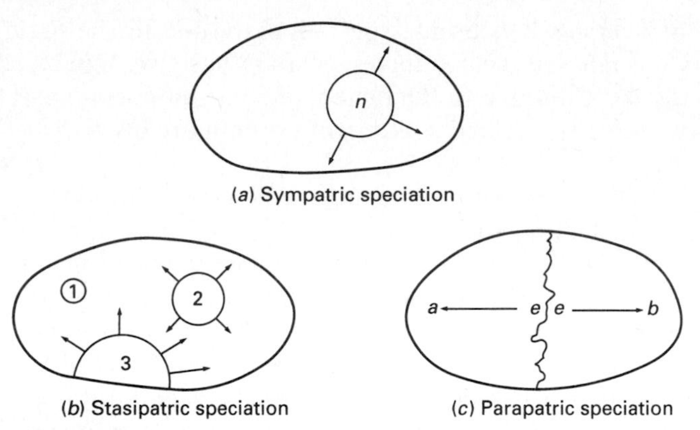

(*a*) Sympatric speciation

(*b*) Stasipatric speciation

(*c*) Parapatric speciation

FIGURE 5-1
Models of gradual nonallopatric speciation. (*a*) *Sympatric speciation.* A new species originates within the range of the parental species through development of different mate preferences and/or ecological segregation. (*b*) *Stasipatric speciation.* New species populations originate within the range of the parental species through chromosomal mutation and subsequent displacement of the parental species. (*c*) *Parapatric speciation.* An intrinsic species barrier within the originally contiguous species *ab* evolves along the ecological escarpment *ee,* resulting in the two species *a* and *b.*

2 Speciation may result from a gradual genetic reconstruction of populations. By and large the various possibilities under this heading can be assigned to six postulated classes:

a *Sympatric speciation:* This can be defined as the origin of a new, reproductively isolated species population within the dispersal area of the offspring of the parental deme (Bush 1975) (Figure 5-1*a*). How frequent sympatric speciation is—indeed, whether it occurs at all (Futuyma and Mayer 1980; Paterson 1982; Barigozzi 1982)—is still controversial. The most probable cases involve host-specific plant feeders and host-specific parasites. Since a switch to a new host is most easily accomplished in a small founder population, the origin of most host-specific species may well be due to peripatric speciation (see below) not requiring sympatric speciation.

b *Parapatric speciation:* In this mode of speciation, it has been postulated (Endler 1977) that isolating mechanisms build up in a cline, along an ecological escarpment, until the two adjacent populations finally are reproductively isolated (Figure 5-1*c*). All the observed cases of more or less drastic belts of intergradation between two subspecies or semispecies known to us are far better explained as zones of secondary contact of two formerly isolated populations.

Nevo (1982), in an analysis of such "tension zones," found no evidence whatsoever for a reenforcement of isolating mechanisms; neither have Butlin (1986), Hewitt (1989), and several other recent authors. For these and other reasons, the occurrence of parapatric speciation is unlikely (Mayr and O'Hara 1986).

c *Geographic or allopatric speciation:* This term designates cases in which reproductive isolation is gradually (at a minimum over several generations) acquired in spatially isolated populations. Within allopatric speciation two subtypes can be distinguished:

- **(1)** *Traditional allopatric speciation (dichopatric speciation):* According to this theory, a reasonably large distributional area is divided by a newly arising barrier (geological, geographic, vegetational), which secondarily splits the previously continuous range into two isolated groups of populations (Figure 5-2).
- **(2)** *Peripatric speciation by primary isolation:* Here a new population is founded outside the continuous species range by a single colonist (a fertilized female) or a small founder group and remains isolated long enough to acquire the genetic basis for reproductive isolation (Figure 5-3).

d *Speciation in time:* Those who believe that speciation in time can occur postulate that a species (phyletic lineage) may change genetically in the course of time to such a degree that the descendants will be reproductively isolated from their own ancestral population if the two could meet. If no appreciable morphological change has occurred during the same period, there will be sibling species in time. The practicing paleotaxonomist has no choice but to ignore such a possibility. However, if morphological change has occurred over time, the taxonomist must decide how much change indicates the attainment of species level. As with geographically isolated populations, it is not possible to provide reliable proof of the species status of *allochronic* populations.

All six of these possible modes of speciation can create situations that make the delimitation of species taxa difficult. This is true even for instantaneous speciation, because individuals with a new karyotype may be reproductively isolated but phenotypically identical. In each of the other five processes of speciation taxonomists assume the existence of incipient species, that is, of populations in the process of genetic reconstruction and a gradual acquisition of isolating mechanisms. The decision whether to designate such incipient species as full species or subspecies is of necessity sometimes arbitrary. The application of the biological species concept helps one make a decision in most difficult cases. For a more detailed treatment of speciation, see Mayr (1970).

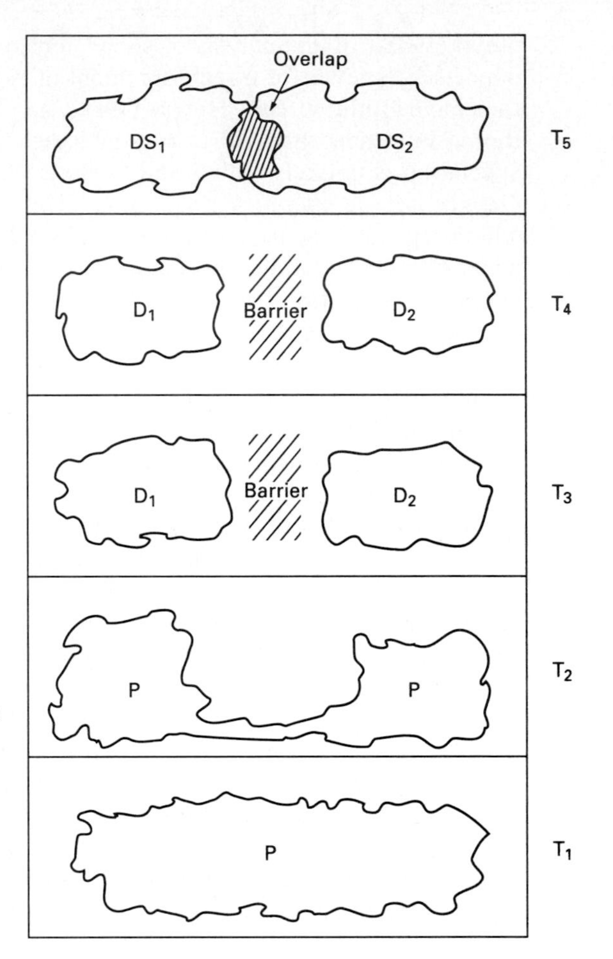

FIGURE 5-2
Dichopatric speciation (speciation by splitting). The parental species P is separated at time 3 (T_3) by a barrier into two daughter populations D_1 and D_2. During their isolation (times T_3 and T_4) these separated populations evolve into two independent species, DS_1 and DS_2, which at time T_5 can overlap without interbreeding.

DECISIONS IN DIFFICULT CASES

The working taxonomist generally encounters two categories of practical problems. Decisions must be made about which individuals encountered in a given locality belong to a single interbreeding population (hence are conspecific) or belong to sympatric species and about which populations, different in space and time, which belong to the same species (see above).

Analysis of Sympatric Samples

On the whole, three very different kinds of situations are responsible for difficulties with sympatric samples:

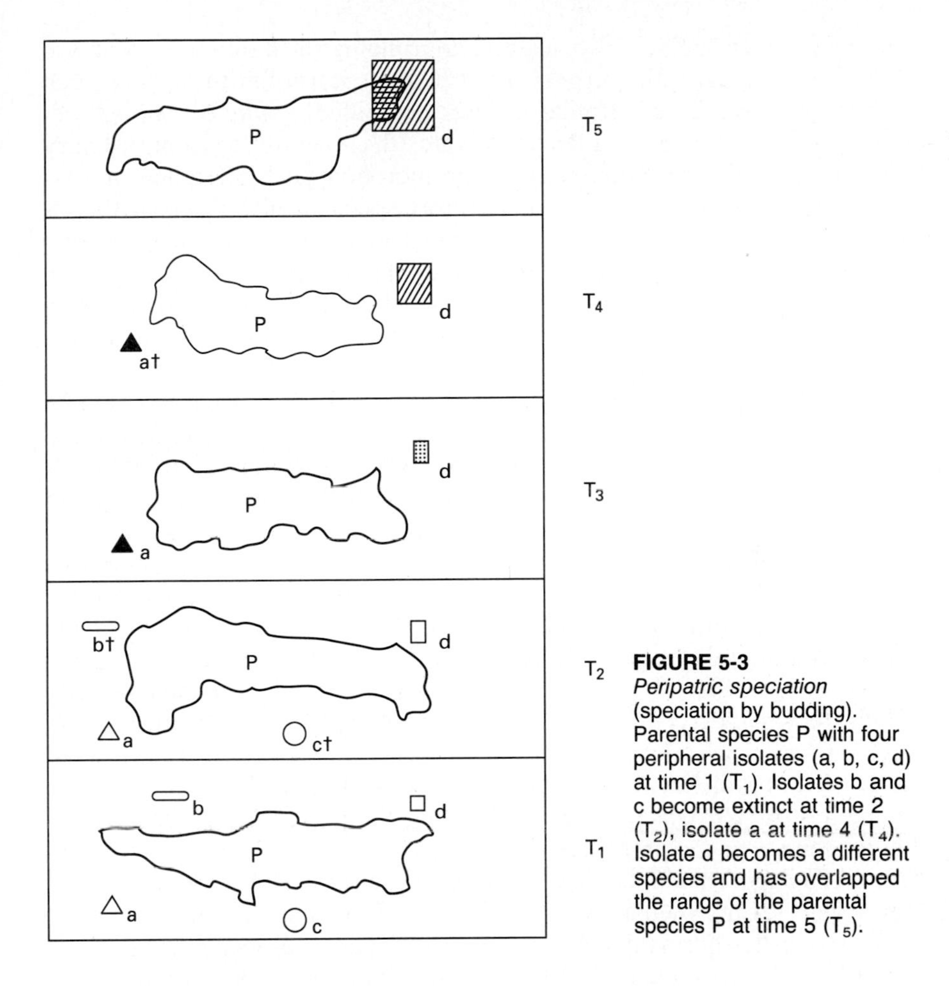

FIGURE 5-3
Peripatric speciation (speciation by budding). Parental species P with four peripheral isolates (a, b, c, d) at time 1 (T_1). Isolates b and c become extinct at time 2 (T_2), isolate a at time 4 (T_4). Isolate d becomes a different species and has overlapped the range of the parental species P at time 5 (T_5).

1 Extreme difference among phena belonging to a single species. The analysis of this situation and methods of resolving it taxonomically are discussed in Chapter 4.

2 Extreme similarity of good biological species (sibling species).

3 Wide variability and phenotypic overlap of two species (discussed in this chapter).

Sibling Species Biological species are reproductively isolated *gene pools* (Chapter 2). When two populations (gene pools) become geographically isolated, they diverge genetically and may eventually acquire isolating mechanisms. As a by-product of the genetic divergence during this process, most species also acquire morphological differences that are suitable for diagnostic purposes, but a few species fail to acquire such differences. Such very similar species are called *cryptic* or *sibling species*. All the available evidence indicates that they differ from ordinary species only by the minuteness of the morphological difference, not by other biological criteria. They are merely species which are near the invisible end of the spectrum of morphological species differences. Sibling species grade imperceptibly into species that are morphologically more and more distinct from one another. However, once discovered and thoroughly studied, they are usually found to have at least a few previously overlooked morphological differences.

Sibling species are not recent species or incipient species. For several pairs of sibling species the date of speciation has been determined, ranging from several million to 18 million years ago (Bullini et al. 1978). If within a single genus both sibling species and morphologically distinct species occur, the latter tend to be older.

Mayr (1963:33–58) has shown how widespread sibling species are in the animal kingdom. Most of them were discovered not during routine taxonomic analysis but during the study of species that are medically (e.g., *Anopheles*), genetically (e.g., *Drosophila, Paramecium*), cytologically, or agriculturally of special importance. It is therefore impossible to indicate what percentage of species are sibling species. In the case of North American crickets, about 50 percent of the species were discovered through differences in their sounds (Walker 1964), and in certain genera of protozoans (e.g., *Paramecium*), the percentage of cryptic species seems to be even higher.

Sibling species can be discovered because they differ in various other attributes even when they are extremely similar in the morphological characteristics normally employed in taxonomic analysis. Mayr (1963:50) listed a number of characteristics that facilitate the recognition of sibling species. Precise measurements sometimes reveal bimodal characteristics, and the two modes can be correlated with additional characters. There are often differences in the number or structure of the chromosomes, a fact which has led to the recognition of sibling species in *Drosophila, Sciara, Chironomus, Prosimulium,* and other dipterans as well as in orthopterans, beetles, and other insects. Various aspects of behavior—such as differences in visual and vocal displays, nest construction, breeding season, migratory behavior, prey selection, and host preference—have perhaps led to the discovery of more sibling species than have any other type of characteristic. Sibling species may differ in their pathogenicity (e.g., *Anopheles*) or in their susceptibility to parasites and suitability to serve as hosts. Various biochemical methods, particularly those used to test protein specificity, are suitable for checking on the probability of a real difference between "stocks" discovered by one of the other methods. The study of enzyme variation by electrophoresis, as well as of other molecular or chromosomal differences, has led to the discovery of numerous sibling species even in relatively well known groups. Lack of success (or sterility) in crossing individuals from different populations sometimes reveals the existence of sibling species, as do drastic differences in life history.

Sibling species are obviously inconvenient to the museum taxonomist because specimens of such species cannot always be identified in preserved material. However, since species are not the creation of museum taxonomists but phenomena of nature, it is impossible to ignore their existence. The museum worker will be unable in many cases to do better than label museum specimens from a group of sibling species by the group name, for example, *Anopheles maculipennis* group. Once a sibling species has been discovered, morphological differences are usually found subsequently that permit the identification of preserved material. Then it is generally possible in a group of sibling species to identify old type specimens to the particular species to which they belong.

The prevalence of sibling species varies greatly among different higher taxa. They are known to be very common in certain families of Diptera, Neuroptera, Orthoptera, and Protozoa but seem relatively uncommon in most genera of Macrolepidoptera and Cerambycidae. They are fortunately rare among vertebrates and most other groups of special interest to the paleontologist. The recognition of a sibling species sometimes depends entirely on the behavior of living animals; such recognition is of course impossible in fossil specimens. Paleontologists must proceed under the assumption that the genera they study do not include sibling species.

One particular class of sibling species—polyploids—raises special difficulties. Polyploids are unable to engage in normal gene exchange with individuals of different ploidy; because of this reproductive isolation, they have the biological characteristics of good species. They may, however, be morphologically indistinguishable, particularly if they are *autopolyploids* (White 1978:278–285). It is up to the specialist to decide whether to give taxonomic recognition to what is biologically surely a different species. In instances of morphological identity and an absence of ecological differences, such recognition is unwise.

Variational Overlap Closely related species are sometimes so variable and their variation is so overlapping that no single character seems to have absolute diagnostic value. A combination of characters usually permits the correct assignment of all seemingly intermediate specimens. As Anderson (1954) perceptively pointed out, mechanical reliance on a biometric analysis or on straight diagnostic characters is often less effective than is a purely intuitive approach based on the totality of characters as revealed by inspection.

A combination of two characters is often sufficient for diagnosis. In such cases it is simplest to plot the two characters on a scatter diagram. This method usually yields several clusters of points when heterogeneous material is involved (Chapter 4). A number of multivariate methods are available when the bivariate scattergram does not yield satisfactory results. Anderson's pictogram is one type of multivariate scattergram (Figure 5-4).

FIGURE 5-4
Pictorialized scatter diagram (pictogram) of 25 individuals from a population of stemless white violets. Two of the characters are indicated along the margins, five by position and length of rays; filled circles indicate individuals with a heavy blotch on the spur petal. (*From Hatheway 1962 after Anderson 1954.*)

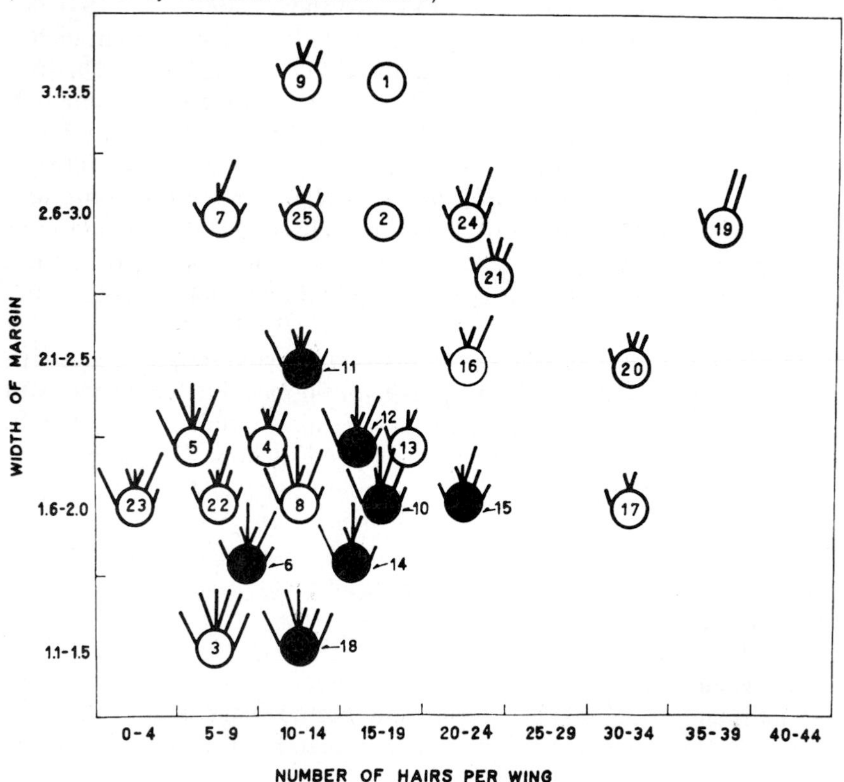

Usually there is no doubt about the validity of species recognized when one uses these methods, but there may be uncertainty about specimens that appear to be intermediate. Calculation of a *character index* is the simplest quick method for placing such specimens. Since this index was first proposed by Anderson (1936) and Meise (1936) to evaluate the hybrid nature of intermediate specimens, it was originally proposed as a *hybrid index*. We prefer the broader term *character index,* for the method is equally suitable for a quantitative treatment of nonhybridizing species with overlapping variability. The method consists in providing a series of states for each character in which the two species differ. The typical condition of a character in species *A* is always designated as 0, the typical condition in species *B* most frequently as 2, and that of intermediate specimens as 1. For 12 characters, essentially typical specimens of species *A* may vary between 0 and 3, typical specimens of species *B* may vary between 20 and 24, and hybrids may vary between 8 and 16.

Two refinements are possible. The first is to allow for more intermediate states in the case of important characters so that specimens may score anywhere between 0 and, say, 6 for such a character. The second is to arrange the scale of values so that the maximum score adds up to 100 (Table 5-1). In this way it is possible to express the similarity of intermediate specimens as percentages. To illustrate the variation, it is advantageous to graph it in the form of histograms. Hatheway (1962) shows how to construct a weighted hybrid index.

A far more precise method than the character index is the multivariate method of *discriminant functions* (Sokal and Rohlf 1981). This method is based on the fact that in the discrimination between two taxa, the contribution made by a character will be greater, the greater the distance between the means and the smaller the combined standard deviations of that character. The method, then, consists in calculating factors (*b*1, *b*2, . . . , *bx*) which, when multiplied by the numerical value of the characters, will result in a maximal value of the function. There are good computer

TABLE 5-1
WEIGHTED DECIMAL CHARACTER INDEX

Character	*Passer domesticus*		*Passer hispaniolensis*	
Crown	Gray	0	Rust-brown	30
Ear coverts	Gray	0	White	10
Sides of nape	Gray	0	Black	15
Flanks	Plain	0	Striped	10
Other		0		35
Total		0		100

Source: After Meise 1936.

programs that calculate the discriminant function from raw data. "Stepwise" programs permit the elimination of characters that do not make a significant contribution to the discrimination.

The method of discriminant functions has been used increasingly in taxonomy in recent years. It is particularly useful in the identification of doubtful specimens and occasionally useful in the establishment of a well-defined class of intermediate hybrids. Kim, Brown, and Cook (1966) point out how to employ discriminant functions to find the diagnostically most useful characters. Although hybridization is on the whole rare among animals, there are exceptional genera in which it is common, such as the crustacean *Bosmina* and some groups of freshwater fishes. Special methods of analysis must be employed in such genera for the analysis of population samples.

Several rather complicated methods of *multivariate analysis* have been tested in recent taxonomic publications, particularly in regard to their application to fossil material. The impression one gets from these papers is that such methods rarely make enough of a contribution to the analysis to justify the considerable labor involved in applying them routinely.

Comparison of Allopatric and Allochronic Samples

The taxonomist has a well-defined objective when comparing samples drawn from different natural populations. The question is never one of whether the compared populations are completely identical: Population geneticists have demonstrated conclusively that no two natural populations in sexually reproducing animals are ever exactly alike. To find a statistically significant difference between several populations is therefore only of minor interest to the taxonomist, who takes it for granted. Even the lowest formal taxon—the subspecies—is normally composed of numerous local populations, some of which differ "significantly" in gene frequencies and in the means of a number of variates.

If no significant differences, statistical or otherwise, between samples from two populations are found, the sampled populations must be referred to the same taxon. If a difference is established, additional evidence is required before it can be decided whether the populations should be considered different taxa and, if so, whether they should be ranked as subspecies or species.

Different Subspecies or Not?

In groups of animals in which polytypic species are generally recognized, it is frequently questioned whether the difference between two populations is sufficiently great to justify their recognition as two different subspecies. The most important prerequisite for making such a decision is a clear understanding of the nature of the subspecies category (Chapter 3 and Mayr 1963:347–351). If the samples are clearly different in one respect or several, they may qualify for recognition as different subspecies (but see below). A problem arises when the ranges of variation overlap. How much overlap between two subspecies is permissible?

The simplest way to determine overlap is to plot the linear overlap of the observed samples (Figure 5-5), but this method is misleading in two ways: It gives only the overlap of the samples (which is always smaller than that of the sampled populations), and it exaggerates the importance of the endpoints of the range, that is, the tails of the overlapping curves. Linear overlap is therefore a very unsatisfactory way of describing the degree of difference between two populations. Is there a more satisfactory way? As far as we know, there is no method that is not open to objections.

The study of subspecies played an important role in the period during which the morphological species concept was replaced by the biological. At that time it was important to point out that populations can be different yet belong to the same species. Evolutionists also had a great interest in subspecies because subspecies were believed to reflect adaptation to the local environment and to represent incipient species. As a result, from the 1920s to the 1950s, specialists in the taxonomy of certain groups outdid each other in describing minutely differing local populations as different subspecies.

There has been a change of attitude in recent decades. Where variation is clinal in continuously distributed populations, no particular benefit accrues from splitting the continuum into several subspecies unless there are pronounced steps in the cline or there is a great difference among the endpoints. Furthermore, only subspecies that are isolates are incipient species, whereas subspecies that are part of a continuum are not. As a result of this downgrading of subspecies, many subspecies that were named from the 1920s to the 1950s have been placed in synonymy.

FIGURE 5-5
Linear overlap of observed samples. *A* = 99–106 mm; *B* = 104–114 mm.

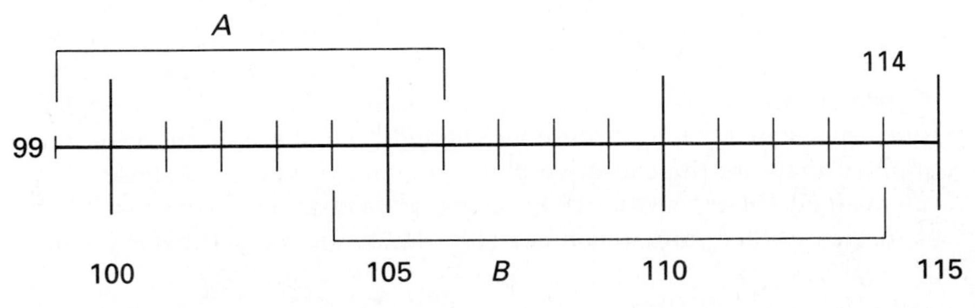

During the period in which subspecies were finely split, elaborate methods were devised (Mayr 1969:188–190) to determine the threshold (*coefficient of difference*) sufficient for subspecies recognition in the case of variational overlap. In the decision whether to recognize subspecies, so many additional considerations enter the picture that extreme accuracy in the determination of the amount of overlap between two populations is not important. Other kinds of information, such as degree of isolation, the presence or absence of clinal variation, the presence or absence of a checkerboard type of distribution, and discordant variation of different characters, may be more important than is the amount of overlap.

The delimitation of subspecies is particularly difficult, if not impossible, if the phenotype of populations is affected by independent causal factors. For instance, many Australian bird populations tend to become increasingly paler from the humid coast toward the arid interior, but at the same time there is a cline of decreasing size from the cool south toward the tropical north. In two species of Adriatic lizards a multivariate analysis based on size, shape, and scale characters revealed the existence of well-defined southern and northern groups. Within each of these groups numerous local populations can be distinguished on the basis of color characters (Thorpe 1980). A rather different set of subspecies has to be recognized in the two species depending on whether coloration or the results of the multivariate analysis are given primacy.

Highton (1962) demonstrated that geographic variation in the salamander *Plethodon jordani* is too discordant to justify the recognition of formal subspecies even though the variation of each individual character shows a definite geographic trend (Figure 5-6). Inger (1961) provided a well-balanced discussion of various criteria to be used in the recognition of subspecies. Degree of difference as expressed in the *coefficient of difference* is only one of them. The standards for subspecies recognition are much more rigorous than they were a generation ago (Chapter 3). Subspecies are rarely recognized in poorly known groups.

Subspecies Borders Most subspecies are geographic isolates or former isolates, and the borders of their ranges are easily established. The delimitation of subspecies that are connected by primary intergradation is difficult, and their recognition is usually unwise. Such subspecies often represent an adaptive response to regional climatic conditions (particularly temperature and humidity) and are no more sharply delimited than are the causative climatic factors. Where substrate races are involved (black lavas, white sands or limestones, red soils, etc.), subspecies borders are sometimes remarkably sharp, particularly when

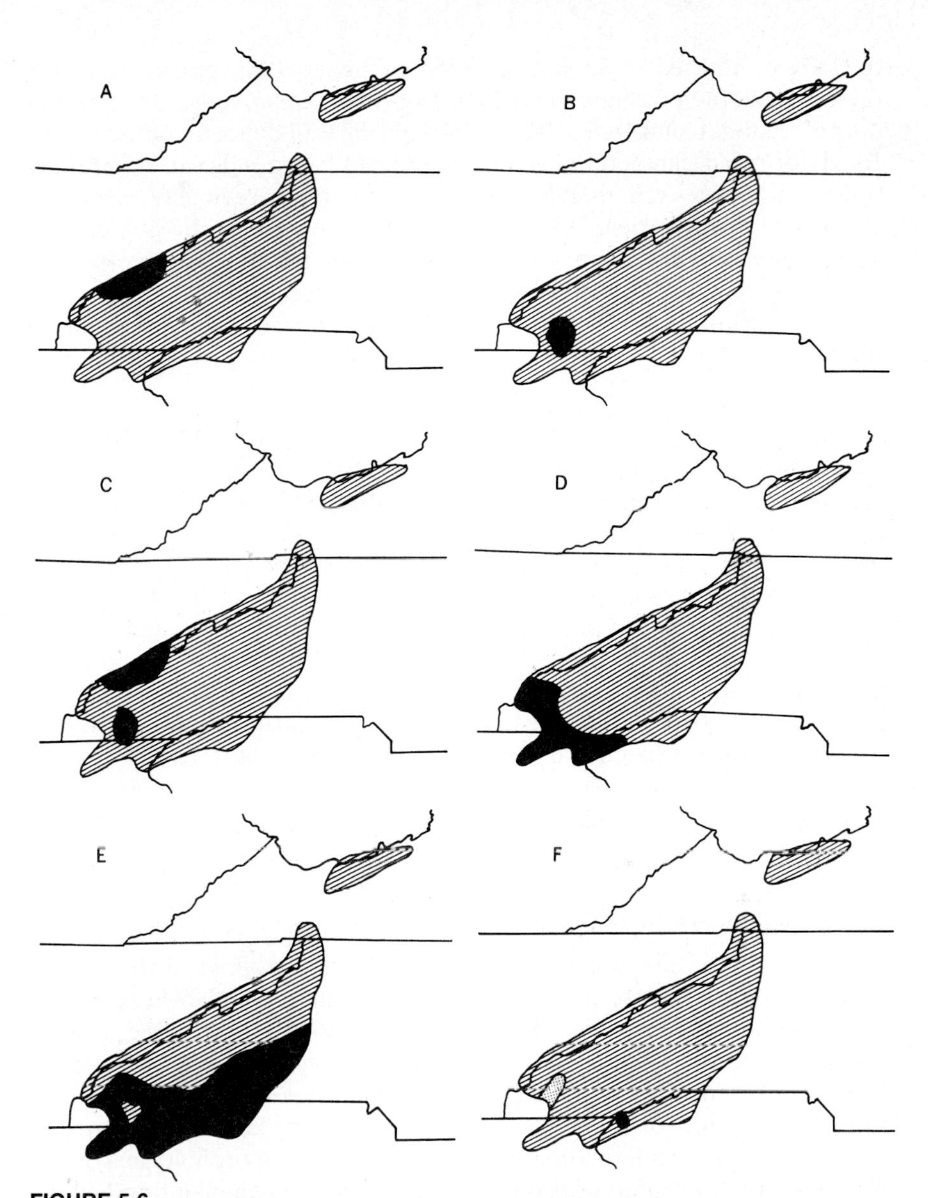

FIGURE 5-6
Discordant geographic variation in the salamander *Plethodon jordani.* Darkened areas represent regions where more than 95 percent have (A) red cheeks, (B) red legs, (C) dorsal red spots in newly hatched young, (D) lateral white spots, (E) a dark belly, (F) small dorsal brassy flecks. An area with small dorsal white spots is indicated by stippling in (F). (*From Highton 1962.*)

they are reinforced by habitat selection. Lidicker (1962) plotted on a map the total character change in a kangaroo rat (*Dipodomys merriami*) per unit of distance and found that it resulted in well-defined contour lines. Bands of rapid character changes, as defined by an index of differentiation, coincided remarkably closely with the previously accepted subspecies boundaries. In other species, however, no well-defined subspecies borders seem to exist, as shown for the marten (*Martes americanus*) and the wolf (*Canis lupus*), two species in which the geographic variation of every character seems to be independent of that of the others.

Polytopic Subspecies When subspecies of a species differ only in a single diagnostic character involving color, size, or pattern, several unrelated and more or less widely separated populations may independently develop an identical phenotype. The evolutionist knows that such populations are not identical genetically, but since the subspecies is not an evolutionary concept, taxonomists sometimes combine such visually identical populations into a single subspecific taxon. A geographically heterogeneous subspecies of this type is called a *polytopic subspecies.* Usually it is preferable not to recognize polytopic subspecies. In the absence of diagnostic differences, there is no legitimate excuse for recognizing several subspecies merely on the basis of locality. It must be remembered that every subspecies is a heterogeneous composite even when it consists of contiguous populations.

Subspecies or Allopatric Species?

Whenever the taxonomist encounters two taxonomically distinct allopatric populations, a decision must be made whether to consider them species or subspecies. Various types of evidence are used in making this decision. Adherents of a typological species concept consider degree of difference per se to constitute sufficient reason for species recognition. Adherents of the biological species concept look for evidence of actual or potential interbreeding. They use degree of morphological difference only to draw inferences on the probability of potential interbreeding.

The word *allopatric* is essentially an antonym of *sympatric* and therefore means "distribution without geographic overlap." There has been a tendency in recent years, following Smith (1965), to make a distinction between *allopatry,* in which populations or taxa are separated by a distributional gap that prevents contact, and *parapatry,* in which contiguous allopatric populations are in contact. Strictly speaking, parapatry is only a special form of allopatry.

Kinds of Parapatric Contact On the basis of the presence or absence of interbreeding and the amount of interbreeding, four kinds of parapatric contact can be distinguished:

1 *A* and *B* intergrade clinally in a (usually fairly wide) zone of contact.
2 *A* and *B* interbreed completely in a (usually rather narrow) zone of contact or hybridization.
3 *A* and *B* meet in a zone of contact where occasional hybrids occur.
4 *A* and *B* meet in a zone of contact but do not interbreed at all.

Populations that qualify under (1) and (2), if sufficiently different, are nearly always treated as subspecies; populations that qualify under (3) and (4) are treated as species.

Primary Intergradation This term refers to situations in which a series of intermediate populations connect two subspecies, with each population being intermediate between adjacent populations and showing approximately the same amount of variability as does any other population of the species. Adjacent populations and subspecies intergrade clinally. In such cases it is usually assumed that the clines are causally related (by selection) to environmental gradients. Whether allopatric populations that intergrade clinally with each other are considered subspecifically different depends on their degree of difference. The modern trend is opposed to subspecific recognition.

Secondary Intergradation This term refers to a zone or belt of hybridization between populations after a breakdown or elimination of a previously existing extrinsic isolation of those populations. This phenomenon is often called *allopatric hybridization*. It occurs when two subspecies meet in a well-defined zone where they form a hybrid population with greatly increased variability, often containing the entire spectrum of character combinations from subspecies *a* to subspecies *b*.

Among North American birds the flickers (*Colaptes*), juncos (*Junco*), and towhees (*Pipilo*) furnish good examples of interbreeding between widely divergent subspecies. For further details and additional examples, see Mayr (1942:263–270, 1963:118) and Short (1969). No taxonomic difficulty arises when the zone of interbreeding is narrow. However, if it is wide and if a well-defined, stabilized hybrid population with intermediate characters develops, it is sometimes convenient and justifiable to recognize the "hybrid" population taxonomically. The taxonomic recognition of a hybrid population is not justified if the population is highly variable and includes a range of phenotypes extending from one parental extreme to the other. If two taxa that previously were recognized as two allopatric species intergrade completely in a zone of secondary contact, this proves that they are not reproductively isolated and should be considered subspecies of a single polytypic species. There must be evidence of random interbreeding in this zone. Some taxonomists use only the terms *interbreeding* and *intergradation* for conspecific populations and restrict the term *hybridization* to the interbreeding of species.

Occasional Hybridization Allopatric forms that hybridize in the zone of contact only occasionally are full species. There are a few cases in which it is difficult to decide whether hybridization is only occasional. Much recent evidence indicates that hybridization has to be fairly complete to restore secondary intergradation. For a discussion of occasional hybridization, see Chapter 2.

Cases where two species remain as distinct species over most of their range but form complete hybrid populations in a few areas are more difficult to evaluate. This happens particularly in regions where human interference in recent years has badly disturbed the natural ecological balance. It is recommended that such forms be treated as full species in spite of the occasional free hybridization seen under the stated conditions (Mayr 1963:119–124; Mayr and Short 1970).

To summarize our analysis of hybridization, sympatric and parapatric species have to be treated as full species if they form only occasional hybrids in their areas of contact or if areas of complete breakdown of the species border are narrowly localized while the two species retain their integrity over most of their ranges. Populations have to be considered conspecific if they interbreed wherever they come in contact and there is no indication of preferred nonrandom mating in the zone of hybridization.

Parapatric Species Allopatric populations that fail to interbreed despite being in contact are full species. Failure to interbreed indicates reproductive isolation and attainment of species rank. The absence of geographic overlap may be due to a number of factors. The zone of contact may separate two very different ecological areas (e.g., savanna and forest). If one of the two neighboring species is specialized for one of these habitats and the second species is specialized for the other, the two species cannot invade each other's range because their ecological requirements are too different.

Another possible reason for nonoverlap of full species is that their ecological requirements are so similar in every respect that they compete with each other. One species is slightly superior on one side of the zone of contact, and the second species is superior on the other side. Such *competitive exclusion* (for a detailed discussion, see Mayr 1963:81–82) will result in strict parapatry of full species. It is important to understand this because allopatry used to be accepted by some authors as an automatic criterion of conspecificity. Vaurie (1955) showed that in 225 species of 7 families of Palearctic songbirds, 22 good species had been treated earlier as subspecies by various authors because of their allopatry. A careful study of the zone of contact usually reveals areas where increased habitat diversity actually permits occasional sympatry of the two species.

Finally, two species may be parapatric because they lack behavioral isolation but their hybrids are completely sterile. As a result, invaders into the range of the other species leave no viable offspring. This is the case with many species of morabine grasshoppers in Australia (Key 1981).

The best evidence for species status among parapatric populations is provided by the sharpness of phenotypic difference in the zone of contact. If two continental species display no evidence of any intergradation in their zone of contact or close approach, they evidently do not exchange genes with each other and must be treated as full species.

Allopatric Populations Separated by a Gap Allopatric populations that are not in contact but are separated from each other by a gap in their distributional range may be either species or subspecies. The most important species criterion—the presence or absence of reproductive isolation—cannot be tested (except experimentally, and even then only with reservations) to determine the status of such populations. This is why the classification of allopatric populations is so often subject to a considerable amount of disagreement among taxonomists. Many solutions for this dilemma have been proposed, but all of them are beset with difficulties.

Some taxonomists insist that all morphologically distinct isolated populations be treated as full species "until it is proved that they are subspecies." This solution is of course impractical, because it is impossible in most of these cases to obtain clear-cut proof one way or the other. Furthermore, this solution overlooks the fact that it is just as serious an error to call a population a species when it is really only a subspecies as it is to do the opposite.

A second solution is to treat all populations that are not connected by intergradation as full species. This procedure is founded on the correct observation that populations that are connected by intergradation are conspecific. The reverse conclusion is then drawn, namely, that populations which are not connected by intergradation are not conspecific. This, however, is in conflict with the rules of logic when it is applied to isolated allopatric populations. Geographic isolation is not an intrinsic isolating mechanism (Mayr 1963:91), and there is no guarantee that the morphological hiatus caused by the temporary halting of gene flow constitutes proof of the evolution of isolating mechanisms. The opposite extreme—considering all related allopatric forms to be conspecific—is equally wrong, as shown above.

Most species of animals are not easy to keep and raise in captivity, but when this is possible, attempts are usually made to cross individuals of populations whose status is in doubt. Such experimental testing of isolating mechanisms must be done very carefully to avoid misinterpretations. An excellent example of the use of experimental systematics is that of Usinger (1966) and associates on the Cimicidae (bedbugs). The studies included cytological examination of hybrids and forced a major reevaluation of species concepts in the family. The fact that bedbugs live well in a laboratory environment and that bedbug mating behavior is promiscuous contributed greatly to the success of the project.

Great care must be taken in the design of such experiments because individuals of different species, such as lions and tigers, often hybridize readily in captivity while never doing so in nature in their area of geographic overlap. Many species of ducks produce fertile hybrids in captivity but virtually never cross in nature.

The minimal conditions for a meaningful test are double-choice experiments in which females are allowed to select either a male of their own population or a male of the population to be tested. Tests in which males are tested against the females of two populations are easier to analyze but less decisive since the females are the choosing sex in most species. If hybrids are produced, they must be tested for fertility and chromosomal incompatibilities. Differences or similarities in molecular characters (e.g., enzyme genes) often provide helpful clues, but where such characters vary geographically or are invariant in related species, they cannot be used. Ecological preferences are part of the isolating mechanisms between species and usually cannot be properly evaluated in the laboratory. For example, the sympatric sibling species *Drosophila pseudoobscura* Frolova and *D. persimilis* Dobzhansky and Epling always hybridize in laboratory populations, but only a few scattered F_1 hybrids have ever been found in nature.

When direct proof is unavailable, it becomes necessary to decide the status of isolated populations by means of inference. Several kinds of probabilistic evidence can be used as the basis for such inferences. All rely on the observation that reproductive isolation is correlated with a certain amount of phenotypic difference, which is fairly constant within a given taxonomic group. The experienced taxonomist can use this evidence to develop a yardstick that can be applied to isolated populations. There are three sets of morphological differences that can be utilized to calibrate such a scale.

1 *Degree of difference between sympatric species:* Within a given genus or within a group of closely related genera, there is usually a fairly well defined amount of morphological difference between valid sympatric species. This difference may be great, as in the case of birds of paradise, or very slight, as in the case of sibling species. This amount of difference between good species can be used to determine the status of isolated populations in the same genus.

2 *Degree of difference between intergrading subspecies in widespread species:* The amount of morphological difference between the most divergent subspecies in species of the same genus indicates how much morphological difference may evolve without the acquisition of reproductive isolation.

3 *Degree of difference between hybridizing populations in related species:* Subspecies or groups of subspecies within a species sometimes become temporarily separated from one another through the development of a geographic barrier but merge again after the breakdown of the barrier. Conspecificity is proved by free interbreeding, which often occurs even after a morphological difference of considerable magnitude has developed. Good examples of such free interbreeding of morphologically strongly differentiated populations can be found in North American birds among some of the juncos (*Junco*) and flickers (*Colaptes*) (Short 1965). These facts must be taken into consideration in the ranking of isolated populations in these genera.

Even after all these criteria have been applied, some doubtful cases remain. *It is preferable to treat allopatric populations of doubtful rank as subspecies,* because the use of trinominal nomenclature conveys two important pieces of information: (1) closest relationship and (2) allopatry. Such information is very valuable, particularly in large genera. Geographic replacement suggests furthermore that either reproductive isolation or ecological compatibility has not yet evolved. Treating such allopatric forms as separate species has few practical advantages. If further analysis shows that such a form has been erroneously reduced to subspecies rank, it can again be restored to full species status.

Categorical concepts may undergo a change during the history of the taxonomy of a particular higher taxon. For instance, when the concept of polytypic species was first applied consistently to bird species (1920s to 1940s), virtually all geographically isolated populations were combined with their nearest relatives as subspecies. The total number of species of birds was thus reduced to less than 8500. After the superspecies concept was adopted, many of the more distinct isolated subspecies were raised (1960s and 1970s) to the rank of allospecies. As a result, the total number of species (including allospecies) of birds has risen again to more than 9000. A similar development seems to be occurring in mammals and other groups of animals.

Difficult Genera or Species Groups In nearly all major groups of animals there are a few genera or species groups that seem to defy the traditional criteria. The North American *Rana pipiens* (leopard frog) complex of about 25 species consists of two groups that are largely sympatric and have minimal interbreeding. The species of which each group is composed are largely allopatric but hybridize in rather narrow contact zones wherever two of them are in contact; however, they are narrowly sympatric when their mating seasons differ. Each species has a characteristic call, but characters vary irregularly, as does the degree of cross-sterility. Molecular analysis has greatly clarified the status of a previously bewildering assemblage of populations and species (Hillis, Frost, and Wright 1983, 1988). Pocket gophers (*Thomomys*) and the crustacean *Daphnia* are other examples of animal genera in which the delimitation of species creates difficulties.

DELIMITATION OF FOSSIL SPECIES TAXA

The paleontologist has to cope with two unique kinds of difficulties in the delimitation of species taxa. These difficulties have been discussed frequently and with considerable detail in the paleontological literature (Imbrie 1957; Sylvester-Bradley 1956, 1958; Simpson 1961:152–155, 163–171). As paleontologists have pointed out, the difficulties they encounter are often exaggerated. Many paleontologists have confused phenon and species, taxon and category, and evidence and inference. A clear understanding of these terms and the underlying concepts removes much of the difficulty.

Evolutionary Continuity

Species are evolving systems, and a vertical delimitation of species in the time dimension should in theory be impossible. Unbroken sequences of fossil populations are, however, extremely rare. Where they exist, the named morphospecies within such a sequence are often so similar that they should not be recognized at all or at most should be ranked as subspecies.

In most fossil sequences there are convenient breaks between horizons that permit a nonarbitrary delimitation of species. This seems to be true even for Brinkmann's (1929) remarkably complete sequence of Jurassic ammonites (*Kosmoceras*) (Raup and Crick 1981). Since much if not most speciation occurs in peripheral isolates, the discovery of strata with intermediate populations (incipient species) is highly improbable and will occur only rarely. Range fluctuations contribute to the appearance of breaks even in cases of gradual change resulting from uncomplicated phyletic evolution in a single vertical column (Figure 5-7). An evolutionary species may be absent from a given locality for shorter or

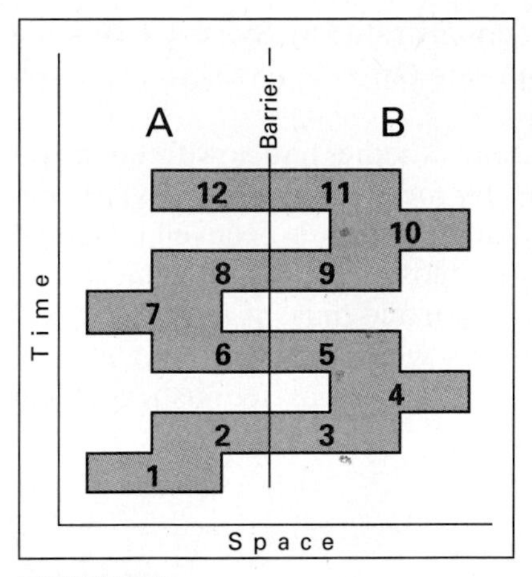

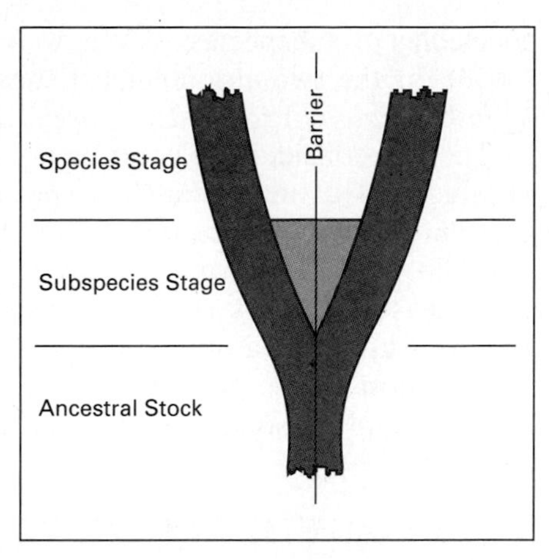

FIGURE 5-7
Left: Geography and time in subspeciation and speciation. Diagram illustrating how geographic fragmentation of successive populations (the numbered rectangles) may accompany vertical differentiation of a phyletic line. The populations rarely remain in one locality for long but migrate. Some migrants are isolated from the parent stock by barriers, ultimately becoming differentiated into geographic races. The faunal succession in any locality (A or B) is never absolutely continuous, even though the gaps may be obscure. The gaps may be produced by migrations, depositional hiatus, or local extermination. *Right:* A population becomes divided by a barrier causing partial isolation with limited gene flow for a time; this is the subspecies stage in speciation. After sufficient genetic differentiation has been reached, interbreeding ceases, gene flow is stopped, and the two branches become separate species. (*From Newell 1947.*)

longer periods, when it recurs, it may have changed sufficiently to be classified as a different species.

Limited Amount of Evidence

Taxonomic evidence supplied by behavior, chromosomes, properties of proteins and other chemical constituents, and other attributes of living populations is not available to the paleontologist. However, most classifications of living species are also made without access to such information; and when the information finally becomes available, it usually confirms the existing classification. As Imbrie (1957) and others have pointed out, the paleontologist has far more to go on than mere morphology. The variation of samples, the associated faunas, the paleoecology, and the geography, horizon, and time level contribute abundant information that facilitates inferences on the delimitation of taxa and their categorical ranking. The analysis, as in much neontological work, often has to begin with a separation of the samples into phena, but a consideration of all the collateral evidence usually allows an unequivocal assignment of these phena to species or subspecies, as was well demonstrated by Sylvester-Bradley (1958) for the two chronological subspecies *Ostrea knorri knorri* and *O. k. lotharingica* (Table 5-2).

There is considerable controversy about whether and how frequently a species evolves into a daughter species by means of gradual phyletic evolution. Even if this occurs far more frequently than is believed by Gould and Eldredge (1977) and Stanley (1979), most new species almost certainly arise by means of peripatric speciation and thus, as far as the fossil record is concerned, quite abruptly. If this were fully substantiated, the delimitation of chronological species would not be as serious a problem as is thought by some paleontologists.

SPECIES DELIMITATION IN PARASITES

Many species of parasites occur only on a single host species. The determination of species status in such cases raises no problems. Difficulties arise when populations on different hosts are slightly different. There are three ways of interpreting these morphological differences.

1 The differences are caused by nongenetic modification resulting from the different physiological environments of the different hosts. Many trematodes can mature in a large number of possible hosts and show great differences. Specimens of the liver fluke *Fasciola hepatica* from a cow, a rabbit, and a guinea pig manifest differences far greater than those usually employed to distinguish species (Stunkard 1957). Large numbers of nominal species in certain genera of cestodes such as *Hymenolepis* may be shown eventually to be nothing but host-induced nongenetic modifications.

2 The differences are indicative of subspecies rank. The differences between host populations of Mallophaga, although constant, are often so slight that these populations are best treated as subspecies (Clay 1958).

TABLE 5-2
FREQUENCY OF 12 PHENA IN TWO SUBSPECIES OF *Ostrea knorri*

Phena	Schönmatt (*knorri*)	Geisingen (*lotharingica*)
D + B + M	49	0
A + G	41	5
K + J + E	10	5
F + H	3	14
C + L	0	45
Total	103	69

Source: From Sylvester-Bradley 1958.

Host separation in this case corresponds to geographic isolation in free-living species, and the amount of gene flow between Mallophaga occurring on different host species is presumably very slight.

3 The differences are indicative of rank as full species. Numerous cases have been described in the parasitological literature in which exceedingly similar parasites could not be transferred from one host to another. There is no opportunity for gene interchange in species that lack an intermediate host. In such a case, even though the morphological difference is comparatively minor, one must assume that the genetic barrier has reached the species level.

PART **B**

MACROTAXONOMY

The sorting of individuals and populations into species is one of the great tasks of taxonomy. The theory and practice of this activity are discussed in Chapters 2 through 5 in Part A. The ordering of species in a rational and practical classification is the second great task of taxonomy. It is the subject matter of Part B and is dealt with in Chapters 6 through 11.

Chapter 6 treats general problems of classification. The evidence used by taxonomists to construct their classifications is provided by taxonomic characters, which are discussed in Chapter 7. In Chapters 8 through 10 the characteristic features of the three major modern schools of macrotaxonomy are analyzed.

Chapter 11 provides a short survey of the most frequently used numerical methods, including methods that require the help of a computer. P. Ashlock was largely responsible for this chapter. Since his death, it has been revised in consultation with D. Maddison. The primary object of this chapter is to introduce the beginner to hand and machine phenetic and cladistic analysis. The literature dealing with numerical methods is increasing every month, and to cover all of it would have required a whole volume. The readers of this chapter must remember that it is only an introduction to a vast field with an enormous literature.

Some readers of this volume might have preferred a simple recipe book describing by what method one can most rapidly arrive at a respectable classification. In the long run this would not be a helpful approach.

Classifiers must understand what they are doing. They must understand the nature of the concepts on which they base their methods and operations. They must be aware of possible pitfalls in the various methods and of the availability of different options.

The present turmoil in views on classification makes this an exciting and stimulating period. Since macrotaxonomy has not yet experienced a synthesis of the various competing schools, it is difficult if not impossible to give clear-cut advice. Many of the problems of classification will probably remain controversial for a long time to come. Readers of this volume must be fully aware of this.

CHAPTER 6

THEORY AND PRACTICE OF BIOLOGICAL CLASSIFICATION

Objects in nature seem to fall into groups or classes, such as birds, butterflies, trees, grasses, bodies of water, and stones, with the members of each group having one key attribute or a number of attributes in common. The traditional definition of *classification* therefore is *the grouping of objects into classes owing to their shared possession of attributes.* Any ordering of the vast diversity of nature, or of any part of it, into such groups is called a classification. Communication would be difficult if such classes were not distinguished and named. Used in all human activities, classifications also exist without benefit of language. Animals prove by their reactions that they classify objects in their environment as food or nonfood, competitors or potential mates, enemies or prey. Michener (1970:5) put it this way: "A bird that classifies a cat as onc of thc rabbits is not likely to maximize its reproductivity."

SOME BASIC PRINCIPLES OF CLASSIFICATION

In view of the ubiquity of classifications and the corresponding need for an understanding of the basic principles of classifying, one is perplexed by the neglect of this subject in the philosophical literature. Some philosophers, such as Nagel in *The Structure of Science* (1961), do not even mention the problems of classification. Others, from ancient (Aristotle) to recent times (Whewell, Mill, and Jevons to Hempel and Popper), have referred to classification, but they have dealt with it typologically and often on the basis of physicalist principles. Most philosophers have failed to see the difference between classification and identification schemes and have relied on logical division (downward classification), a method which leads to identification schemes but not to genuine classifications. Almost all of them have assumed that there is no difference between the classification of inanimate objects and that of organisms.

As we shall see below, a classification of living organisms must take into account certain special constraints (evolutionary history), yet at the same time it should not violate certain elementary rules of classification that are equally applicable to both organisms and inanimate objects. Curiously, we have been unable to find in the philosophical literature a well-articulated list of such rules.

General Rules of Classification

The attempt is made here to fill this gap in the literature by supplying a set of general rules of classification. These elementary rules, here tentatively proposed, are the ones that are in daily use in arranging books in a library, goods in a warehouse, or character traits in psychology. Additional rules, which are supplementary, but not contradictory, are needed for the classification of organisms.

1 Items that are to be classified are assembled into classes that are made as homogeneous as possible.

2 An individual item is included in that class with the members of which it shares the greatest number of attributes.

3 A separate class is established for any item that is too different to be included in one of the previously established classes.

4 The degree of difference among the classes is expressed by arranging them in a hierarchy of nested sets. Each categorical level in the hierarchy expresses a certain level of distinctness.

Surely there must be more general rules than these four, but they will suffice for this provisional listing.

Additional Rules for the Classification of Organisms

Most philosophers and even a number of taxonomists, among whom Gilmour (1940, 1961) and his school have been the most prominent, have chosen not to make a distinction between classifying organisms and classifying inanimate objects. In either case, they state, that classification would be the most natural that served the largest number of purposes (for critiques, see Simpson 1961:25 and Farris 1979*b*:497).

Eventually taxonomists, Darwin most decisively, realized that it is neither useful nor legitimate to classify objects or phenomena strictly on the basis of some defining quality (essence) or in any other arbitrary manner when the grouping of these objects is actually the result of history or another cause (Mayr 1982*a*:238). Such forms of causation exert constraints that severely limit the number of possible meaningful classifications. For instance, classifications of human diseases by means of such arbitrary and artificial criteria as "quick recovery versus slow or no recovery," "with fever or without fever," and "with pain or without pain" were found by physicians not to be very helpful. A far more appropriate system of classification sorts diseases into those *caused* by bacteria or viruses, malignancies, degenerative changes, and inheritance; that is, it sorts them by causal agents. Darwin was the first to see clearly and state emphatically that the grouping of organisms is a result of the common descent of the members of a group (taxon). In order to be "natural," a classification of organisms therefore has to reflect descent. This theoretical basis of all natural biological classifications is a powerful constraint and refutes the claim that the same principles of classification are applicable to organisms and to inanimate objects. From these considerations one arrives at the following definition: *A biological classification is the ordered grouping of organisms according to their similarities and consistent with their inferred descent.*

It is thus evident that there are two independent sets of criteria responsible for the grouping—*similarity* and *sameness of causation* (lines of descent)—on the basis of which organisms can be assembled into taxa. As we shall see, these two sets of criteria are frequently in conflict, and their application is controversial. In fact, the most important difference among the three major current schools of taxonomy lies in the relative weight they give to these two criteria. These differences are the subject of this and the next five chapters.

Special Classifications

Classifications that attempt to reflect a similarity and/or relationship are based on numerous ("greatest possible number of") characters. For practical purposes other, special classifications based on single characteristics are sometimes needed, for instance, diploid versus polyploid plants, annual versus perennial herbs, and inedible (or poisonous) versus edible mushrooms. A limnologist may divide plankton into autotrophs, herbivores, and predators. Plants are traditionally subdivided into trees, shrubs, herbs, and grasses, and this grouping is still useful in some areas of ecology. Such special-purpose classifications have a low information content and usually cannot be used for broader generalizations.

One must remember that the taxonomist classifies populations at all levels, not characters. Characters are only the means by which taxa are recognized. See Chapter 9 for the dissenting views of pattern cladists.

IDENTIFICATION

There is a fundamental difference between a classification and an *identification* scheme. A classification orders a diversity of items into groups or taxa on the basis of principles and criteria that are discussed below. An identification scheme or key permits the placement of an unidentified object (specimen) in one of these taxa. In identification one uses a few characters, ideally a single diagnostic one, that throw a given specimen into one line or another of a key. Computer identification methods work somewhat differently but also focus on the specimen that is to be identified. The procedure of identification is based on deductive reasoning.

All so-called field guides for the recognition of birds, butterflies, flowers, or other organisms are based on identification schemes. Identification keys are also valuable components of monographs and revisions. However, the preparation of these keys must be preceded by a careful taxonomic analysis. Identification keys and classifications serve very different purposes, yet the quality of an identification key depends largely on the quality of the classifications on which it is based. It is wrong to say "I classified a specimen" when one means "I identified a specimen."

A classification is a filing system, and the unique name of a species or higher taxon is like the index number of a file. It is the key to the entire literature on that taxon. Many of the developments in taxonomy in the last 200 years have led to a clearer separation of the procedures of identification and those of classification. Prior to Linnaeus, virtually all so-called classifications were actually identification schemes; so were the artificial downward classifications of Linnaeus (Chapter 1). They were based on a method of logical division in which the organic world as a whole was considered a summum genus, which at the first step of the identification was divided into two "species." With the last of a series of dichotomous divisions, one arrived at the species of the specimen that was to be identified. This method produced artificial groupings; the Aptera (arthropods without wings) contained such distantly related taxa as fleas, collembolans, arachnids, and certain crustaceans.

CRITERIA OF ZOOLOGICAL CLASSIFICATION

The procedure of classifying animals consists of combining "related" species into groups called taxa. A *taxon* is defined as *a monophyletic group of populations or taxa that can be recognized by their sharing of a definite set of characters; such a group must be sufficiently distinct to receive a name and to be ranked in a definite taxonomic category.* Although one demands that a taxon be recognizable by sharing a definite set of attributes, this set may be a set of one. Many taxa are recognizable by a single diagnostic character, such as the possession of a notochord in the chordates. It is assumed that all the members of a taxon are derived from the nearest common ancestor.

How does the taxonomist find the characters that group species into higher taxa? Similarity was the exclusive yardstick in the delimitation of taxa in pre-Darwinian days. Darwin established an entirely new theoretical foundation for biological classification: the genealogical principle of common descent. He emphasized correctly that the development of groups of species is a result of evolution and that the delimitation of taxa must be based on this recognition. Only groups of species related by common descent should be recognized as taxa. Difficulties inherent in the definition of common descent are discussed in Chapters 9 and 10.

Hennig (1950) rightly praised the "process of reciprocal illumination," in which a tentative classification produces a deeper understanding of the information content of the characters, possible homoplasies, and other information that permits the production of an improved revised classification. This, of course, corresponds to the hypothetico-deductive approach (Hull 1967). In the construction of a classification, the first step ordinarily consists in recognizing, by inspection, seemingly "natural" groups on the basis of similarity. The "naturalness" of these groups is subsequently tested by all available methods, including the presence of appropriate synapomorphies, the elimination of conflict with the fossil record, and the unmasking of homoplasies and other factors affecting judgment on relationship. The primacy of a first phenetic approach was recognized by Hennig (1966*a*) and has been stressed by Simpson (1961), Wagner (1980), and many other taxonomists.

CLASSIFICATION AND PHYLOGENY

In spite of Darwin's clear statement that classifications should be based on genealogy, his advice was difficult to follow, primarily because of a continuing uncertainty about how to recognize the descendants of a common ancestor. The first to implement Darwin's recommendation was Ernst Haeckel (1866), who claimed to have based his classifications on the phylogeny of the groups concerned. His opponents objected to this method by asking, How can we know the phylogeny? Is the truth not exactly the reverse of Haeckel's claim? Is not the phylogeny inferred from the findings made during the establishment of classifications?

Actually, neither side was entirely correct in its claims. As Mayr (1969:77–78) stated:

> Neither is phylogeny based on classification nor is classification based on phylogeny. Both are based on a study of natural groups found in nature, groups having character combinations one would expect in the descendants of a common ancestor. Both sciences are based on the same comparisons of organisms and their characteristics and on a careful evaluation of the established similarities and differences.

This method is not circular, as was shown by Hull (1967). The independence in principle of phylogeny and classification was perceptively recognized by Felsenstein, who emphasized the idea that "the reconstruction of phylogeny is a logically distinct task from making classifications, and can be discussed separately" (1983*b*:316). Sober (1988:264) stated that "it is essential to see that classification and the reconstruction of genealogical relationships are separate undertakings." However, this view is not shared by some taxonomists, and even those who accept it are not certain how to reconcile these two independent objectives.

In spite of Darwin's conceptual clarification, most botanists and many zoologists continued to question whether we could ever be certain about the actual phylogeny of any group of organisms. Eventually this became a semantic argument. The study of homologies, the use of additional (especially molecular) characters, and the application of the careful distinction made by Hennig between ancestral (plesiomorphic) and derived (apomorphic) characters undoubtedly permit, at least in zoology, quite robust inferences on the most probable phylogeny in most cases. Such a reconstruction lacks the certainty of a physical law, but most revisions of older phylogenies based on modern character analysis have affected only minor details, not the major outline. More important, each segment of a phylogeny corresponds to a separate hypothesis that is open to testing and can be corrected if refuted.

What is the evidence that permits inferences on phylogeny? There are, in general, three classes of evidence.

Taxonomic Characters

Almost any attribute of an organism is useful as a taxonomic character if it differs from the homologous feature in some members of another taxon. Classical taxonomy employed structural characters almost exclusively, although Aristotle had recognized that groups of animals might be characterized by "their way of living, their actions, and their habits" (*Hist. Anim.* 1.1.487 all). Morphological characters have been increasingly supplemented in recent times by a large array of nonmorphological, particularly molecular, characters (Chapter 7).

An entirely new frontier of research was opened up through studies of the phyletic evolution of macromolecules: thousands of kinds of proteins as well as nuclear, ribosomal, and mitochondrial nucleic acids and other organic molecules. Each of these variables permits the construction of an independent phylogeny. By consolidating the phylogenies of different characters, one arrives at the "most probable" phylogeny. That phylogeny is most probable which is supported by the greatest number of independent sets of characters. For a more detailed study of taxonomic characters and an analysis of the principles of weighting, see Chapter 7.

Fossils

When very large gaps were found between higher taxa, reconstruction of the common ancestor through the study of homologies often failed in the past to produce satisfactory results, particularly as long as a study was limited to morphological characters. In such cases the discovery of a fossil was often very helpful, particularly in vertebrate classification. *Archaeopteryx* as a missing link between birds and reptiles, the mammal-like reptiles, and types such as *Seymouria* (between amphibians and reptiles) and *Ichthyostega* (between fishes and amphibians) proved very illuminating. A single occasional fossil, the only representative of the vast number of unknown members of intervening phyletic lines, is of course not to be accepted literally as the true ancestor of the later group. Such a fossil merely indicates in the roughest outline what the characters of those ancestors might have been.

At the same time, fossils sometimes create difficulties in the construction of classifications. When fossil organisms are somewhat aberrant or belong to extinct groups, the interpretation of structure is often controversial. In a group of so-called carpoid echinoderms that were designated by Jefferies (1986) as calcichordates, specialists still argue which is the anterior end and which is the upper surface. The place of this taxon in the classification of the Deuterostomia depends entirely on the solution of the biological problem of the locomotion and feeding of this group, because this information determines which structures are homologous. Similar problems exist for some of the Precambrian fossils.

Geographic Distribution

The pattern of distribution often reveals clues to relationship (Chapter 7).

THE THREE SCHOOLS OF MACROTAXONOMY

Macrotaxonomy made remarkably little progress in the 100 years after the publication of the *Origin* (1859). Although it was generally agreed that taxa should be related by descent, no one developed a specific methodology for the determination of common descent. Many taxa were still based on single key characters, and others were based simply on the "overall similarity" of the included species. Most taxonomists during this period concentrated on the species level, and their work eventually culminated in the new systematics. The first well-thought-out statement of the principles of traditional *evolutionary taxonomy* was made by Simpson in 1945; he followed this up in 1961 with a comprehensive treatment.

A more active interest in the problems of macrotaxonomy did not develop until the 1950s and 1960s. It has led to heated controversies which are still unresolved. The nature of the disagreement among the schools and the difference in their methods can be best understood when presented as a disagreement about how classification can reflect both similarity and descent. Since the criteria of these two indications of relationship are frequently in conflict, four procedures have been proposed to resolve the conflict.

1 To give primacy to one of the two sets of criteria, hoping that the result will simultaneously satisfy the other set:
- **a** *Phenetics,* which gives primacy to similarity
- **b** *Cladistics,* which gives primacy to the branching points of descent

2 To consider the two sets of criteria equally but sequentially:
- **c** *Evolutionary taxonomy* (taxa provisionally delimited by similarity and subsequently tested by *monophyly*)
- **d** Some schools of cladistics (construction of a cladogram, with taxa delimited by cutting branches of approximately equal length)

Each of these approaches has advantages and disadvantages, and the controversy concerning their relative merits has not ended. The principles and methods of the three major schools are discussed in detail in Chapters 8 through 10. However, because this chapter frequently refers to these methods, a short characterization of each of the three schools is needed here.

Phenetics

This school of taxonomists on the whole simply applies the elementary rules of classification specified above to the classification of organisms. Degree of overall similarity, as displayed by a large number of characters, is the overriding criterion of relationship for pheneticists. They believe that taxa delimited and ranked on the basis of this principle are natural groups with respect to which a greater number of general propositions can be made; those propositions are considered to be more important than those involving groupings arrived at in accordance with other taxonomic philosophies (Gilmour 1940).

Pheneticists also believe that their approach is not in conflict with Darwin's theory of common descent. Similarity, they believe, is to be expected among the descendants of a common ancestor. Therefore, bringing the most similar species and higher taxa together ought to produce a phylogenetic classification automatically. The phenetic method, however, does not have any procedure for testing this assumption.

The difficulties for the phenetic approach caused by homoplasy, mosaic evolution, absence of criteria for character selection, and other factors are discussed in Chapter 8.

Cladistics

Until 1950 no one had articulated an adequate method for testing the monophyly of taxa. Although most animal taxonomists silently agreed with Darwin that classifications should reflect phylogeny and that taxa should not be polyphyletic, no methodology had been published. The fact that so many pre-1950 classifications were largely or completely confirmed by subsequent cladistic analysis shows, however, that many taxonomists used such criteria. Apparently they grouped species that shared the greatest number of homologous characters.

In 1950 Willi Hennig (1913–1976) published his *Grundzüge einer Theorie der phylogenetischen Systematik,* in which he articulated a set of principles that he claimed would permit the establishment of an unambiguous genealogical classification. His most basic criteria were these: Only taxa based exclusively on the possession of synapomorphies, i.e., shared derived characters, should be recognized, while ancestral (plesiomorphic) characters should be ignored. Furthermore, every taxon should be "monophyletic" (=holophyletic), consisting of the stem species and all its descendants, including all "ex-groups." Finally, all sister groups should be accorded the same categorical rank.

The outstanding feature of Hennig's system is its endeavor to exclude reliance on mere similarity. His cladistic analysis consists of the partitioning of characters into derived and ancestral ones. The recognition of taxa and their ranking are entirely controlled by the inferred branching pattern of the phylogenetic tree (*cladogram*). The cladogram is then converted into a hierarchy of categories (construction of a cladistic classification) (Chapter 9).

There has been a characteristic difference between botanists and zoologists in regard to the choice of methods of classification. Owing to the seemingly greater frequency of homoplasy in plants and the seeming scarcity of diagnostic morphological characters for higher taxa, most botanists despaired of ever being able to construct a phylogenetic classification of plants (Stevens 1986). Furthermore, most botanists worked at the species level; a phylogenetic tree and the resulting classification seemed beyond their reach. They were satisfied, therefore, to use phenetic methods in the hope that this would produce a classification that reflected genealogical descent reasonably well. In numerical botany we find, therefore, that phenetic methods predominate. In zoology, by contrast, as soon as suitable computer methods for cladogram construction were available, they tended to replace phenetic methods.

Evolutionary Classification

In Chapter XIII of the *Origin* (1859) Darwin presented a carefully worked out theory of classification. He used two criteria in the construction of a classification: common descent and amount of evolutionary divergence. A classification must be "genealogical," he said; that is, it must reflect the common ancestry of the members of each taxon. However, Darwin also stressed that the unequal amounts of subsequent divergence of descendant groups must be duly considered in the ranking of taxa. "Genealogy by itself does not give classification" because "the *amount* of difference in the several branches or groups, though allied in the same degree in blood to their common progenitor, may differ greatly, being due to the different degrees of modification which they have undergone" (Darwin 1859:420, discussed in Mayr 1985). For Darwin there was no reason why sister groups should receive the same categorical rank, and many modern taxonomists agree. Although birds and crocodilians are sister groups, one would not give the widely divergent birds the same categorical rank as the crocodilians, which in most respects are still reptiles.

The acceptance of the principle of dual determination of a classification creates a serious dilemma for the taxonomist. At the present time there is no well-tested method for the simultaneous consideration of similarity and descent in the process of making a classification (see Chapter 10 for some early attempts). Therefore, the two factors have to be treated sequentially. A cladist might favor determining descent first and then ranking the resulting holophyletic taxa in a Linnaean classification. The difficulties of this approach are discussed in Chapter 9. The evolutionary taxonomist (Simpson 1961; Mayr 1969) prefers to establish reasonably homogeneous taxa and then test them for monophyly and remove all alien clements. It is obvious that some subjectivity is unavoidable in this procedure, particularly as a result of mosaic evolution, but evolutionary taxonomists still consider it preferable not to base classifications on heterogeneous holophyletic taxa.

It is evident from the recent controversies about virtually every method used in numerical taxonomy (Chapter 11) that none of these methods is free of weaknesses. There may be likelihood but never certainty that the best dendrogram achieved using a given method is a correct representation of the phylogeny. This situation seems to cause a good deal of distress to those with a strong mathematical background. It is less disturbing to those who realize that the reconstruction of a phylogeny and the construction of a classification are independent operations (Hull 1970). Even when it is impossible to decide by numerical methods which of several possible trees is the best one, it is nevertheless often obvious which would be best suited to form the basis of a useful classification. The selection of the evolutionarily most important characters and a due consideration of mosaic evolution and homoplasy are usually more important for the construction of a useful classification than is the choice of the numerical method used for the construction of the tree.

Comparisons

The different schools also differ in the extent to which they realize that a "good" classification must serve both practical and theoretical needs and that if these two needs are in conflict, a somewhat arbitrary compromise may be required.

One major difficulty in comparing the methodology and philosophy of the three major schools is their internal diversity. This is least pronounced among the pheneticists, although different authors may prefer different algorithms and although branching may or may not be taken into consideration. Differences among evolutionary taxonomists generally involve microtaxonomic procedures that have little effect on classification. The diversity would seem to be greatest among the cladists. Leading current Hennigians (Wiley 1981) now retain only the most basic of Hennig's original rules (holophyly and rejection of the use of plesiomorphic characters). Pattern cladists no longer aim to depict the evolution of taxa in their cladograms, but only a pattern of characters. Finally, the followers of Farris deviate from Hennig's classical method in considering distance in ranking, although they still insist that all taxa be holophyletic. All generalizations about the methods of the three schools discussed in Chapters 8 through 10 are subject to exceptions for some representatives of each school.

All three schools present the results of their analyses in the form of treelike diagrams, so-called *dendrograms*. Despite the pronounced differences in the dendrograms of the three schools, their main differences are in their philosophies. A cladist wants all the groups to be holophyletic; a pheneticist classifies by overall similarity without regard for phylogeny; an evolutionary taxonomist wants taxa to be monophyletic (traditionally defined) but also to reflect the information content of plesiomorphic characters in the delimitation and ranking of taxa.

Although the principles and methods of the three schools seem to be too different for any synthesis, it would be regrettable if this split in taxonomy continued. However, as long as all three groups believe that their method is the best, there is not much hope for a synthesis, for such a synthesis would presumably require abandonment of the key components of the methods of each school. Phenetics, for instance, might have to adopt the weighting of characters and the testing of taxa for monophyly (traditionally defined). Cladistics might have to give up delimiting taxa by holophyly and use kinds of information for ranking and classification different from those used for the construction of cladograms. Finally, evolutionary taxonomists would have to adopt cladistic analysis (some already have), work out more concrete methods for the delimitation and ranking of taxa, and come to a consensus on the definition of monophyly. These are for the time being impossible hopes.

IS A CLASSIFICATION A THEORY?

A protracted controversy has been going on in the taxonomic literature in recent years as to whether a classification is a theory. The answer depends on one's definition of theory. Classifications certainly do not correspond to the causal law explanations that philosophers call universal process theories. Biological classifications instead are historical narrative explanations based on the evolutionary theory of common descent. They share the explanatory and predictive properties of that theory. No classification is a single, universal proposition; instead, all are composed of a large number of individual propositions, each of which is a hypothesis that is more or less independent of the others, subject to the process of testing and attempted refutation.

The classification of the vertebrates, for instance, contains hypotheses regarding the relationships of thousands of genera, families, and orders. Each claim of relationship is a separate testable item. One can ask whether the homologies from which the nearest common ancestor is inferred are valid; one can ask whether the shared characteristics of two groups are genuine synapomorphies or independently acquired homoplasies or result from the retention of ancestral characters; one can test whether the chronology of the geological record substantiates or contradicts the location of branching points; one can determine by a character analysis whether one of two sister groups differs (by the acquisition of more autapomorphies) more widely from the inferred ancestor than from the other sister group; and one can check for evolutionary discordance, including mosaic evolution (Michener 1977). One can test separately every relationship reflected in a classification, and several methods are available for some tests.

Using the methods of evolutionary classification, one can sometimes base several different classifications on the same phylogram, and none of these classifications can be falsified in the strict sense of the word. The choice of which classification to adopt in that case has to be somewhat arbitrary, depending on one's conclusions about the information content and practicality of the competing classifications. This does not mean that the alternatives have been falsified.

Nevertheless, the definition of a biological classification given earlier in this chapter indicates that classifications have many attributes of a genuine theory. First, they have explanatory value, at least when they are specifically based on the theory of evolution. Such classifications explain the similarity of the members of a taxon as being due to common descent. By accepting evolution and by constructing classifications on the basis of homology and synapomorphy, one transfers the explanatory power of evolution to any classification based on it. Furthermore, a good classification, like a theory, permits one to make "predictions" (see below).

Finally, like most theories, all classifications are tentative and thus are subject to revision. The assignment of every species, genus, or family to a particular position in a classification can be tested by means of a character analysis. If the classification fails the test, it must be modified. Classifications thus have many of the traditional properties of theories. However, because classifications are based on historical narratives, they are unlike theories relating to physical laws in that they cannot be proved or, as wholes, falsified. Nevertheless, when well constructed, they reflect a pattern found in nature that is due to common descent and divergent radiation. Ultimately, however, a given classification is a human convention, even though every step in the analysis has theorylike qualities.

THE FUNCTIONS OF A CLASSIFICATION

In pre-Linnaean days, the only purpose of a classification was to serve as an identification key. Today a classification serves multiple purposes, both practical and theoretical (Warburton 1967; Bock 1973; Ashlock 1979).

A Classification Is an Index to Stored Information

Wherever there is diversity, as in books in a library or goods in a warehouse, there is need for a classified index to facilitate access to individual items. Even though the systematic classification stores only a very small part of this information, it is the key to the storage system of biological science. This system consists of museum collections and the vast scientific literature published in the form of books and scientific journals. The quality of any classification may be judged by its ability to facilitate the storing of information in relatively homogeneous divisions and permit, with the help of an efficient labeling system (nomenclature), rapid discovery and retrieval of this information.

A classification provides information on the structure of the included taxa: whether they are monotypic or rich in species, whether they possess numerous close relatives or are highly isolated, and other information important for evolutionists, biogeographers, and other biologists.

A Good Classification Has Heuristic Properties

It permits one to make predictions concerning the attributes of other members of the taxon (including those still undiscovered) and previously unused characters. An individual can be identified as a lark on the basis of just a few diagnostic characters, but, knowing that it is a lark, one can make probabilistic statements about its skeleton, internal organs, physiology, reproduction, and behavior without examining it.

A classification permits extrapolation from known to previously unstudied characters. An analysis of a few species that are strategically scattered through the natural system may provide us with much of the needed information about the distribution of a new enzyme, hormone, or metabolic pathway. Many animals cannot be kept in the laboratory, and others do not reproduce in captivity. Again, a sound system will permit all sorts of inferences from the genetically well known types, and the systematist can fill many gaps in our knowledge that specialists in the experimental branches of biology cannot. The goodness of a classification is documented by the degree to which such predictions are validated.

However, such predictions are not universal. All taxa vary, and a given prediction may not be valid for a given species because the taxon is polythetic or because the particular species or genus within the higher taxon has a unique specialization. The predictions made possible by classifications thus are probabilistic; they are not facts.

Classifications Permit the Making of Generalizations

The validity of generalizations derived from comparisons depends on the goodness of the classification on which the comparisons are based. The comparative method is most reliable when obviously homogeneous classes of phenomena are compared. Therefore, it is the objective of classification to construct groups that are homogeneous in the sense that they consist of descendants of the nearest common ancestor. Disturbing heterogeneity, however, can be avoided in classifications only if the degree of divergence of different phyletic lines is taken into consideration.

A Classification Has Explanatory Powers

If one accepts the principle of phylogenetic classification, one accepts the prescription that every taxon must be monophyletic (traditionally defined), that is, that all members of a taxon must have descended from the nearest common ancestor. Consequently, when we conclude after analysis that dogs and cats, together with some other families, constitute a natural mammalian order, we have implicitly advanced the theory that they have descended from a common ancestor. This tentative theory explains the shared attributes of members of a taxon as well as the gaps that separate different taxa. To be sure, it is not classification as such that supplies the explanation but rather the theory of evolution by common descent when applied to a classification.

It might be useful at this point to summarize our conclusions concerning the relationship between phylogeny and classification: It is not true that classification gives phylogeny but rather that an analysis of characters permits inferences on phylogeny that are used in the construction of a classification.

WHAT IS THE BEST CLASSIFICATION?

As soon as it is admitted that taxa are the product of evolution, it can be demonstrated that certain taxonomic groupings are inadmissible. Placing the whales with the fishes and grouping the bats with the birds are well-known examples of such misclassifications. As Simpson (1959*a*:294) rightly stated, "a classification can be wrong if it contradicts its own principles, but it can never be right in the sense of being the only one possible for a given set of principles." Nevertheless, from the days of Cesalpino, Ray, and Tournefort, it has been the stated ambition of taxonomists to construct the "best" classification of the particular group of organisms with which they are concerned. Even today, members of each of the three major schools of macrotaxonomy claim that their classifications are the best. The classifications of pheneticists attempt to bring the "most similar" organisms together in the belief that similarity, being a product of evolution, best reflects relationship. For a cladist, that classification is best which reflects most accurately the branching pattern of the phylogenetic tree, in other words, the genealogy of the various recognized taxa. A traditional evolutionary taxonomist might claim that his or her classification is the best because it combines a regard for genealogy with a regard for general principles of classification. In a given case it is often impossible to decide which is the best classification. Some authors therefore offer several alternative classifications. However, it is usually not too difficult to conclude that a certain classification is wrong because it is definitely contradicted by the evidence.

Instead of claiming to have constructed the best classification, some authors have claimed that theirs is the "most natural" classification. That classification was traditionally considered most natural which best reflected degree of relationship. However, there are several concepts of relationship, and their determination is controversial in all cases. Prior to Darwin, the parallel variance of characters was considered the best test for the naturalness of a classification. As stated by Whewell (1840), "the maxim by which all systems professing to be natural must be tested is this: that the arrangement obtained from one set of characters coincides with the arrangement obtained from another set."

On this basis the pheneticists, using overall similarity, consider their classifications to be the most natural. However, there are several other ways to define *natural* (Wiley 1981:71). For instance, if one agreed that a classification should reflect the causes responsible for the ordered groupings of organisms—common descent and a sharing of the selection pressures to which all members of a single phyletic lineage have been exposed in the past—a natural classification would have to reflect phylogeny. In that case the association of characters would be merely a by-product and indicator of common descent. The true clue to common descent would be the shared possession of uniquely derived characters. Even if one accepts this criterion, one still has two options concerning the most natural classification, one exclusively based on cladogenesis (cladistics) and one based on both cladogenesis and an evaluation of all characters (evolutionary classification). In view of these contradictions and ambiguities, few modern authors claim any longer that their classification is the most natural one.

THE STEPS IN THE TAXONOMIC PROCEDURE

It is often assumed that a classification is the product of a single procedural step, but this is not correct. The construction of a classification usually requires a sequence of steps, some of which are taken silently for granted while others are ignored or their necessity is denied by some taxonomists. Three major steps are involved, and the procedure of classifying can be best understood if these steps are dealt with separately (Mayr 1969:199).

1 Assembling phena into populations and populations into species. This step—the procedure for assembling individuals and populations into species and their delimitation against each other—was described in the section on microtaxonomy in Chapter 5.
2 Gathering species into species groups and formal taxa.
3 Ranking taxa in a hierarchy of categories.

The second step—the grouping of species into species groups and formal taxa—was traditionally guided by similarity. The same methods were used in biological classification that were used in the classification of inanimate objects. Overall similarity was the traditional criterion of relationship. After the acceptance of the theory of common descent, when such taxa were tested for monophyly (particularly in recent years, when cladistic analysis was employed), some were found to be heterogeneous. This was particularly true in cases where the classifier failed to guard against the pitfalls of convergence. However, most classical taxa that had been based on similarity, such as birds, mammals, reptiles, beetles, and cats, were found to be "natural." Traditional phenetic classifications have on the whole been so successful simply because, other things being equal, species that are descendants of a common ancestor tend to be more similar to one another than they are to species with which they do not share immediate common descent. To overcome the subjectivity of the phenetic approach, cladists have adopted a radically different method of delimiting taxa (see below).

Polythetic Taxa

Some seemingly natural taxa cannot be characterized by a few clear-cut diagnostic characters. The members of such *polythetic* taxa resemble each other—and differ from those of other taxa—by the possession of a considerable set of characters but do "not resemble each other in respect to every single one" (Beckner 1959:23). Beckner used the preoccupied term *polytypic* for such taxa, a term replaced with *polythetic* by Sneath (1962). Since not all members of a polythetic taxon necessarily share the same set of synapomorphic characters, the recognition of polythetic taxa is rejected in principle by most cladists. However, cladists may make exceptions in the case of reversals (loss of structures) and a limited amount of autapomorphy.

THE DELIMITATION OF TAXA

The methodology of grouping species into species groups and formal taxa of increasingly higher rank has undergone great changes during the history of taxonomy. A group of species that is sufficiently distinct to justify its recognition by a separate name is a higher taxon. Such a taxon is not a class of objects brought together by a definition, but a "historical entity" (Wiley 1980) with some of the properties of particulars as defined by philosophers (Chapter 1). Each taxon is ranked in a hierarchical classification in a definite category, from genus (and subgenus) to kingdom (see below).

Pheneticists consider higher taxa to be products of phenetic operations and thus to be human constructs. They consider these taxa to be classes. Cladists and evolutionary taxonomists consider taxa to be particulars, that is, to have discrete reality in nature. Penguins, birds, beetles, marsupials, and nearly all higher taxa have all the defining characteristics of particulars rather than of classes and are thus facts of nature rather than human constructs (Ghiselin 1980; Wiley 1980). They "exist in nature independent of man's ability to perceive them." This is equally true for the monophyletic taxa of the evolutionary taxonomist and the holophyletic taxa of the cladist.

It is the task of the taxonomist to bring together species that form a "natural" taxon that is well demarcated against the species of its sister taxon. Many of the disagreements among specialists stem from differences of opinion about how this should be done.

The traditional approach was to sort species according to similarity, but similarity (or its absence) can be misleading when groups have become similar by convergence (legless lizards and snakes) or drastically different by entering very different adaptive zones (whales and bats). The acceptance of the theory of common descent led to the development of a different criterion of naturalness: community of descent. Since legless lizards descended not from snakes but from four-legged lizards, they were classified with lizards, and since whales descended from terrestrial mammals, they were classified with the mammals rather than with the fishes.

The question of which species are sufficiently similar to be placed into a natural taxon was replaced by the question: How can one recognize groups formed by descent from a common ancestor?

Although the fact is often overlooked, this shift to a new concept was involved in a drastic change in the status of the taxon, from "artificial" to "discovered." The taxon delimited on the basis of similarity was always more or less an artificial entity. What is similar is decided by means of subjective evaluation and thus is somewhat arbitrary. Such taxa are "made," as taxonomists often frankly admitted. How often does one find in the older literature statements such as "I made a new genus, a new order, etc."? Such subjectivity is, at least in principle, eliminated by the adoption of the objective criterion of community of descent. Taxa based on this principle are discovered, not made. We say "in principle" because in practice there is still one major difficulty: Taxa are products of evolution, but evolution is not an orderly process. It allows for all sorts of intermediacy between higher taxa and, in particular, much conflict between criteria of similarity (as evaluated by the practicing classifier) and degree of relationship.

There is a strong difference between the methods of cladists and those of evolutionary taxonomists in regard to the delimitation of taxa. Since cladists "make" taxa by applying the principle of holophyly, their taxa are unambiguously delimited. The taxa of evolutionary taxonomists are arrived at in a two-step process. At the first step similarity classes are established. They are tested for monophyly at the second step, during which all nonconforming species are eliminated. The delimited taxon is both monophyletic (traditionally defined) and relatively homogeneous.

HIERARCHICAL CLASSIFICATION

If the classification system of animals consisted only of a list of species names in alphabetical sequence, it would be virtually useless as the basis of an information storage system (Warburton 1967); it would not provide a key for finding any particular item of information. Furthermore, it would not correspond to the situation in nature, for organic diversity does not consist of an evenly spaced array of equally different species. Rather, species are arranged in groups (genera) of related species; these genera in turn are grouped in families that represent the next higher rank, and so on up to the level of kingdom. In other words, species in nature are ranked, according to their comprehensiveness, in a hierarchy of nested categories, the so-called *Linnaean hierarchy*.

Within the animal kingdom the highest category in regular use is the phylum and the lowest is the species. The hierarchy is constructed by assigning a definite rank (such as family) to each taxon, subordinating the lower categories to the higher ones. The Linnaean hierarchy thus consists of a nested set of taxa of different categorical rank. For instance, the species of doglike animals of the genus *Canis* are grouped with other canidlike genera (e.g., *Vulpes*) in the family Canidae; the members of this family and those of other carnivorous families such as the Felidae and Mustelidae are grouped in the order Carnivora, and so on to larger and larger groups of ever higher rank. The process continues until all living organisms are grouped into five kingdoms, of which the Animalia is only one; others are the Plantae, Fungi, Protista, and Monera (prokaryotes). Rank in this hierarchy, other things being equal, indicates degree of similarity and recency of common origin. The lower the rank of a taxon, the more similar the included species usually are and the more recent their nearest common ancestor is.

Hierarchical arrangements are used in many classifications, not only those of organisms. The classification of the language families is hierarchical, and even most artificial classifications, such as those of books in a library, are hierarchical. A hierarchical arrangement has many virtues, telling us, for instance, that a taxon of lower rank also has the attributes of the higher taxa under which it is ranked. Once we know that a given organism is a cardinal, we know that it also has the attributes of songbirds, birds, vertebrates, chordates, and animals.

The Linnaean Hierarchy

Linnaeus, the first taxonomist to establish a definite hierarchy of taxonomic categories, recognized within the animal kingdom only five categories: *classis, ordo, genus, species,* and *varietas*. When the number of known animals grew, making finer divisions necessary, two additional categories soon became obligatory: the *family* (between genus and order) and the *phylum* (between class and kingdom). The varietas, used by Linnaeus as an optional category for infraspecific variants, was eventually discarded or replaced by the *subspecies*.

The listed categories form the basic taxonomic hierarchy of animals. Any given species thus belongs to seven obligatory categories, as follows:

	Wolf	**Honeybee**
Kingdom	Animalia	Animalia
Phylum	Chordata	Arthropoda
Class	Mammalia	Insecta
Order	Carnivora	Hymenoptera
Family	Canidae	Apidae
Genus	*Canis*	*Apis*
Species	*lupus*	*mellifera*

These five higher categories (genus, family, order, class, and phylum) permit the placing of a species of animal with a fair degree of accuracy. However, as the number of known species increased, and with it our knowledge of the degrees of relationship of these species, the need arose, particularly in species-rich groups, for a more precise indication of the taxonomic position of a given species. This was accomplished by splitting the original seven categories and inserting additional ones. Most of these additional categories are formed by combining the original category names with the prefixes *super-* and *sub-*. Thus there are superorders and suborders, superfamilies and subfamilies, etc. The most frequently used additional new category name, especially in entomology, is the term *tribe,* which indicates a category between genus and family. Vertebrate paleontologists also routinely use the category *cohort* between order and class. Some authors use terms for additional subdivisions, such as *cladus, legio,* and *sectio*. Some use *infraclass* below the subclass and *infraorder* below the suborder. Some cladists have proposed numerous additional designations of categories for the fine dichotomous branching of cladograms.

The generally accepted categories are the following:

Kingdom
Phylum
Subphylum
Superclass
Class
Subclass
Cohort
Superorder
Order
Suborder
Superfamily (-*oidea*)
Family (-*idae*)
Subfamily (-*inae*)
Tribe (-*ini*)
Genus
Subgenus
[Superspecies]
Species
Subspecies

The standardized endings in zoology for the names of tribes, subfamilies, families, and superfamilies are indicated in parentheses. Standardized endings for the categories above the family group have not been adopted in zoology. The brackets indicate that the category superspecies is used only in some branches of zoology (e.g., ornithology). On the whole, the same categories are accepted in botany, but the standardized endings are generally different.

The rank of a taxon in a Linnaean classification is indicated by three conventions: (1) the category into which it is placed (e.g., family), (2) the ending of the taxon name (e.g., *-idae*), and (3) the appropriate indentation in a sequential listing of higher taxa.

The Linnaean hierarchy, with its need for arbitrary ranking, has often been attacked as an unscientific system of classification. Alternative methods such as numerical schemes have been proposed but have not found favor among taxonomists for two reasons. First, assigning definite numerical values to taxa demands a far greater knowledge of their relationships than can be inferred from the available evidence. Second, the assignment of such values would freeze the system into a state of finality that would preclude further improvements. The very subjectivity of the Linnaean hierarchy gives it the flexibility required by the incompleteness of our knowledge of taxonomic relationships. It allows the proposal of alternative models of relationship, permitting different authors to test which particular balance between splitting and lumping stores the greatest amount of information. Like any other scientific theory, it will forever be provisional.

HIGHER CATEGORIES AND HIGHER TAXA

A higher category is a class into which are placed all the higher taxa that are ranked at the same level in a hierarchic classification. The category selected for a given taxon indicates its rank in the hierarchy. As explained in Chapter 1, taxa are based on zoological realities, while categories are based on concepts. In the Linnaean hierarchy there is no difference between the species category and the higher categories. In other respects, the concept of the species and the concepts of the higher categories are quite different.

The category species is "self-operationally" defined by the testing of isolating mechanisms in nature in nondimensional situations (Chapter 2), but definitions for categories above the species are arbitrary. The species category signifies singularity, distinctness, and difference, while the higher categories have the function of grouping and ordering by emphasizing affinities among groups of species—they are collective concepts. Even though a nonarbitrary criterion does not exist for the higher categories or for the rank they signify, these categories do have an objective basis. A taxon correctly delimited and placed in a higher category is natural provided that it is consistent with the theory of common descent. Higher taxa are often, if not usually, well delimited and separated by a pronounced gap from other taxa of the same rank.

A further difference between the species and higher categories is that comparative data furnish the evidence used for the delimitation of higher taxa and their ranking into categories, while interbreeding or its absence is the criterion used for ranking at the species level; i.e., the species is a relational concept (Mayr 1957) whereas the higher categories are not.

Darwin supplied the scientific interpretation of categorical ranking when he stated (1859:422) that "the natural system is genealogical in its arrangement, like a pedigree; but the degrees of modification which the different groups have undergone have to be expressed by ranking them under different so-called genera, sub-families, families, sections, orders, and classes." In the course of geological history, the descendants of an aberrant species may evolve into a different genus, the genus into a different family, and so forth. The origin of higher categories is thus exactly the opposite of that envisioned by some scholastic philosophers [and by Goldschmidt (1952)]. *Higher categorical rank of taxa evolves through evolution from a lower rank; lower rank does not evolve through subdivision of higher categories.*

Most taxa above the family level are sharply delimited. Mollusks, penguins, beetles, and indeed most higher taxa are separated from their nearest relatives by a decided gap, far more distinct than the gaps that separate most genera and families. Nevertheless, it remains true that the higher categories in which we place these taxa are ill defined. Category means categorical rank, and no yardstick has been found for the nonarbitrary ranking of taxa. There is hardly a higher taxon that is not ranked higher by some specialists and lower by others. It is by the arbitrariness of definition that all higher categories differ from the species category. Different criteria and operations for ranking are employed by the three schools of macrotaxonomy (Chapters 8 through 10).

The Genus

The genus is the lowest obligatory higher category and the lowest of all categories established strictly by comparative data (Cain 1956). For the modern taxonomist the genus is not different in concept from the family, the order, or other higher categories. For Linnaeus, who based his theory of classification on the principles of Aristotelian logic, the genus occupied a very special place (Cain 1958). This fact would have only historical interest if Linnaeus had not incorporated Aristotelian logic into the binominal system of nomenclature.

A convenient pragmatic definition of the genus is as follows: *The genus is the obligatory taxonomic category directly above that of the species in the Linnaean hierarchy.* A genus (generic taxon) is a monophyletic group composed of one or more species that are separated from other generic taxa by a decided gap. For practical reasons, it is recommended that the size of the gap be in inverse ratio to the size of the taxon. In other words, the more species in a species group, the smaller the gap needed to recognize that group as a separate genus; the smaller the species group, the larger the gap needed to recognize it. One of the functions of the genus from Linnaeus's time on has been to relieve the memory (to facilitate information retrieval), and the inverse ratio recommendation prevents the recognition of a burdensome number of monotypic genera. Delimiting species groups of optimal size as genera is an operation that requires experience, good judgment, and common sense. For a discussion of helpful criteria, see Chapter 10.

Generic Characters Taxonomic characters that prove generic distinctness do not exist. Taxonomic literature would have been spared countless generic synonyms if taxonomists had always remembered Linnaeus's (1737) dictum: "It is the genus that gives the characters, and not the characters that make the genus." This is still generally valid, even though we have abandoned the Aristotelian logic on which Linnaeus based this statement. The soundest genera are based on an overall appreciation and weighting of the various considerations cited in this chapter (Michener 1957; Inger 1958).

The species included in a genus usually have many features in common, and recognition of a higher taxon is generally based on the presence of a correlated character complex. This may include some rather minute and inconspicuous characters, but as Darwin said (1859:417), "The importance, for classification, of trifling characters, mainly depends on their being correlated with several other characters of more or less importance. The value indeed of an aggregate of characters is very evident in natural history." So important has this principle been considered by taxonomists that it led to much generic splitting when a species was found to lack one character or another of the correlated complex. Instead of revising the diagnosis of the genus and if necessary recognizing a polythetic genus, the taxonomist named a new one.

Many genera that are clearly natural groups cannot be diagnosed unequivocally by means of a single character. This is the case when a character diagnostic for the majority of the species is modified or lost in at least one species of the genus (*polythetic taxa*). This situation exists in many genera and even in families of birds.

Meaning of the Genus When we assign generic rank to a group of species, we wish to express a number of characteristics of all species included in that genus. A genus taxon is a phylogenetic unit, which means that the included species are descended from the nearest common ancestor. Very often it is also true that the genus is an ecological unit consisting of species adapted for a particular mode of life. The genus niche is obviously broader than the species niche, but both kinds of niche exist. Lack (1947) has convincingly shown the adaptive significance of genera in the Galápagos finches (Figures 6-1 and 6-2). Of course, as emphasized by Dubois (1982), the species of a genus possess considerable genetic identity. Species belonging to the same genus sometimes are able to produce hybrids. Dubois has gone so far as to demand that all species capable of hybridizing with each other should be placed in the same genus.

Recognition of a generic taxon corresponds to the stating of a theorylike proposition; like any scientific theory, it must have explanatory, heuristic, and predictive value. If there are several alternative ways of delimiting a genus, we must be guided by the same principles used in the recognition of any scientific theory. "Where alternatives are available, we stand by the theory or concept that is most useful—the one that generalizes the most observations, and permits the most reliable predictions" (Inger 1958:383).

The Family

As with all other higher categories, a nonarbitrary definition of the family category is not possible. What a layperson would call a "kind" of animal is often a family: the ladybird beetles (Coccinellidae), the long-horned beetles (Cerambycidae), the woodpeckers (Picidae), the swallows

FIGURE 6-1
One suggested tree for the adaptive radiation of Darwin's finches on the Galápagos Islands. (*From Lack 1947.*) For a few recent modifications, see Grant (1986).

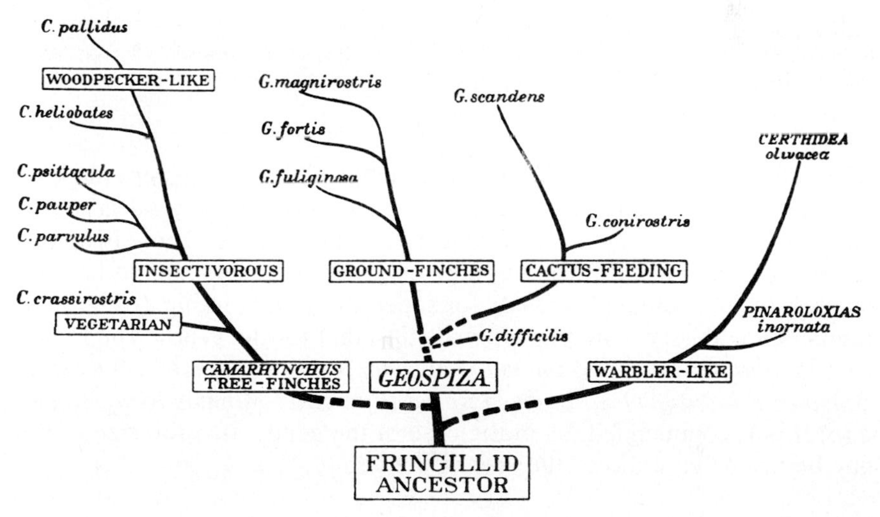

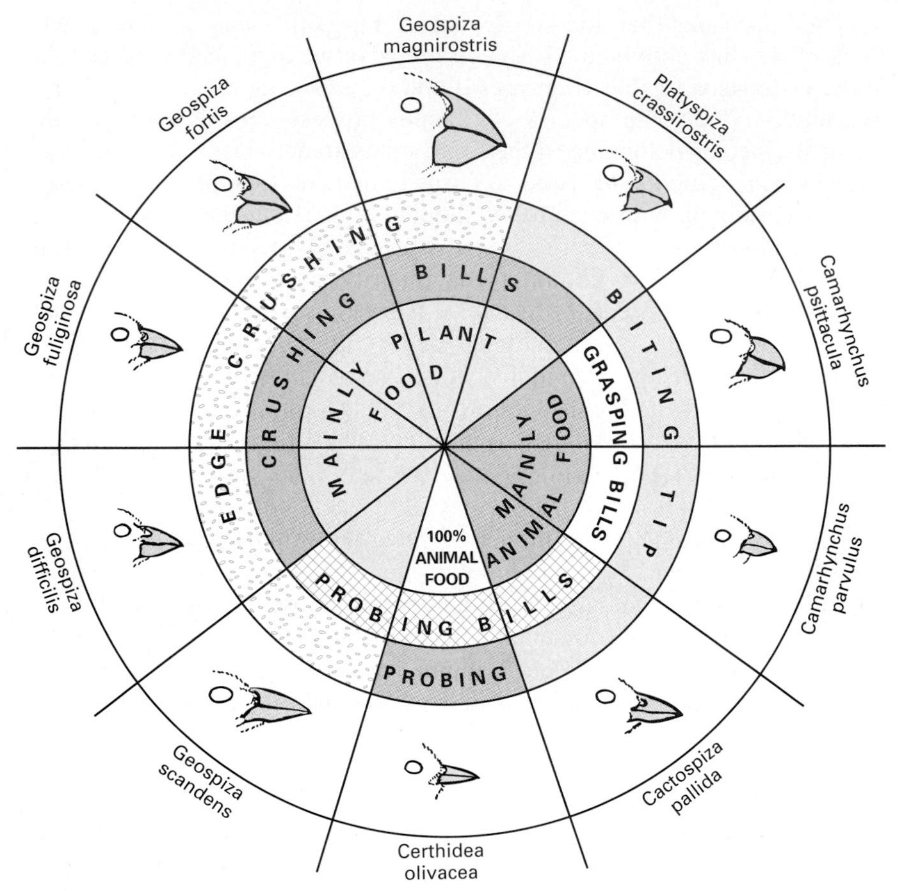

FIGURE 6-2
Niche occupation, feeding habits, and bill structure in 10 species of Darwin's finches from Santa Cruz (Indefatigable Island, Galápagos). (*From Bowman 1961.*)

(Hirundinidae), and so forth. How distinctive a group of genera must be in order to be considered a family varies from one zoological group to another for the various reasons indicated in this chapter. There is no criterion that will indicate whether a given group of genera should be considered a tribe, subfamily, family, or superfamily. A working definition of the family category would be equivalent to that for the genus: *The family is a taxonomic category for a taxon composed of a single genus or a group of related genera; it is separated from other families by a decided gap.* It is recommended, as in the case of the genus, that the size of the gap be in inverse ratio to the size of the family.

Like the genus, but perhaps to an even greater degree, the family tends to be distinguished by certain adaptive characters that fit it for a particular adaptive zone, e.g., the woodpeckers or the family Picidae, the leaf beetles or the family Chrysomelidae. The more distinct the adaptive zone, the wider the gap from other families. Families are older than the genera they contain and more often have a worldwide distribution. An entomologist who knows the 422 families of British insects can go to Africa or even Australia and recognize nearly all the same families that occupy similar niches.

Thus the family is a very useful category. A British entomologist would have to learn only 422 names to place a total of about 4800 genera and more than 20,000 species. It is especially useful to the general zoologist because each family usually presents a general facies that is recognizable at a glance, and most or all its species occupy a similar niche in their particular communities, as, for instance, do most of the thousands of species of Cerambycidae.

At a given locality the various families, like the various species, are generally distinct. Decided gaps between families are the rule rather than the exception, and little or no difficulty is encountered in "keying out" families in local faunal works. Unfortunately, the situation becomes much more complicated when a worldwide study is undertaken. Families are often found to form different distinctive groups on each continent, and types that bridge the gap between families are sometimes found. Relict groups existing at the family level can defy efforts to attain a clear-cut classification. Thus, in many insect groups (scale insects, aphids, water striders, etc.) a choice has had to be made between enlarging the family concept beyond the limits of local convenience or recognizing connecting exotic types as separate families and using a superfamily category for the group as a whole. In entomology there appears to be a trend, not necessarily desirable, toward the employment of superfamilies. By contrast, some large traditional families of birds, such as the flycatchers (Muscicapidae), warblers (Sylviidae), and shrikes (Laniidae), were found to have been based on homoplasy. The Australasian components of these unnatural families had to be recognized as separate families or transferred to other Australian families (Sibley and Ahlquist 1983).

Linnaeus did not recognize the family as a category, but it is significant that most of his animal genera have since been elevated to family rank. From this we may infer that his generic concept was compatible with our modern family concept, with the difference between the genus and family being merely one of degree. With only 312 genera of animals known in 1758, Linnaeus had no need for an intermediate category between genus and order. However, the number of newly discovered animal types increased so rapidly after 1758 that early nineteenth-century naturalists gradually developed and universally applied the family concept to designate an intermediate level between genus and order.

The number of families continues to grow because of the advance in knowledge about existing animals and the discovery of new types. By the end of the nineteenth century about 1700 families of animals were recognized (Perrier 1893–1932). The latest count (per Mayr) is 5600 families of Metazoa and 580 families of Protozoa, totaling approximately 6200 families. Thus, we now know more families of animals than Linnaeus knew species (4162).

Orders, Classes, and Phyla

The higher taxa above the family level in the recent fauna are on the whole very well defined and are even less often connected by intermediates than families are. There are, however, two exceptions to this broad statement. First, a drastic reevaluation of the significance of taxonomic characters in certain groups of lower invertebrates, for instance, sponges and turbellarians, has led to a drastic reclassification at the ordinal level. Even where a complete consensus exists in the delimitation of taxa, there may be a strong disagreement about the ranking. Instead of recognizing more suborders and superfamilies, certain authors have raised almost all taxa in rank, producing a great imbalance among the respective portions of the system (Chapter 10).

The taxa ranked in these highest categories represent the main branches of the phylogenetic tree. They are characterized by a basic structural pattern laid down early in evolutionary history, the special adaptive significance of which can now be perceived only dimly if at all. Superimposed on this pattern are seemingly endless modifications resulting from a series of *adaptive radiations* that have taken place within the classes and phyla. In general, then, taxa in the higher categories are definable in terms of a basic structural pattern, but except for certain highly specialized groups, such as the order Siphonaptera (fleas), the order Chiroptera (bats), and the order Impennes (penguins), the higher taxa are not primarily or even predominantly distinguished by special adaptations. The taxa contained in the higher categories are in most cases widely distributed in space and time.

As with genera and families, there has also been an increase in the number of recognized taxa above the family rank. According to recent tabulations, there are approximately 29 phyla, 144 classes, and 772 orders of recent animals.

THE PROCESS OF RANKING

The process of classifying is not complete when species are assembled into species groups and genera, for the genera still must be combined into families and families must be combined into still higher taxa until the complete Linnaean hierarchy has been constructed. This step—the determination of the appropriate rank for each recognized taxon—is the issue on which there is the least agreement among contemporary taxonomists.

As long as the diversity of nature was believed to reflect a *scala naturae* from the most primitive to the "most perfect," higher categories did not make much sense. Cuvier and von Baer, however, recognized major types or archetypes of animals rather than a *scala naturae* and adopted a hierarchical arrangement of taxa from phylum down to genus and species. Although this hierarchical system was followed by most authors in the first half of the nineteenth century, one great uncertainty remained: the explanation of this hierarchy (Mayr 1982*a*).

The reasons why a hierarchy was needed were quite clear to Darwin. He saw that organic diversity could have originated only through speciation and that chance and adaptive processes were responsible for the gradual evolution of higher and still higher taxa separated by gaps caused by divergent evolution and extinction. It was the evaluation of this multiplicity of processes—speciation (branching), adaptational divergence, and extinction—that caused dissension among taxonomists. Could existing classifications be reconciled with the new phylogenetic thinking? One must remember that by 1859 a categorical ranking (no matter how erroneous in hindsight) was available for most groups of animals. Could this pre-Darwinian classification, based as it was entirely on *affinity* (similarity), be converted into one that appropriately reflected Darwin's emphasis on genealogy? On the basis of what principles should one rank higher taxa?

The three current major schools of macrotaxonomy differ in their views on this subject. The pheneticist uses a measure of overall similarity to group and rank organisms. Cladists who follow Hennig (1966*a*:156) introduce a new rank at each branching point of the cladogram and give sister groups identical categorical rank (for recent departures from these conventions, see Chapter 9). For the classical taxonomist, ranking results from the degrees of difference found among taxa; much divergence from the ancestral condition requires that a taxon be given a higher rank.

Classifications proposed by cladists are on the whole rather more elaborate than those of evolutionary taxonomists, because cladists want their classifications to reflect as minutely as possible the actual branching pattern of the genealogy (Wiley 1981:199–238); gaps are consciously ignored. Evolutionary taxonomists tend to emphasize major groupings and the existence of major gaps.

RELATIONSHIP AND SIMILARITY

In the taxonomic literature there has been an unfortunate equivocation about the meaning of the term *relationship,* a situation that goes back almost 200 years. In some recent arguments between pheneticists and cladists, both acclaimed the capacity of their respective methods to bring out "relationships." The pheneticists actually meant overall similarity, while the cladists meant genealogy. The term *relationship* has different meanings even when it is used to reflect genealogy. For the evolutionary taxonomist it means both ancestor-descendant relationship and collateral relationship among sister lineages, while for the cladist it applies only vertically within a holophyletic lineage.

Relationship between two taxa is most often indicated by degree of similarity, but there are many exceptions. False similarities can be unmasked only through a careful analysis of taxonomic characters. In such an analysis one must distinguish between different potential causes of similarity. These are best grouped under the terms *homology, convergence,* and *parallelism.*

Homology

Relationship among species and higher taxa is indicated by the existence of homologous characters, but there is considerable uncertainty about what homology is and how it can be established. Morphologists in the first half of the nineteenth century (Geoffroy, Oken, the Quinarians) ascribed similarity among groups of organisms to two causes, affinity and analogy (Mayr 1982*a*:202, 209, 458). Owen, in the pre-Darwinian period, clearly distinguished between similarity due to performing the same function (*analogous* characters) and similarity due to belonging to the same type of organism (*homologous* characters) (Mayr 1982*a*:464). When Darwin discovered common descent as the cause of homology, it became possible to adopt a more rigorous definition than the ones suggested by these forerunners, and yet, 125 years after the publication of the *Origin,* there is still considerable argument over the definition of homology. The problem is how to avoid a definition that is circular.

Simpson's analysis of the problem (1961:68–93) is particularly enlightening. He points out that what is involved is the relation between a definition and the evidence that the definition has been met in a particular instance. Such a relation is well illustrated by the phenomenon of identical twins. Two siblings are not identical twins because they are so similar; rather, they are so similar because they are identical for having been derived from a single egg cell. The establishment of an unambiguous definition of homology is thus the first step in the analysis. The biologically most meaningful definition would seem to be as follows: *A feature in two or more taxa is homologous when it is derived from the same (or a corresponding) feature of their nearest common ancestor.*

This definition applies equally to structural, physiological, molecular, and even behavioral features of organisms. One must be careful in the application of this principle. The wings of birds and those of bats are homologous as tetrapod forelimbs, but they are not homologous as wings because they were not derived from the wings of a common ancestor. Hence, in order to establish homology, the evidence which leads to a particular inference of homology must be critically analyzed and stated. For instance, the metamerism of arthropods (and annelids) evidently evolved independently from that of chordates. It is therefore erroneous to treat metameric phenomena of the two phyla as homologous. Also, homology must be clearly demarcated against other phenomena, such as homonomy, analogy, parallelism, and convergence, with all of which it is frequently confused (Bock 1977; Holmes 1980).

Any claim of homology based on our definition is a hypothesis which must be substantiated. As Simpson (1961:88) stated, "What is the evidence that the definition of homologous is met in this particular case?"

Many types of evidence have been proposed by comparative anatomists and other evolutionary biologists. The most detailed presentation of this evidence is that of Remane (1952:31–84), who distinguished the following homology criteria:

1 *Position in relation to neighboring structures or organs:* This criterion is particularly useful for the determination of skull bones in vertebrates but is in most cases inapplicable to nonmorphological features.

2 *Quality as expressed particularly in function:* The liver, heart, or other organs of vertebrates are of the same quality and, as shown by other criteria, are clearly homologous. However, this criterion is on the whole rather weak because a similar quality and function can often be attained by means of convergence. For instance, eyes have evolved in the animal kingdom independently at least 40 times (Salvini-Plawen and Mayr 1977).

3 *Connecting of two dissimilar stages by connecting intermediate stages:* Evidence that the thyroid of mammals is homologous with the endostyle of *Amphioxus* is provided by an intermediate organ found in the cyclostomes.

4 *Similarity in the ontogeny:* Structures which in the adult condition appear highly dissimilar may have diverged through a set of ontogenetic stages from a very similar embryonic structure. Even when one rejects the extreme interpretation of recapitulation, it remains true that related taxa are often far more similar to each other as embryos than they are as adults. An analysis of embryonic stages may reveal homologies that would not be apparent in a comparison of adults.

5 *Existence of intermediate conditions in fossil ancestors:* This category of evidence was not mentioned by Remane but has been used very successfully by paleontologists. Evidently, ancestral groups often have characters that are intermediate between those of derived groups and those of a still earlier ancestor. This evidence for homology is of course available only in groups with an adequate fossil record.

6 *Features compatible in their distribution with the delimitation of monophyletic taxa (Patterson's principle of congruence):* "A true homology will circumscribe a group that is congruent with those specified by other homologies" (Patterson 1988:606). It is not quite clear whether circularity is involved in this criterion.

It would be misleading to base the definition of homology on one of these criteria, as many authors have done. The definition of homology must be based on the underlying concept of homology, and that is quite clearly derivation from a corresponding feature in the common ancestor.

The objection has been raised that the recognition of homology involves circular reasoning. This is an unwarranted argument. Claims of homology are clearly based on legitimate hypothetico-deductive reasoning.

Homologous features usually show similarity. They usually also perform the same or similar functions, particularly in close relatives. However, it is misleading to refer to any kind of similarity as homology, as is done by some molecular biologists. A protest against such a confusing misuse of the word has been published by Reeck et al. (1987); this problem has also been discussed by Patterson (1988). It is precisely the function of the word *homology* to discriminate between fortuitous similarity and convergent similarity due to common ancestry. In view of these two very different potential causes of similarity, it is misleading to include similarity in a definition of homology. In fact, some homologous features are exceedingly dissimilar in form and function, such as the articulating bones of the reptilian jaw, which evolved into two of the middle-ear ossicles of the mammals. The detection of the homology of such dissimilar features is one of the most gratifying triumphs in comparative research.

All homologies are inferred. They have at first the character of hypotheses. The more the phylogeny postulated by them is also supported by homologies of other characters, the more the validity of such a homology is strengthened. Actually, real problems arise only among rather distant relatives, particularly those which are separated by a major gap. Relationship in such cases is generally tested by the analysis of homology in several entirely different systems of features. Molecules, for instance, evolve very much as macroscopic structures do, and the phylogeny of molecular changes can often be reconstructed with considerable certainty (Fitch and Margoliash 1967). When the molecular and the morphological phylogenies confirm each other, the validity of both postulated homologies is greatly strengthened.

As with most problems in science, obvious hypotheses are accepted provisionally unless they lead to logical contradictions. The establishment of homologies ranges from the simple comparison of features of closely related species, where the matter need hardly be given a second thought, to the frustratingly difficult comparison of dissimilar features in higher taxa.

Serial Homology This term was coined by Owen (1866) for the morphological correspondence among repetitive or serial structures within a single organism, such as consecutive vertebrae, teeth, body segments, extremities, and any other metameric or meristic features. Metamerism is clearly a phenomenon that is so completely different from homology among different taxa that it is best referred to by Bronn's term *homonomy* (see Remane 1952:84–87 for a thorough discussion).

Analogy For the sake of completeness, the term *analogy* must be mentioned here; it is often, but rather erroneously, called an antonym to homology. *Analogy is phenotypic similarity; it is not related to community of descent but is due to similarity of function.* The wings of birds and those of insects are analogous structures. Convergent characters are analogous insofar as the similarity can be attributed to function, as is usually the case. Most analogous characters are so evidently not homologous that they create no problem for the taxonomist.

Homoplasy There are three kinds of changes of characters in evolution that mimic homology. They are usually grouped together under the term *homoplasy*. Homoplasy is similarity in a character independently acquired by two taxa. The term is particularly useful for similar taxa in an intermediate degree of relationship that makes it impossible to determine whether the similarity is due to a common intrinsic potential (parallelism) or to entirely unconnected evolutionary developments (convergence). The attempt to determine whether an observed similarity is a genuine homology or a homoplasy ought to be an indispensable component of every taxonomic analysis. Unfortunately, it is altogether ignored in unweighted phenetic procedures and often insufficiently considered in the construction of shortest trees (Chapter 9).

Convergence in Characters

The acquisition of a similar character by two taxa whose common ancestor lacked that character is called *convergence*. Although some taxonomists feel that convergence is the greatest obstacle to proper phylogenetic classification, the difficulty it causes has been exaggerated. Even though convergence involving particular characters, entire organs, or general shape and proportions occurs quite frequently, convergence between the entire phenotypes of unrelated organisms is rare if not nonexistent. We are not aware of a single case among higher recent organisms. No flying mammal, reptile, or arthropod could ever be mistaken for a bird, nor can a whale, a seal, or a manatee be mistaken for a fish. Convergence is a problem only among fairly close relatives in rather uniform groups, particularly when classificatory schemes are based on few characters. For a discussion of convergence, see Cain (1982).

Convergence almost invariably involves an adaptation for similar niche utilization. The lagomorphs (hares and their relatives) were long mistakenly included in the rodents because of their similar gnawing incisors. When several kinds of only distantly related seed-eating songbirds were combined in the family Fringillidae, it was because of their cone-shaped bills. Other familiar examples include the distantly related but similar-appearing families of water beetles that share a streamlined form, the mantids (Mantodea) and mantispids (Neuroptera) with their strikingly similar forelegs, and the superficially similar ectoparasites of vertebrates, which belong to at least six different insect orders. Extreme convergence in response to adaptive need occurs also in many physiological properties, for instance, blood pigments in marine invertebrates and neurotoxins in certain fishes and salamanders. Most cases of convergence are restricted to a single character or functional character complex. The utilization of additional characters will lead to the unmasking of unnatural convergent assemblages.

Similar convergent characters can produce seriously mistaken ideas about relationships if they are misinterpreted as being uniquely derived. An excellent example of convergence is the eye of vertebrates and the eye of cephalopod mollusks. The gross morphology of the two kinds of eye is strikingly similar, and the general optical principle which permits the reception of detailed images is the same. This correspondence was exploited by those who opposed Darwin's theory of natural selection. They claimed that such similarity in so complex a structure could hardly have come about through "accident." Therefore, the existence of similar complex organs in such different animals as human beings and squids could be explained only by special creation. Darwin (1859:186) admitted the difficulties of accounting for the eye but suggested that if numerous gradations between simple and complex eyes could be found, then clearly both kinds of eye could have been formed through natural selection.

Salvini-Plawen and Mayr (1977) reviewed the evolution of photoreceptors and eyes and concluded that subsequently found evidence unavailable to Darwin has proved him right. Vertebrate and cephalopod eyes differ in almost every micromorphological and embryological feature. The superficial similarity of the eye in the two phyla is indeed an example of character convergence. There are, after all, only a limited number of physical designs that can work as efficient imaging eyes. It is not surprising that two very different groups of animals have come to similar solutions to the problem of seeing well.

Convergent characters are most often found when different animals become adapted to similar niches. For example, loons and grebes, which are both diving birds, agree in numerous structural characters, particularly of the leg, yet are only very distantly related to each other. Many marsupial adaptive types (wolves, mice, moles, badgers, etc.) are remarkably similar to analogous placental types; the similarity is due to selection for similar modes of life.

Parallel Characters Similar characters derived independently by related taxa with a similar genetic background cause systematists the most trouble. These characters range from rather distinctive features found sporadically in a restricted group of organisms to simple characters that constantly recur. As an example of the first kind, Hennig (1966*a*) cited the stalked eyes of Diptera. Confined to six families of flies, stalked eyes vary considerably in frequency of occurrence in these families. They are found in all the approximately 150 members of the Diopsidae but in only one member of the family Tephritidae, which has 4700 species. The fewer than 200 species of stalk-eyed flies are confined to one higher taxon, the acalypterate Cyclorrhapha, a group containing more than 5000 species. The stalk-eyed flies, then, constitute less than 4 percent of a taxon that in turn accounts for less than 5 percent of the well over 100,000 species of flies now known. It is clear that the stalk-eyed condition has evolved independently several times. The fact that the condition is confined to so small a group of related families also clearly indicates that something in the genetic background of the Acalyptera which permits development of stalked eyes is missing in the remaining Diptera.

There is at present an unresolved difference of opinion in the interpretation of the relation between parallelism and homology. On the strictly phenotypical level, characters that evolved as a result of parallelism are not homologous because they are not "derived from the same [phenotypic] feature of their nearest common ancestor" (see p. 143). This is the interpretation that is most congenial to taxonomists who are simply concerned with the construction of a character matrix. Geneticists and evolutionists, however, understand that the propensity for the parallel evolution of a derived character among related species is a property of the genotype of the common ancestor. Hence, the parallel characters are "derived from the same [genotypic] feature of their nearest common ancestor." Therefore, some evolutionary taxonomists consider characters due to parallelism to be homologous (Mayr 1974*a*; Hecht and Edwards 1977); the synapomorphy is the potential to develop the character (see also Chapter 10 under parallelophyly).

Reversed Characters Still under the influence of the *scala naturae,* phylogenists have tended to consider morphological change an inexorably advancing process. Character analysis, however, remarkably often shows that what appear to be primitive (ancestral) characters are actually reversals (pseudoprimitiveness). In these cases it would be a mistake to treat such characters as plesiomorphic. There is much evolutionary reversal owing to the loss of specializations or other derived (apomorphic) characters. Recent cladistic analyses have revealed that such reversals are far more frequent than was formerly believed. Seemingly, "plesiomorphous features can and do reappear often after protracted periods of time and over large taxonomic distances" (Stiassny 1986:414). They generally affect only single characters or character complexes and can usually be discovered by means of an appropriate analysis. However, *Dollo's rule,* according to which a more or less complex structure that has been lost is not reacquired in the same complexity, has few if any exceptions (Chapter 9).

DIFFICULTIES ENCOUNTERED IN MACROTAXONOMY

Evolution is a capricious process, and taxonomists attempting to follow it, no matter which school of macrotaxonomy they adhere to, encounter difficulties that do not allow for easy solutions. In some genera (e.g., *Drosophila*) abundant speciation occurs without much morphological change, while in others every act of speciation produces an obvious morphological restructuring. This pluralism results in the so-called hollow curve encountered in all groups of organisms (Chapter 10). Several other evolutionary phenomena raise problems in the construction of the Linnaean hierarchy.

Similarity, as we have just shown, may be the result of a number of different types of evolutionary causation; it is obligatory to discriminate rigorously between instances of genuine homology and incidences of various forms of homoplasy. To distinguish between homology and parallelism and to recognize reversal often requires very careful analysis. The following other difficult situations that are not always considered by taxonomists also require careful analysis.

Mosaic Evolution

Perhaps the greatest difficulty encountered by the taxonomist is created by the discordant evolution (noncongruence) of different sets of characters. Entirely different classifications, for instance, may result from the use of characters of different stages of the life cycle, such as larval versus adult characters. In the study of species of a group of bees, Michener (1977) obtained four different classifications when he sorted these species into similarity classes on the basis of the characters of (1) larvae, (2) pupae, (3) external morphology of adults, and (4) male genitalia. When a new set of characters becomes available, its use often leads to a new delimitation of taxa or a change in rank.

Different characters, or, strictly speaking, different components of the genotype, may show drastically different rates of evolutionary change. This difference is found particularly often in comparisons of molecular and morphological characters. The anthropoid genus *Pan* (chimpanzee) is more similar to *Homo* in certain molecular characters than some species of *Drosophila* are to one another, but as we all know, the human being differs from even this closest relative among the anthropoid apes very drastically in certain traditional characters (central nervous system and its capacities) and in the occupation of a highly distinct adaptive zone. Recognition of this fact led Julian Huxley (1958) to propose assigning to *Homo* the rank of a separate kingdom, Psychozoa.

In species with metamorphosis—holometabolic insects, most Amphibia, and many parasites—the different stages in the life cycle may differ drastically in morphology. Usually larvae and adults in these taxa occupy very different niches, and the different stages may have acquired strikingly different specializations. Also, one stage (sometimes the adult, sometimes the larval) may be evolutionarily stable while the other stage exhibits major adaptive modifications. The different rates of *anagenetic change* in different phases of the life cycle inevitably lead to problems not only in the determination of "overall similarity" (phenetics) but also for the ranking criteria of the evolutionary taxonomist. The taxonomist will try to combine information from all stages of the life cycle but may favor the classification that is most practical. When there is a need for ready identification, as in insects and amphibians, the classification will be biased toward the stage that is most commonly encountered, usually the adult stage. A drastic discordance between stages of the life cycle is, however, rare.

It must be realized that numerical methods are quite powerless in the face of manifestations of mosaic evolution. Invariably each different mixture of characters will produce a different dendrogram. And there is no method that will tell us which is the correct tree, indeed even which is the "best" tree.

Fossils

The recent fauna consists of the endpoints of the countless branches of the evolutionary tree. Higher taxa are separated from each other by gaps caused by divergent evolution and extinction. Entirely new taxonomic problems arise, however, when this "horizontal" classification is expanded by including the extinct fauna. These problems are unavoidable, since all organisms, including extinct ones, are related to each other by descent and should be included in a single reference system (Schoch 1986).

Recent fossils, that is, those of the last 25 million to 40 million years, usually can be accommodated in classifications based on the recent fauna. Major difficulties in the placement of older fossils result in two situations. First, a fossil may belong to an extinct lineage that at one time filled a gap that now widely separates two taxa. Should this fossil be recognized as a new major taxon or be attached as an aberrant member to one of the two recent taxa? The most famous transitional group is the clade leading from the mammal-like reptiles to the Mammalia. Among recent groups the elephant louse, intermediate between biting lice and sucking lice, is a good example.

Second, ancestral taxa virtually never have the exact character combination one would expect in a common ancestor. Whether due to mosaic evolution, the acquisition and subsequent loss of temporary specializations (evolutionary reversals), or both, these peculiar combinations explain why Osborn (1936) considered none of 51 fossil species of mastodons to be ancestral to the two descendant lineages he recognized. In fact, the actual common ancestor might not have developed the characteristics of its later descendants, but its genotype had the propensity to develop these characteristics (Mayr 1974*a*). This warns us not to take reconstructed phylogenies too literally. They roughly indicate the path evolution has taken, but we may never be completely certain about some of the details.

THE IMPROVEMENT OF EXISTING CLASSIFICATIONS

The complete reclassification of a major higher taxon may be the most important achievement of a taxonomist, but the taxonomist's daily routine consists of minor additions to or modifications of existing classifications. The following are the most frequent activities of the taxonomist:

1 The assignment of a newly discovered species to the proper genus by answering these questions:
 a Can it be included in an established genus?
 b Does it require a new genus and possibly a new higher taxon?

2 The transfer of an incorrectly placed taxon to its proper position, e.g., moving a species to a different genus or a genus to a different subfamily or family.

3 The splitting of a taxon into several taxa of the same rank (genus, family, etc.) either by cleaving a heterogeneous assemblage of species into several smaller and more homogeneous ones or by removing an alien element from an otherwise homogeneous taxon. When one breaks up too large a taxon, certain rules must be observed in the naming and ranking of the resulting new taxa.
 a The rank of the original taxon is to be maintained if at all possible. Finer discrimination can be achieved by means of the elaboration of subtaxa. For instance, it is usually less desirable to raise a heterogeneous family to the rank of superfamily and then to raise the previously recognized subfamilies to the rank of families than it is to develop a finer subdivision of the subfamilies into tribes and genus groups.
 b In ranking, no taxon should fall out of step with its sister groups. The classification of fossil humans by certain anthropologists who recognized more than 30 genera of fossil hominids is an illustration of an unbalanced classification.
 c A minimal number of names is desirable. If one adopts informal groupings such as species group (instead of a new genus or subgenus) and genus group (instead of a new family, subfamily, or tribe), the same information can be conveyed without burdening the memory and disturbing the balance of the hierarchy of categories.
 d An inconveniently large taxon should be subdivided only if it can be "cleaved," that is, if it can be divided into taxa of approximately equal size. Splitting off a number of monotypic genera from a genus with 500 species would only impede information retrieval. This advice does not proscribe the removal of clearly alien elements from currently recognized taxa.

4 The raising in rank of an existing taxon, e.g., a genus to a subfamily or a subfamily to a family.
5 The fusion of several taxa of the same rank and the synonymizing of the taxa with junior names.
6 The reduction in rank of a taxon, for instance, that of a genus to a subgenus or that of a family to a subfamily. Such a reduction in rank may lead to a considerable simplification of a classification.

Such a reduction is necessary in many groups of animals. For instance, there is little doubt that both birds and fishes are badly oversplit and that natural taxa in these groups are ranked in categories higher than necessary. Even the specialists concerned admit that there is little justification for having 412 families of fishes and 171 families of birds. But which of these families should be reduced to subfamilies? There is no easy answer.

Modern molecular methods sometimes permit the placing of previously puzzling groups. In 1969 Mayr said (p. 248) that

> there is a well-defined group of songbirds in the Old World tropics, the drongos or crow shrikes. In spite of their wide distribution and numerical abundance, they consist of only 2 genera and about 20 species. Up to now, not a single good morphological character diagnostic for this group has been found, and yet in general habitus and in behavior they stand reasonably well apart from all other songbirds. Ornithologists would be perfectly willing to consider the drongos a subfamily or perhaps a tribe of some other songbird family, but as yet no character is known that would help in finding that family. Among the families that have been suggested are the Campephagidae (cuckoo-shrikes), the shrikes (Laniidae), the Muscicapidae [including the Monarchidae] (flycatchers), the Paradisaeidae (birds of paradise), and the Sturnidae (starlings). In desperation, ornithologists finally raised the drongos to the rank of a family, the Dicruridae, while perfectly willing and ready to reduce this rank as soon as additional information becomes available.

With the help of the method of DNA-DNA hybridization, Sibley and Ahlquist (1983) have since shown that the Monarchidae are the nearest relatives of the drongos and rank them as a subfamily of that family.

7 The creation of a new higher taxon not by raising the rank of a taxon (e.g., a family to superfamily rank) but by making an entirely new grouping of taxa of the next lower rank. The proposal of a new superfamily for a number of existing families or a new order for a series of families illustrates this procedure.
8 The search for the nearest relative of an isolated taxon and, if this is successful, the study of the question whether a new taxon of higher rank should be created for the newly established group of relatives. For instance, behavioral and anatomic research has indicated that the

Tubinares (shearwaters, etc.) are the nearest relatives of the penguins (Impennes), which formerly were considered a rather isolated taxon. Should one establish a superorder for these two orders?

Stability

During such minor improvement activities a determined effort must be made to disturb the stability of the currently prevailing classification as little as possible and to maintain, if not improve, its information retrieval qualities. *The usefulness of a classification as a communication system stands in direct relation to its stability, which is one of the basic prerequisites of any such system.* The names for the higher taxa serve as convenient labels for the purpose of information retrieval. Terms such as *Coleoptera* and *Papilionidae* must mean the same thing to zoologists all over the world to have maximum usefulness. This is even more true for the genus, which is included in the scientific name. The overriding need for stability dictates that accepted taxa and their names be maintained in all cases except when they are strongly contradicted by the evidence.

When a well-established taxon is found to be somewhat heterogeneous, it is often inadvisable to split it into several taxa of the same rank if the components are each other's nearest relatives. Murphy and Ehrlich (1984) have shown in the case of butterflies the absurdities to which the pulverization of well-established higher taxa can lead.

The currently adopted zoological system contains numerous taxa suspected of being polyphyletic owing to convergence. Among the birds, particularly among the songbirds (titmice, warblers, babblers, flycatchers, shrikes, finches, etc.), there are many such groups. As long as the nearest relatives of the components of such unnatural groups remain unknown, it is far better to retain these groups provisionally for ease of reference. Such provisional classifications must be abandoned, however, as soon as the true relationship of the components is established. This is increasingly often possible through the use of molecular methods.

In publishing the classification that has resulted from one's taxonomic studies, one must present it either as a printed list, a diagram, or both. Both methods of presentation raise problems.

The Printed Sequence The technology of printing requires a linear, one-dimensional sequence for any printed classification. One species will have to come first and another species last, while all others will have to be listed sequentially between the first and the last. How can one determine the simplest, most convenient sequence of species?

When the classification of a group is still obscure and catalogues consist merely of lists of nominal species, an alphabetical sequence is often most useful for information retrieval. However, an alphabetical listing lacks the heuristic value of a classification arranged according to inferred phylogenetic relationship. By not placing closest relatives near each other, such a listing makes it difficult to undertake evolutionary studies. Finally, there is no stability because it necessitates a change in the sequence every time the name (synonymy) or rank (e.g., shift from species to subspecies) of a taxon changes (Mayr 1965*b*). Species in all better-known groups should be listed according to their relationship with each other. However, this raises various difficulties.

The multidimensional *phylogenetic tree* with the dimensions of time, space (longitude and latitude), and adaptational divergence must be converted into a single linear sequence. To do this, the taxonomist must make some inevitable compromises between various considerations. Most important among these considerations are the following three:

1 *Continuity:* Each species is to be listed as near as possible to its closest relatives.

2 *Progression:* Each series of species or higher taxa should begin with the one closest to the ancestral condition ("the most primitive one"), to be followed by derived taxa that deviate increasingly from the ancestral state.

3 *Stability:* One should not change previously accepted sequences unless they are proved unequivocally wrong. A classification is a reference system, and adopting undocumented "experimental" changes can drastically reduce its usefulness, particularly in a comparison of faunal lists.

These three principles are often in conflict with one another, particularly principles 1 and 2. It is sometimes possible to establish a well-defined morphological sequence without being able to state which end of the sequence is the more primitive. In other cases, there is a dual progression from a group of primitive species toward two specialized extremes (Figure 6-3). Instead of dividing the closely related primitive species into two groups, each leading to one of the extremes, as might be demanded by the progression principle, one sometimes achieves greater continuity if one starts at one specialized end and establishes a single sequence by first descending to the most primitive species and then ascending again to the other extreme. This avoids a more or less arbitrary split through the middle of the group of primitive species.

Because of mosaic evolution, most groups show several trends of specialization at the same time. For such groups, the decision regarding which specialization is considered most advanced may be entirely arbitrary. Among birds, for instance, we find four particularly conspicuous specializations of the wing:

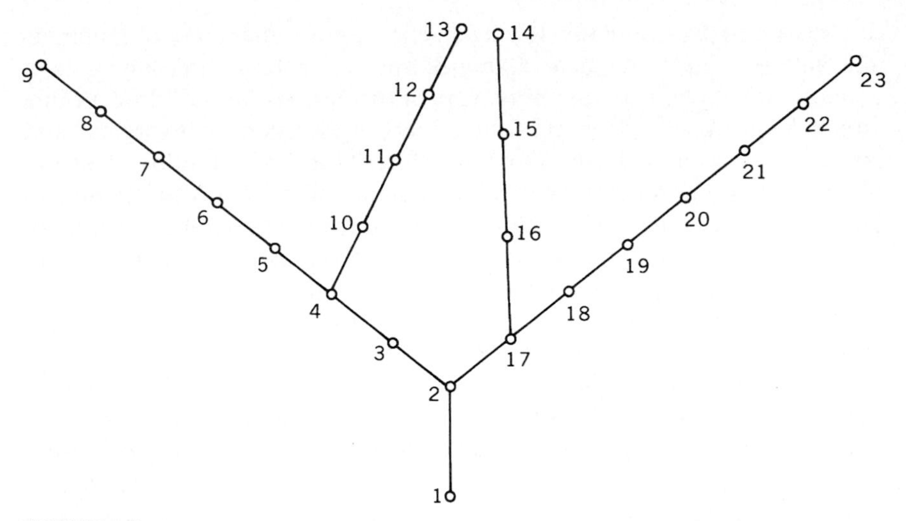

FIGURE 6-3
Different possible ways of arranging 23 species of a dendrogram in a linear sequence. Other options are 1–13, 17, 16, 15, 14, 18–23 and 1–4, 10–13, 5–9, 17–23, 16, 15, 14.

1 Increased flight efficiency:
 a Primaries reach functional peak (swifts, hummingbirds).
 b Secondaries reach functional peak (albatrosses).
2 Loss of flight:
 a Wing has changed to swimming paddle (penguins).
 b Locomotory function of wing has been entirely lost (flightless birds such as ratites).

The decision about which of these four groups of taxa are considered "high" or "low" in the system has to be based on other characters. Most of the so-called primitive animals, for instance, the monotremes among the mammals, are highly specialized in certain respects. Consideration of the totality of characters usually permits a decision about the most logical or convenient sequence, but in highly uniform groups such as birds, no clearly superior sequence is obvious. The taxon that is highest in wing development may be lowest in the development of the central nervous system and vice versa. Even a consideration of molecular characters is often unable to resolve such a dilemma. Cladists make the practical recommendation to list first any taxa that have more ancestral (primitive) characters than do their sister taxa. (See Chapter 9 for phyletic sequencing.) However, owing to mosaic evolution, different sets of characters may show conflicting trends.

Graphic Representation Living species are the endpoints of innumerable phyletic lines. A linear listing of species, genera, and higher taxa cannot begin to convey an impression of the various lines of descent in a complex phylogeny or express the effects of divergent, convergent, and parallel evolution or those of highly different rates of evolution. This deficiency of the printed sequence has induced taxonomists to attempt to depict various aspects of relationship in the form of diagrams, which are usually treelike. Paleontologists have used diagrams with particular emphasis on the age and prevalence of each taxon.

Each of the three modern schools of macrotaxonomy prefers a different kind of treelike diagram (dendrogram). The *phenogram* of the pheneticist is an unrooted tree. It is merely a graphic representation of degree of phenetic difference; it is not a phylogenetic tree. The *cladogram* of the cladist is a branching diagram of taxa as inferred from synapomorphies. It reflects events of cladogenesis. Taxa are delimited by holophyly. The *phylogram* of the evolutionary taxonomist is a phylogenetic dendrogram in which an attempt is made to represent the taxa by the totality of their characters, not only their diagnostic (synapomorphic) ones, and by changing the lengths and angles of internodes to reflect differing rates of evolution.

For further details, see Chapters 8 through 10 and Mayr (1969:62–63, 251–253) and Wiley (1981:93–114).

Phylogenetic Trees

Voss (1952) and O'Hara (1989) have recounted the history of attempts, beginning with Pallas, to design phylogenetic trees. However, the construction of such trees did not begin in earnest until Haeckel (1866) published several picturesque trees (Figure 6-4). The trees published before and after Haeckel are of a highly heterogeneous nature. At worst they were more or less dichotomous identification keys, but in most cases they were phenograms depicting similarity. Quite remarkably, however, many of the so-called phylogenetic trees constructed as long as 100 years ago anticipated to an extraordinary degree the best cladograms recently constructed by a rigorous application of cladistic analysis.

THE IMPORTANCE OF SOUND CLASSIFICATIONS

A sound classification is the indispensable basis of much biological research. It is a prerequisite for the application of the comparative method. Consistent with Simpson's (1961:7) definition of systematics as "the scientific study of the kinds and diversity of organisms and of any and all

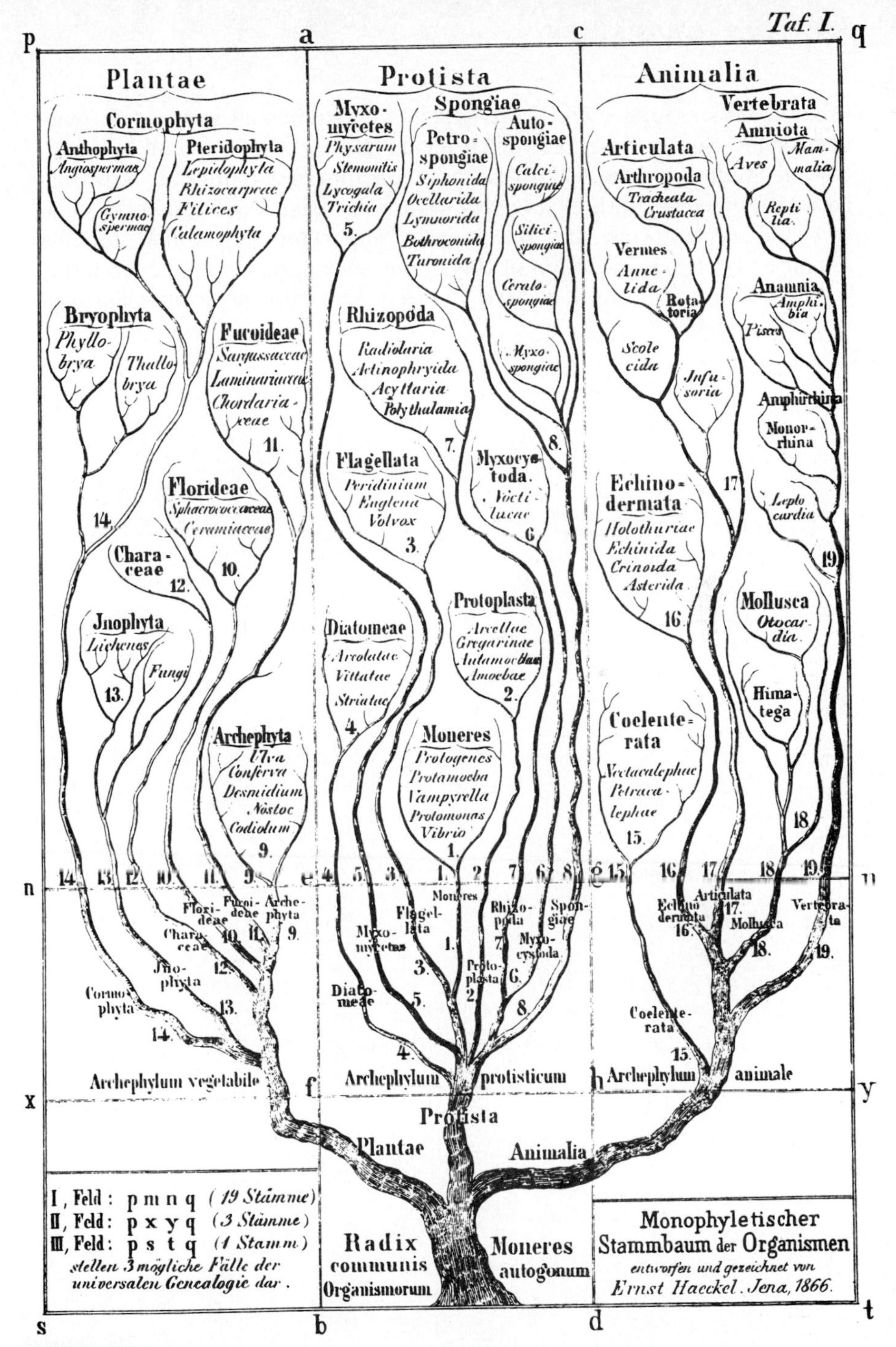

FIGURE 6-4
Early example of a phylogenetic tree as conceived by Haeckel (1866).

relationships among them,'' the systematist studies all aspects of living organisms. Such studies are often meaningless without a sound classification. Studies of species formation, the factors of evolution, and the history of faunas are unthinkable unless they are based on sound classifications. Classifications are particularly important in applied biology (Chapter 1). The recognition of this importance explains why even today so many biologists are dedicated to the task of improving the classification of animals.

CHAPTER **7**

TAXONOMIC CHARACTERS

Taxonomic characters provide the evidence from which relationship between taxa is inferred. The more characters two taxa have in common and the more similar they appear to the taxonomist, the more closely related they are considered to be (Chapter 6). In taxonomic practice, however, it is actually more productive to look for differences between taxa. The definition of a taxonomic character is based on this experience: *A taxonomic character is any attribute by which a member of a taxon differs or may differ from a member of another taxon.*

Any difference between two individuals is a character, but not all characters are useful for taxonomic purposes. Knowing where useful characters are to be found and establishing their specific value is perhaps the most important skill of a systematist, requiring not only a good theoretical knowledge but experience as well.

Defining a taxonomic character as "any attribute of an organism" is incorrect. Differences between phena (Chapter 4) are not taxonomic characters. Therefore, features by which individuals of the same population differ—that is, differences between sexes, age classes, and other polymorphisms—are not taxonomic characters. However, differences between corresponding phena of different species are taxonomic characters.

CHARACTERS AND SIGNIFIERS

The word *character* is not the exclusive property of systematics. The letters in a type alphabet are called characters, and the moral attitudes of a person constitute that person's character. The word's meaning in systematics is derived from one of its most common uses: *A character is a property, attribute, or feature that distinguishes one thing, individual, or group from another.* Black feathers are a character of Australian swans, and white feathers are a character of European swans. This use of the word predates Linnaeus.

In the late 1950s, numerical taxonomists encountered a difficulty with the word. To be handled by a computer, all characters must be assembled into a table called a data matrix (Chapter 11). Typically, the columns of the table are for taxa, and each row of cells contains the numerical codes for characters. The problem was what to call the rows.

To solve this problem, these workers unfortunately transferred the meaning of the term *character* from the difference between taxa (e.g., black versus white feathers in swans) to the structure that varies (e.g., feather color) in order to have a term to designate the rows in their tabulations. The datum in each cell, traditionally a character, came to be called a character state. As explained by Sokal and Michener (1958*a*:1410):

> Our use of the word "character" will require some elaboration. In its commonest taxonomic usage, a character is any feature of one kind of organism that differentiates it from another kind. Thus the red abdomen of one bee is a character distinguishing it from another bee with the abdomen black. In this paper we use the word in a second connotation only; that is, as a feature which varies from one kind of organism to another. Now, to use the above example, abdominal color is the character, which occurs in two "states" or alternatives, red and black.

This proposal had the unfortunate result that the term *character* is now used in taxonomy in two very different senses. In computer taxonomy it is used in the transferred sense of Sokal and Michener, while in ordinary taxonomic discussions it is used in the classical sense of the word. This equivocation has caused considerable confusion in the taxonomic literature.

Problems caused by the dual use of this word are frequently found in studies done by numerical taxonomists. For example, in phenetics (Chapters 8 and 11), some formulas used to measure difference between taxa produce results that also reflect how many actual characters are used in a study. This makes it difficult to compare a study using few characters with a study using many. Pheneticists suggested that each similarity value be divided by the number of "characters" in each study to find an average distance. What they intended was to divide by the number of distinct characters, but the correction factor actually employed in their formulas and computer programs was the number of variables. The result of this confusion of terms was that the newly made and supposedly comparable phenograms were nothing of the sort. Numerous errors of this type can be found in the recent numerical literature.

To end this confusion Ashlock (1985) proposed restoring the term *character* to its traditional meaning and introducing a new term, the *signifier,* for a feature that varies from one organism to another. The feather, then, would be the signifier (sign bearer), and "feather white" and "feather black" would be the character. As desirable as such a restoration of the traditional meaning of the word *character* would be, the transferred meaning has been so firmly established after 30 years that all our consultants strongly advised against the adoption of the term *signifier.* Thus *character,* particularly in regard to numerical methods, refers to a variable feature, and *character state* is defined as an attribute by which a member of a taxon differs from a member of another taxon.

KINDS OF CHARACTERS

Almost any attribute of an organism may be useful as a taxonomic character if it differs from the equivalent feature in members of another taxon. However, proper classifying work is possible only when adequate material from many species is simultaneously available for comparison. Museums provide this opportunity, and this is why the taxonomist prefers characters that can be easily observed in preserved specimens (e.g., morphological characters) (Table 7-1).

In each group of organisms, whether birds, butterflies, sea urchins, or snails, different taxonomic characters exist. It is part of the training of a taxonomist to become familiar with the characters that are most useful in the particular taxon in which that taxonomist plans to specialize. Monographs and handbooks usually give detailed descriptions of the characters used. In his revision of the North African scorpions, for instance, Vachon (1952) devoted 27 pages to a detailed description and illustration of the taxonomic characters of that group.

Morphological Characters

Features of external morphology vary according to kinds of animals. They range from such superficial features as plumage and pelage characters of birds and mammals through scale counts of fishes and reptiles to the highly conservative and phylogenetically significant sutures and sclerites of the arthropod body. Internal anatomy provides many taxonomic characters in practically all groups of higher animals. The extent to which such characters are used routinely varies from group to group, generally in inverse proportion to the abundance and usefulness of easily

TABLE 7-1
KINDS OF TAXONOMIC CHARACTERS

1 Morphological characters
- **a** General external morphology
- **b** Special structures (e.g., genitalia)
- **c** Internal morphology (anatomy)
- **d** Embryology
- **e** Karyology and other cytological differences

2 Physiological characters
- **a** Metabolic factors
- **b** Body secretions
- **c** Genic sterility factors

3 Molecular characters
- **a** Immunological distance
- **b** Electrophoretic differences
- **c** Amino acid sequences of proteins
- **d** DNA hybridization
- **e** DNA and RNA sequences
- **f** Restriction endonuclease analyses
- **g** Other molecular differences

4 Behavioral characters
- **a** Courtship and other ethological isolating mechanisms
- **b** Other behavior patterns

5 Ecological characters
- **a** Habitats and hosts
- **b** Food
- **c** Seasonal variations
- **d** Parasites
- **e** Host reactions

6 Geographic characters
- **a** General biogeographic distribution patterns
- **b** Sympatric-allopatric relationship of populations

observed external characters. In the preparation of mammal skins, the skull (with teeth) is routinely preserved and used in classification, while reptiles, amphibians, and fish are normally preserved in alcohol and are always available for dissection. On the whole, aspects of internal anatomy supply characters for the classification of the higher taxa more often than they do for discrimination at the species level. Fossils consist almost entirely of preserved hard parts; in the case of Mesozoic mammals, for instance, they consist largely of teeth.

Even in this traditional area great advances have been made in recent decades. Descriptions have become more detailed and better standardized. Careful microscopic analysis of lower invertebrates has revealed an abundance of characters even in such seemingly nondescript forms as nematodes. The development of new silver impregnation techniques has revealed a wealth of characters even among protozoans, particularly ciliates. Scanning electron microscopy has added enormously to our knowledge of the morphology of insects, arachnids, and other small organisms.

New organs and structures are steadily added to those which show taxonomically important differences. The spermatozoa of many taxa, for instance, have a highly peculiar and specific morphology and may serve as useful indicators of relationship.

Hard Parts and the Work of Animals It would be senseless to worry about whether the items referred to as hard parts (shells, external skeletons, etc.) are morphological, physiological, or behavioral characters. Much of the classification of invertebrates is based on characters of exoskeletons and shells. Similarly, among the protozoans, tests, shells, thecal plates, cysts, and other hard parts are vital in the classification of foraminiferans, radiolarians, testaceous rhizopods, flagellates, and other organisms. The orientation types of calcite crystals in the skeletons of echinoderms agree well with their classification in families and orders (Raup 1962).

The description of many taxa of dinosaurs has been based on fossil tracks. In the classification of gall insects, the gall sometimes yields as good a clue to relationship as do the insects themselves. The form of the mines is an important taxonomic character in mining insects, and it even sheds light on their history since these mines are sometimes well preserved in fossil leaves (Figure 7-1). However, since 1930 it has not been permissible to base the name of new species exclusively "on the work of an animal" [Articles 12b (8) and 23f (iii) of the International Code of Zoological Nomenclature].

Coloration Color pattern and other aspects of coloration are among the most easily recognized and thus most convenient characters in certain groups of animals. Every species of bird can be recognized by its coloration except for a few genera with sibling species (e.g., *Collocalia, Empidonax*). The same is true of certain reef fishes and butterflies. Even where coloration is not completely diagnostic, it often helps narrow down drastically the number of species to be considered. In groups in which subspecies are recognized routinely, such as mammals, birds, butterflies, and some wasps, color again plays an important role; many subspecies are identified entirely on the basis of coloration.

The quality of color is not easy to describe in words. In original revisionary work, therefore, it is preferable not to rely on descriptions but to base one's judgment on the comparison of specimens if possible. There are, however, various ways in which greater precision can be given to color determination and description (Chapter 12).

FIGURE 7-1
Diagnostic mine patterns caused by six species of leaf miners of the genus *Phytomyza* on the leaves of *Angelica* (50, 51) and *Aquilegia* (54). The letters *a, b,* and *c* refer to different species of *Phytomyza* on the same host plant. (*From Hering 1957.*)

Genitalic Structures For reasons that are not yet fully understood (Mayr 1963:103; Eberhard 1985), the genitalia of many animals, particularly arthropods, not only show a great deal of structural detail but are also highly species-specific (Figure 7-2). Because they are three-dimensional structures, genitalia have to be carefully prepared to be strictly comparable. In many groups of insects and spiders genitalic structures are more important for species diagnosis than is any other character. However, even here it has been found that a single species may have a good deal of variation or that two related species may have

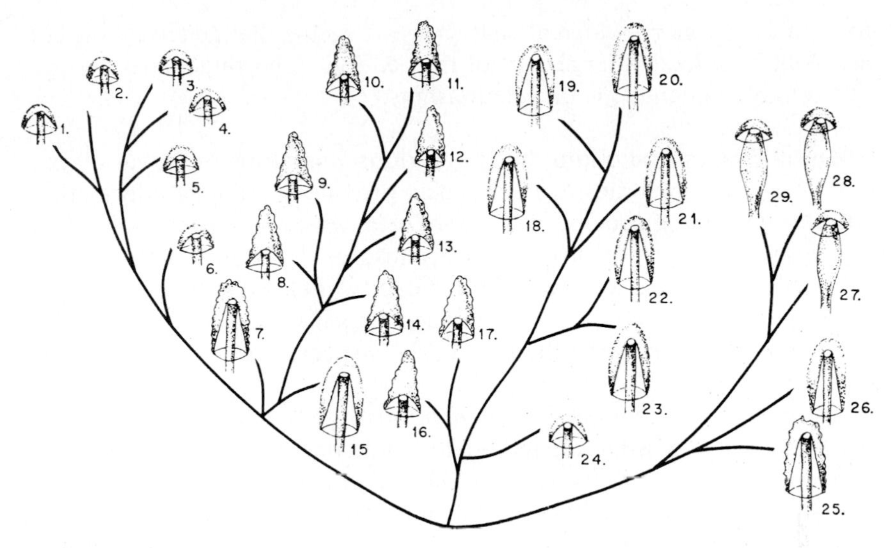

FIGURE 7-2
Types of spermathecae in the *Drosophila repleta* group placed on a phylogenetic tree constructed on the basis of chromosomal characters. (*After Throckmorton 1962.*)

indistinguishable structures. In most vertebrates the genitalia are soft, but the gonopodium of some fishes, the hemipenis of snakes, and the baculum of mammals may supply good taxonomic information.

Other Characters Morphological characters of adult specimens are still used more frequently than are any others, but they are supplemented to an increasing extent by other characters, as listed and discussed below (Blair 1962; Munroe 1960). This is particularly true for "difficult" species, genera, and families in which the evidence from morphology has been equivocal or contradictory. The increasing utilization of new characters is justified because (1) morphology reflects only part of the genotype and may not reflect genetic relationship accurately, (2) morphology in certain taxa does not supply sufficient characters, and (3) any character may be misleading because of special adaptations. The introduction of new kinds of taxonomic characters has been one feature of the so-called new systematics. These characters (molecules, chromosomes, behavior, etc.) supplement but do not displace the use of morphological characters.

A particularly important reason for the utilization of new characters is that they serve as a check on the conventional morphological characters. When discrepancies occur between a classification that is based on morphology and one that is not, still other sets of taxonomic characters must be used. Fortunately, the newer characters usually confirm classifications based on morphological characters. It seems that general morphology usually reflects a large part of the genotype and thus generally permits reliable conclusions on relationships.

Larval Stages and Embryology Various immature or larval stages, embryos, and sometimes even eggs may provide taxonomic information. The various sibling species of the *Anopheles maculipennis* complex (malaria mosquitoes) were discovered because of differences in egg structure. The classification of the Aleyrodidae (whiteflies) is based primarily on the pupae. Discovery of the leptocephaluslike larval stage of the fish order Heteromi confirmed the previously suspected relationship with the eels.

In groups with complete *metamorphosis,* entirely different sets of characters evolve in larvae and adults, and conclusions drawn from the characters of one stage form a very useful check on conclusions drawn from those of another. This was shown for frogs (Anura) by Orton (1957) and Inger (1967). A careful comparison of larval and adult characters of digger wasps (Sphecidae) permitted Evans (1964) to show that some features of the adult structure (emarginate eyes, loss of wing veins) had been overvalued. A proper evaluation of larval characters led to an improved classification. Similar findings have been made in other groups of insects (Van Emden 1957). Larval and adult characters are visible manifestations of the same *genotype*. There can be different identification schemes for larvae and adults, but there can be only one classification for a given group of organisms, and this classification must be based on the proper weighting of both adult and larval characters. Sometimes it is the adults that have acquired specialized adaptations, and sometimes it is the larvae; only a biological analysis can lead to the proper evaluation. In groups such as the sponges (Porifera), which show extremely few morphological characters, consideration of the embryology has been a great help in classification (Lévi 1956).

Genetic Characters There has been much confusion about the meaning of the term *genetic characters*. Because, broadly speaking, all characters except nongenetic modifications of the phenotype are genetic characters, the term is virtually meaningless. Isolating mechanisms and monogenic characters have sometimes been designated as genetic characters, but this restriction is unjustified.

Sterility On the whole, only closely related species can hybridize successfully. In groups of related forms the presence or absence of cross-sterility provides important information, but it must be employed carefully. Ducks, for instance, show much cross-fertility not only of species but even of related genera, yet crosses of two such closely related species as the wood duck (*Aix sponsa*) and the mandarin duck (*A. galericulata*) are completely sterile as a result of a chromosomal rearrangement. In birds there is much fertility among closely related species, but it is hard to generalize about insects. Geographically remote populations of the same species may be nearly intersterile in some insect groups, while in other groups some *congeneric* species are fully fertile. A few exceptions notwithstanding, the relative degree of fertility in species crosses is on the whole a very sensitive measure of relationship. Dubois (1982) suggested that any two species whose genotypes are still sufficiently similar to permit the production of hybrids (sterile or not) ought to be included in a single genus. Conversely, the inability to produce hybrids does not, of course, constitute proof of generic difference.

Chromosomes Botanists have made use of the abundant information provided by chromosome patterns far longer than have zoologists. Improvements in cytological techniques over the past several decades now permit chromosomal studies even in such difficult groups as mammals. Birds, lepidopterans, and other groups with many small chromosomes are harder to work with than are the Orthoptera with their few, large chromosomes and those groups of Diptera in which giant salivary chromosomes are especially suitable for cytotaxonomic studies. White (1973, 1978) provided useful summaries, but they have to be continuously updated because of recent advances. A series, *Animal Cytogenetics,* edited by John (1974 et seq.), includes several fascicles which review recent work on invertebrate (especially insect) and vertebrate groups.

The taxonomic character most frequently used by early cytogeneticists was chromosome number. Since then a rich array of additional chromosomal characteristics has been discovered which can be used to determine degrees of relationship. Chromosomes may have a single arm owing to the nearly terminal position of the centromere (acrocentric chromosomes) or two arms (of equal or more or less unequal length) owing to a more central position of the centromere (metacentric chromosomes); alternatively, they may have a diffuse centromere (holocentric chromosomes), a condition widespread among the Hemiptera and a few other orders of insects. The number of chromosome arms (*nombre fondamental*) in a group is often more constant than is the ratio of acrocentric to metacentric chromosomes; this is due to the frequency of Robertsonian fusion and fissions, in which two acrocentric rod chromosomes are fused into one metacentric one or the reverse. Such fusions and fissions may be reversed, resulting in homoplasy.

Species and higher taxa of animals may differ by an extraordinary number of chromosomal phenomena, including paracentric inversions, pericentric inversions, translocations, Robertsonian changes, sex-determining mechanisms, the presence of diffuse or concentrated heterochromatin, and the presence of supernumerary chromosomes (White 1973). The fine structure of chromosomes can be brought out by appropriate staining methods that show G bands, C bands, or Q bands. Not only do these banding patterns reveal a wealth of detail, they are also highly conservative in some groups, convincingly documenting relationship (Baker, Qumsieh, and Hood 1987). Bickham and Carr (1983) have demonstrated the importance of chromosomal information for the determination of turtle phylogeny. Robertsonian changes (fusions and fissions) are not necessarily unique events in phylogeny. Even this chromosomal phenomenon is subject to homoplasy.

The phylogeny of horses, asses, and zebras (genus *Equus*) has been greatly clarified by means of a comparison of their chromosomes. The dendrogram of the karyotypes agrees remarkably well with the dendrogram based on morphology. What no one understands, however, is why there are such drastic numerical differences among closest relatives in this taxon. The chromosomal number (2n) ranges from 66 in *Equus przewalskyi* to 32 in *E. zebra*. *E. przewalskyi* has 40 acrocentric and 24 metacentric autosomes (and a pair of sex chromosomes), while *E. hemionus onager* has only 8 acrocentrics and 46 metacentrics. Fusions and translocations must have been rampant within this genus. By contrast, in the camel family the two old world species and the four South American species (llama, etc.) have 74 very similar chromosomes, documenting great chromosomal conservatism. In numerous orders and families of animals, for instance, bats and turtles, the study of chromosomes has led to a considerable improvement in our understanding of relationship. The greatest value of chromosomal information is that it can serve as a check on morphological, molecular, and other types of information.

Chromosomes are particularly useful on two different levels. At the lower level, they aid in the comparison of closely related species. Sibling species are often far more different in chromosomes than they are in external morphology (Mayr 1963, Chapter 3). In higher taxa, chromosomal patterns may be of extreme importance in establishing phyletic lines. Most chromosomal changes are unique events; once the new pattern is established, it is characteristic of all descendants of the ancestral population. Changes in sex determination; rearrangements of chromosomes and centromeres through fusions, fissions, or translocations; and the acquisition of supernumeraries often supply unequivocal clues to relationship. For instance, the similarity in the spermatogenesis of Mallophaga and that of Anoplura (true lice) strongly supports the belief in a close relationship of these taxa. The giant salivary gland chromosomes of certain larval dipterans permit the construction of very precise dendrograms of related species.

Polyploidy is rare in animals compared with plants, but there are numerous other ways in which an increase or decrease in chromosome number may occur. Curiously, the most frequent evolutionary trend in many groups is from high to low chromosome number through chromosomal fusion. Even though the chromosomes represent the genetic material, it is not true that the amount of chromosomal change reflects the amount of genetic change. Close relatives may show considerable rearrangement; indeed, many species are polymorphic for various types of chromosomal rearrangements. However, cases are known in which a considerable degree of genic change is not reflected or is only lightly reflected in the chromosomal pattern, as in some Hawaiian *Drosophila* (Carson 1967).

Physiological Characters

This group of characters is hard to define. All structures are the product of growth, that is, of physiological processes, and are thus ultimately physiological characters. Also, physiology is regulated by enzymes and other macromolecules and thus is not separable from biochemical characters. Physiological characters generally include growth constants, temperature tolerances, and the various other processes studied by the comparative physiologist. Species differences are abundant in these characters, but since they are not present in preserved material and their study usually requires special apparatus, they are rarely used by the taxonomist. Some representative cases are discussed by Mayr (1963:60–65), and fuller treatments are given in textbooks of comparative physiology (Prosser and Brown 1973; Schmidt-Nielsen 1979).

Molecular Characters

The major molecular inventions were made by the earliest living organisms. Even the most primitive prokaryotes have on the whole the same kinds of macromolecules and metabolic processes that are found in the highest animals and plants. Still, there is enormous specificity at every taxonomic level, and this specificity is being increasingly exploited by taxonomists.

Serology provided the earliest widely used method of comparing proteins. This method is based on the principle that the proteins of one organism will show a stronger antibody reaction to the proteins of a closely related organism than they will to those of a more distantly related organism. Unfortunately, this method has various technical difficulties. Though used for more than 60 years, it has not contributed as much to a clarification of otherwise ambiguous cases as had been hoped. A summary of some of the achievements of this method can be found in Leone (1964). As a result of improvements in these techniques, there has recently been a revival of interest in the quantitative study of antigenic reactions. An extension of this work is the method of microcomplement fixation (Champion et al. 1974). The study of blood-group genes (immunogenetics) has shed light on relationships among species of pigeons (Irwin 1947 and later authors of his school) and has been used in the study of primates.

Much recent work has been devoted to developing a taxonomy of specific chemical components and macromolecules. The early technique of diffusion chromatography used capillary action to separate molecules on filter paper. In this method a sample (e.g., a body fluid or macerated tissue) is spotted on the paper, which is dipped in a solvent. Movement of the solvent carries different molecules at different rates, so that different samples produce different patterns of spots that represent various components of the samples. Diffusion chromatography has largely been replaced by *electrophoresis,* which uses an electrical potential instead of capillary action and hydrolized starch or polyacrylamide instead of filter paper. Improved resolution has been obtained by processing samples a second time at 90 degrees to the first run to produce a two-dimensional separation. Most components are colorless and must be stained to show differences or must be viewed under ultraviolet light, which may be absorbed or cause fluorescence. Improvements are constantly being introduced, and the newest literature must be consulted for the latest techniques and instrumentation (Brewer 1970; Harris and Hopkinson 1976).

A large number of new molecular techniques have been introduced since the early days of serology and electrophoresis. They make use of proteins, nuclear DNA, mitochondrial DNA, and ribosomal nucleic acids. A study of the restriction site mapping of ribosomal DNA showed the value of this method in establishing relationships in the genus *Rana* (Hillis and Davis 1986). A comparison of 18S ribosomal RNA permitted a provisional reclassification of the invertebrate phyla (Field et al. 1988), but subsequent findings led to considerable modifications. Ribosomal RNA has been particularly useful for an understanding of the relationship of the lower *eukaryotes* and the various branches of the *prokaryotes*.

The molecular method that provides the greatest amount of information is the *sequencing* of proteins and nucleic acids. The number of differences in the base pairs of two nucleic acids reflects the number of mutational steps that separate them (except for transposons and other insertions). By comparing the sequence of base pairs of two homologous

DNAs or RNAs, one can determine the number of mutational differences (Goodman 1982). The same technique can be applied to amino acid replacements in proteins, but this provides much less information. The complete amino acid sequence is now known for many macromolecules (Dayhoff 1973, 1976, 1979), and the degree of relationship of two species can be inferred from the amount of difference. The close relationship of humans and chimpanzees, for instance, is indicated by the fact that their hemoglobins and fibrinopeptides are still identical (King and Wilson 1975). Different macromolecules change at different rates—histones very slowly, cytochrome C quite slowly, globins at an average rate, and fibrinopeptides very rapidly.

A given macromolecule usually changes in a particular phyletic lineage at a rather constant rate. This observation led Zuckerkandl and Pauling (1962) to propose the idea of a *molecular clock*. The clock is calibrated by estimating the age of certain branching points between higher taxa with the help of the fossil record or biogeographic evidence and then calculating from these dates and the known difference in the amino acids the average number of amino acid replacements per 10 million years. The resulting time frame permits the construction of dendrograms that depict the inferred phylogeny (Fitch and Margoliash 1967).

Use of molecular evolution permits inferences on the probable time of the splitting of phyletic lines and sometimes even on the geography of the splitting event. For instance, the immunological distance of the hemolymph proteins of the Hawaiian drosophilids from those of their North American relatives is consistent with a colonization of the islands about 40 million years ago, that is, before any of the currently existing islands had emerged (the older islands are now submerged sea mounts) (Beverley and Wilson 1985). This confirms the previous reconstruction of the history of the Hawaiian biota by Zimmermann (1948).

The assumption of a completely constant molecular clock is, of course, unrealistic (Britten 1986). The same molecule may evolve at quite different rates in different phyletic lines. The less affected by selection certain molecular changes are, the more likely it is that they will occur at a constant rate; this seems to be the case with many enzyme genes, as revealed by electrophoresis (Kimura 1983). Regulatory genes, introns, transposons, and gene families may experience considerable deviations from a molecular clock constancy. The molecular clock thus is only a first approximation, and its results must be fine-tuned through the use of additional methods.

Chronologies can also sometimes be established through an analysis of particularly stable molecular changes. For instance, in the echinoderm phylum a certain gene regulatory mechanism (alpha subtype histone maternal mtRNA) originated in the late Triassic and is now found in all the orders of echinoids that originated after that date (about 200 million years ago) (Raff et al. 1984). In doing such calculations one must always remember not only that different molecules may evolve at different rates but also that within a lineage the rate may change drastically. For instance, nucleotide substitutions for seven mammalian proteins seem to have occurred in the primates at a distinctly slower rate than was the case in most other organisms (Fitch 1976). It is now known that several molecules change as much as five times as fast in some phyletic lineages as they do in others.

The ideal of the pheneticists has always been to determine the total overall similarity of two taxa. This is what the technique of DNA hybridization attempts to achieve at the molecular level (Sibley and Ahlquist 1983, 1985). The method is fairly complex. The DNA is freed of all protein and RNA, and then the purified DNA is sheared into fragments with an average length of 500 base pairs. The DNA is then melted at about 100°C so that the two strands of double-stranded DNA separate. Then the repeated sequences are removed so that only the pieces of a single-copy DNA are left. When the DNA of two species is mixed (one DNA is labeled with a radioactive isotope) and the mixture is gradually cooled, homologous single-stranded pieces of the two DNAs will pair while the nonmatching pieces will remain in solution. This indicates what percentage of the DNA pairs and what percentage has become sufficiently different during evolution to pair no longer. A delta (Δ) value can be determined which reflects the degree of difference. (This is a greatly abbreviated account of a considerably more elaborate procedure.) Sibley at first postulated the rate of DNA evolution (i.e., nucleotide substitution) to be the same in all lineages of birds and stated that it took about 5 million years of divergence for the genomes of two species to become different in 1 percent of their base pairs. On the basis of these assumptions he constructed a dendrogram of all avian families. The discovery of considerable inequalities of rates has necessitated certain modifications (Britten 1986). Since Sibley's dendrogram is based on a phenetic method, it is really a phenogram, even though it is quite different from Sokal's overall morphological similarity. Mitochondrial DNA is of considerable help in the comparison of populations of a single species, for such DNA is inherited directly from the maternal parent and is not subject to sexual recombination, as is nuclear DNA (Avise and Lansman 1983).

Different molecular techniques are useful at different taxonomic levels. The study of mitochondrial DNA and that of electrophoretically discovered enzyme differences are most useful for the comparison of populations and closely related species (Barrowclough 1983; Patton and Yang 1977; Selander and Whittam 1983), while protein and nucleic acid sequencing, immunological methods, and DNA hybridization are most informative in the study of distantly related species and higher taxa, that is, in the construction of phylogenetic trees. One of the first molecular trees, that of Fitch and Margoliash (1967), was so close to the traditional phylogenetic tree of the vertebrates that it made taxonomists aware of the importance of molecular methods. Unfortunately, the methods that are most precise and informative, such as protein and DNA sequencing, are also the most time-consuming.

The thoroughly documented demonstration by Davis (1964), based on morphological characters, that the giant panda is a bear and is not related to the raccoons was doubted by a few scientists until Sarich and O'Brien confirmed the validity of Davis's conclusion by using several molecular methods (Mayr 1986). In a similar manner the findings of comparative anatomists have been confirmed again and again in recent years by molecular methods. However, there are exceptions. The cheetah (*Aconyx*) has always been considered by morphologists to be the most aberrant of all the cats, but molecular studies have revealed that it actually is a rather close relative of the lion-tiger group, with other genera of the cat family being far more distant (Collier and O'Brien 1985).

The study of structural characters has revealed that each organ or organ system may have its own specific rate of evolutionary change (mosaic evolution) (Chapter 6). Much evidence indicates that this principle is equally valid for molecular characters. A comparison of the human being (*Homo*) with the chimpanzee (*Pan*) shows, for instance, that there has been little evolutionary divergence in the hemoglobins and some other molecules since they branched from each other, even though the hominid line has since entered an entirely new adaptive zone (Figure 7-3). When one uses taxonomic characters to draw inferences about classification, one must always balance the potentially conflicting information derived from different character domains. One must also understand the subtle difference between evolutionary phenomena at the molecular level and the organismic level (Mayr 1964; Simpson 1964).

Molecular taxonomy permits the analysis of a wholly new set of characters that are seemingly independent of more traditional characters (Barrowclough 1985). While molecular characters are subject to the same problems of convergence, parallelism, and reversal that any character set may have, it is highly unlikely that morphological and molecular characters will be affected in the same manner. Where the molecular characters are superior is in the extraordinary detail they can supply. The genome of a mammal consists of about 2 billion nucleotide base pairs. In many cases where the morphological evidence is ambiguous, molecular methods have produced totally unequivocal conclusions. The study of evolving

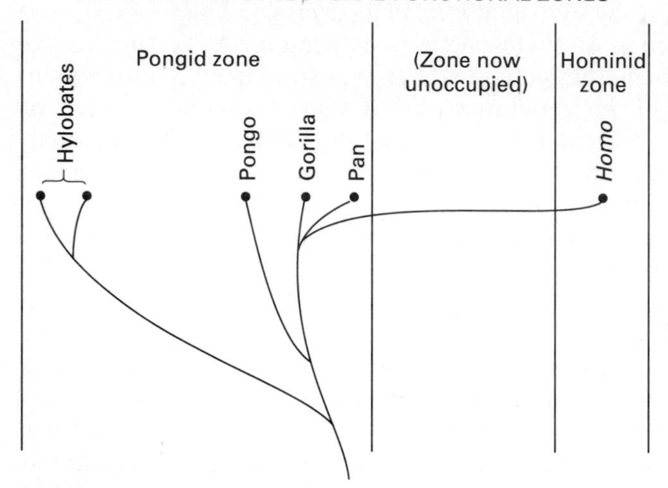

FIGURE 7-3
Recency of descent versus degree of adaptive divergence. Dendrogram of inferred affinities of recent hominoids in relation to their radiation into adaptive and structural-functional zones. (*After Simpson 1963.*)

molecules superimposed on the background of classical taxonomy is bound to reveal discrepancies and inaccuracies which will lead to improvements in classification (Hillis 1987).

Molecular methods are not a panacea. In almost every case (particularly among more distant relatives) where taxa were classified with the help of different molecular methods, different classifications were arrived at. This is not surprising in view of the universality of mosaic evolution. Such conflicts pinpoint the branches in a phylogeny that are in need of further analysis. Instead of concentrating on such trouble spots, some molecular biologists have wasted time and valuable resources by confirming totally uncontroversial findings of taxonomists.

The most important aspect of some of the molecular methods is that they do not determine relationship by the presence or absence of shared characters but permit one to calculate the distance (degree of difference) between two taxa. The special numerical methods required for such calculations are described in Chapter 11. These methods are so new and there are so many competing ones that it is still uncertain which the best methods are and in what circumstances a distance determination is superior to a character analysis.

The increasing need for suitable material for taxonomic analysis has induced many museums to establish *molecular collections,* that is, collections of material suitable for molecular analysis. Such material consists of whole specimens, tissues, or body fluids preserved in such a way that they are suitable for the study of proteins, nucleic acids, or other molecular constituents or for chromosome analysis. For information on methods for the preservation and use of such material, see Chapter 13.

Behavior

Behavior is undoubtedly one of the most important sources of taxonomic characters. Indeed, behavioral characters are often clearly superior to morphological characters in the study of closely related species, particularly sibling species (Mayr 1963). However, there are two major technical drawbacks. Behavior cannot be studied in preserved material, and it is intermittent even in a living animal. Some types of behavior occur only during the breeding season; others, only during part of the 24-hour period. The comparative study of the behavior of related species has become an autonomous discipline known as comparative ethology. It has already made major contributions to the improvement of classifications of birds, bees, wasps, orthopterans, frogs, fishes, and other groups.

The reason for the importance of behavior is obvious. Behavioral characteristics are the most important isolating mechanisms in most animals, and new adaptations are often initiated by changes in behavior. The rapidly expanding literature on behavioral systematics has in part been summarized in a number of reviews, including those of Mayr (1958), Alexander (1962), and Wickler (1967). Exemplary studies include those of Evans (1957, 1966) on the digger wasps (Sphecidae), Spieth (1952) on the genus *Drosophila,* Tinbergen (1959) and Moynihan (1959) on gulls (Laridae), Eickwort and Sakagami (1979) on halictid bees, and Gordl and DeBach (1978) on chalcid wasps. Flash patterns are species-specific in most genera of fireflies (Figure 7-4) and have been used by systematists to unmask sibling species (Barber 1951; Lloyd 1983). However, in the genus *Photuris* males may mimic the signal of sympatric species of other genera (*Photinus, Pyractomena,* etc.). When one encounters several flash patterns in the repertory of a single mimicking species (Lloyd 1985), one must not conclude that one has discovered several sibling species.

A great technical advance in the study of behavior has been the development of devices for accurate sound recording and the translation of sounds into graphic patterns [the sonagraph (Figure 7-5)]. More than 40 species of North American crickets were either discovered or rediscovered by B. B. Fulton and his followers as a result of a careful analysis of their songs. The classification of species in several avian genera (for instance, *Myiarchus, Empidonax, Tyrannus*) has been greatly helped by an analysis of sound recordings. A comparison of the calls of frogs and

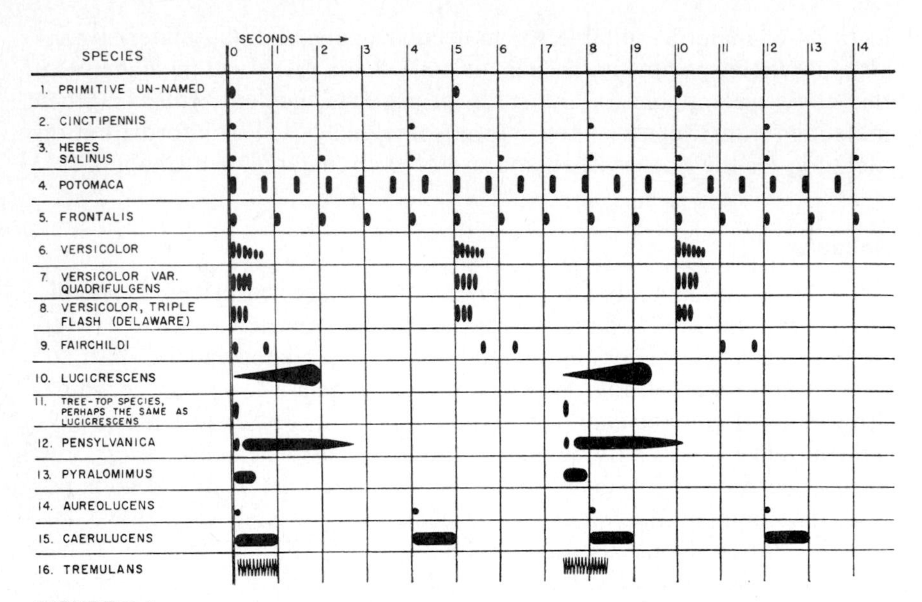

FIGURE 7-4
Patterns of light flashes in North American fireflies (*Photuris*). Height and length of the marks indicate intensity and pattern of the flashes. (*From Barber 1951.*)

toads not only has led to the discovery of previously unrecognized sibling species but also has shed light on the relationship between previously established species. Important studies of comparative sound analysis in anurans were done by Barrio, Blair, Bogert, Littlejohn, Main, and Mecham. The acoustics of animal behavior has been summarized in a number of recent volumes (Tavolga and Lanyon 1960; Busnel 1963; Sebeok 1977; Claridge 1985). The importance of bird song in avian systematics has been reviewed by Payne (1986).

In addition to courtship and acoustic behavior, various other kinds of behavioral elements have taxonomic value. For example, the pattern of the webbing constructed by various mites and caterpillars may be used at various levels in the classification. The two bee genera *Anthidium* and *Dianthidium* were slow to be recognized on morphological grounds, yet all known species of *Anthidium* construct nests from cottony plant fibers while those of *Dianthidium* construct nests from resinous plant exudations and sand or small pebbles.

The use of extraneous materials in the construction of nests or of larval or pupal cases provides characters at various levels in the classification of caddisworms and bagworms. The egg cases of praying mantids have a species-specific form.

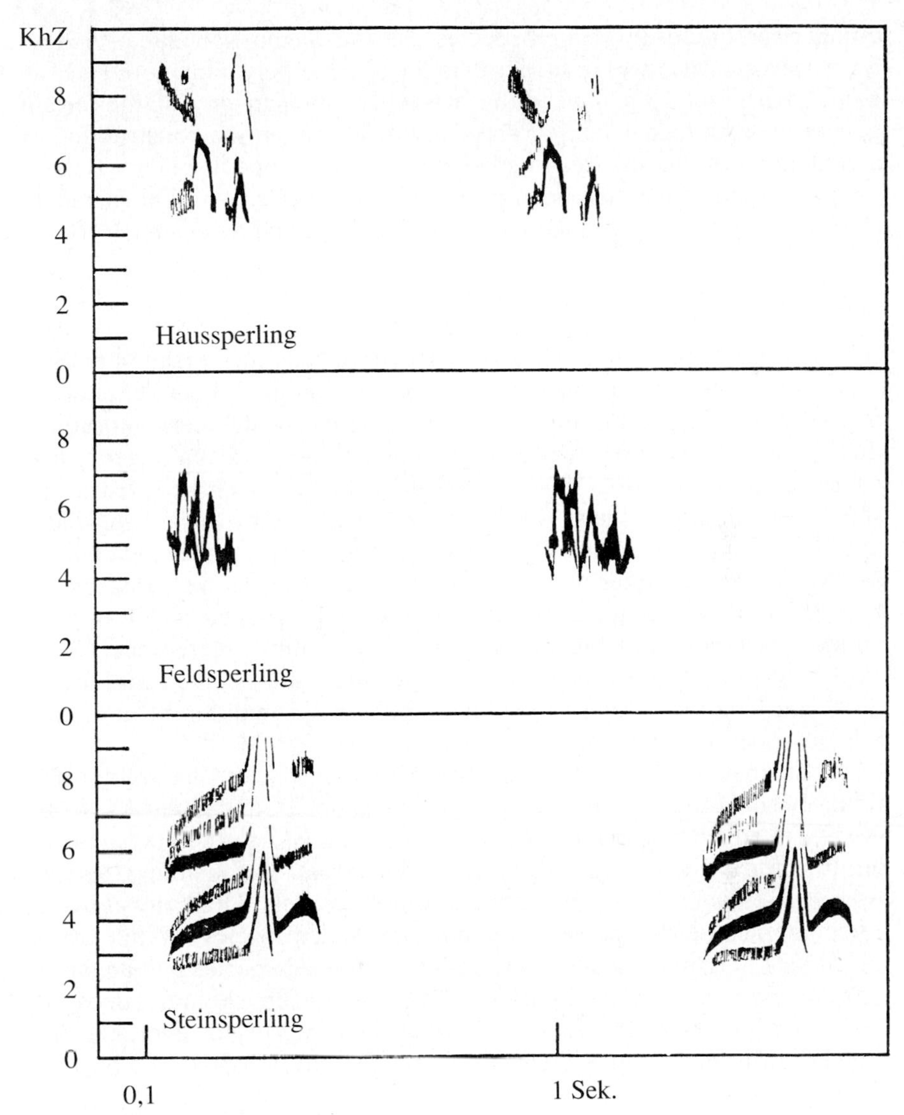

FIGURE 7-5
Different vocalizations of related species of sparrows. (*After Thielcke 1964.*)

Behavior patterns that are characteristic of higher taxa are far more rare. Examples are the use of mud in nest building by barn swallows and crag martins (*Hirundo*), certain "comfort movements" (scratching, stretching, bathing, etc.) in birds (McKinney 1965), and grooming movements in insects. Attention up to now has been directed so strongly at the comparison of closely related species that the diagnostic value of behavior patterns at the level of higher taxa has gone largely unexplored. However, Eberhard (1982) showed that web structure is diagnostic for certain genera or even higher taxa of spiders; this permitted the construction of a cladogram of the orb-weaving behavior of the araneoid spiders. Behavior cladograms have also been published for other groups, for instance, the nest architecture of halictine bees (Eickwort and Sakagami 1979).

Ecological Characters

It is now well established that every species has its own niche in nature, tending to differ from its nearest relatives in food preference; habitat selection; breeding season; tolerance to various physical factors; altitudinal distribution; resistance to predators, competitors, and pathogens; and other ecological factors. When two closely related species coexist in the same general area, they avoid fatal competition by means of these species-specific niche characteristics (principle of competitive exclusion; see Mayr 1963, Chapter 4). A number of sibling species were discovered as a result of discrepancies in food preference (host specificity)—for example, the apple and blueberry maggot—or habitat preference (Mayr 1963, Chapter 3). Many aspects of the life cycle, such as life span, fecundity, and length or time of breeding season, may be different in closely related species (Mayr 1963).

Niche specificity is quite pronounced even in species that are not particularly substrate-specific, such as birds, mammals, and mollusks. Kohn (1959) found that every species of the genus *Conus* in the Hawaiian Islands differs ecologically from related species. Two sibling species (*ebraeus* and *chaldaeus*) feed on nereid polychaetes, but of 199 *ebraeus,* 136 contained nereid species *a* and none contained species *b,* while of 106 *chaldaeus,* 5 contained species *a* and 98 contained species *b*. The larvae of *Drosophila mulleri* and *D. aldrichi* live simultaneously in the decaying pulp of the fruits of the cactus *Opuntia lindheimeri*. However, the two species differ markedly in their preference for certain yeasts and bacteria (Wagner 1944).

Niche specialization is even more pronounced in animals that are substrate-specific, including host-specific plant feeders among insects and mites and host-specific parasites. Some of this information can be found in recent books on coevolution (Nitecki 1982; Futuyma and Slatkin 1983). There is also the excellent treatment by Dethier (1947), who emphasized the sensory aspects of host selection. Many new species of insects were discovered by comparing populations of the "same" species occurring on different plant hosts. Some enthusiasts, however, carried this principle too far and made any occurrence on a different host an excuse for the naming of a new species. In the North American Cryphalini (bark beetles) alone, 53 of the species described by Hopkins turned out to be synonyms. Downey (1962) and Kohn and Orians (1962) have given useful summaries of some of the relevant literature. Host specificity of external parasites has been discussed by Clay (1949), Hopkins (1949), and Holland (1964).

Ecological differences also occur among populations of the same species. Indeed, most widespread species exhibit differences in the ecology of local populations, particularly of peripherally isolated populations (Mayr 1963:312ff, 355).

Again, as with behavioral characters, differences at the species level have been studied far more extensively than have ecological differences between higher taxa, many of which are simply taken for granted. The fact that whales occupy a different adaptive zone than do bats is too obvious to be mentioned, yet a close study shows that even most genera, when well founded, occupy definably different niches or adaptive zones. Lack (1947) showed, for instance, that each genus of Galápagos finches is characterized by its utilization of the environment. *Geospiza* is a ground finch (chief food, seeds), *Camarhynchus* is a tree finch (chief food, insects), and *Certhidea* is a warbler finch (chief food, small insects) (Grant 1986).

Parasites and Symbionts In several instances sibling species were discovered because their parasites were different. A previously undiscriminated species of *Octopus* was discovered in California because it had its own set of mesozoan parasites. A new species of termites was discovered because its nests contained a different set of termitophile staphylinid beetles than did the nests of a previously known species. Parasites are also important in contributing to our knowledge of the relationship of higher taxa. Parasites evolve together with their hosts and are in some cases more conservative than the hosts. Unfortunately, they shift to new hosts more frequently than is sometimes admitted, and so evidence based on parasites must be evaluated very carefully (Baer 1957). For instance, flamingos (Phoenicopteri) exhibit characteristics which they share with both storks and geese. Their bird lice (Mallophaga) are of the same genera as those which occur on geese. At first this might suggest a close relationship, but it actually indicates merely a comparatively recent transfer of the lice from geese to flamingos. If the bird lice had been derived from a common ancestor, we would expect related but slightly different parasites in the two orders of birds. Neither anatomic nor molecular evidence indicates a close relationship of flamingos and geese. Human beings (*Homo*) and the African apes (*Pan* and *Gorilla*) share more external and internal parasites with each other than the African apes do with the orang (*Pongo*). This strengthens the case for a close relationship between *Homo* and *Pan* (established on other grounds).

The fact that intracellular symbionts supply important taxonomic characters was discovered by Buchner (1966*a*) and his school. For instance, the most primitive tribes of the coccids (Steingeliini, etc.) have no symbionts, but once a coccid taxon has acquired one, this symbiont (with all its highly specific adaptations) will be found in the derived phyletic lines of coccids. Repeatedly, unnatural taxa of coccids could be unmasked because they had heterogeneous complements of symbionts (Buchner 1966*b*). The same is true of symbionts in other groups of insects. The protozoan faunas in the intestines of termites evolved together with their hosts and are potentially useful indicators of relationship in cases of ambiguity in termite classification (Kirby 1950*b*).

Geographic Characters

Each taxon, from the species to the highest taxon, has a geographic range. Since this range changes as the taxon evolves, an intimate relation exists between systematics and biogeography. Most great biogeographers were systematists. Patterns of distribution often provide decisive clues for clarifying a confused taxonomic picture and for testing taxonomic hypotheses. The distributions of most taxa display patterns that permit conclusions about their past history. Because geographic range provides such useful taxonomic characters, every taxonomist needs to become familiar with the literature in the field of biogeography. The following works should be mentioned: Briggs (1974), Brown and Gibson (1983), Carlquist (1974), Darlington (1957, 1965), Mayr (1976), Pielou (1979), Raven and Axelrod (1974), Simpson (1965), and Udvardy (1969).

Geographic characters are important on two levels. In microtaxonomy the sympatric-allopatric relationship of populations is often decisive in the determination of species status. Two sympatric populations in reproductive condition can never be conspecific. The mapping of populations, subspecies, and allopatric species is indispensable in the delimitation of polytypic species and superspecies.

In macrotaxonomy an understanding of the relationship of higher taxa is often helped by an analysis of their distributions. Two questions are particularly important: (1) What is the distribution of the nearest relative (the sister group) of the studied taxon? and (2) If the taxon has a disjunct range, what is the probable cause of the range disjunction?

Since Darwin, we know that there are potentially two causes for disjunction:

1 Primary isolation (establishment of a founder population through *dispersal*)
2 Secondary isolation, that is, the fracture of a previously continuous range by a new distributional barrier across the previously continuous range (*vicariance*)

A knowledge of the Earth's history and the geological record often permits one to decide whether the isolation is primary or secondary and, more important, for how long a time the two isolates have been separated. The fact that the mockingbirds on the volcanic Galápagos Islands had flown across from the adjacent South American continent was perfectly clear to Darwin. The fact that the fauna and flora of the volcanic Hawaiian Islands had gotten there by transoceanic colonization has also been long evident; the same conclusion was inevitable for all volcanic oceanic islands in any ocean. Some continental islands, however, for instance, Madagascar and New Zealand, have a mixture of old vicariant elements that are remnants of the biota of the continents or plates to which these islands had formerly been attached and of more recent elements that reached them by transoceanic colonization.

Depending on the dispersal facility of a taxon, either primary or secondary isolation is the more important cause of range disjunctions. Some groups, such as earthworms (except when passively transported) and primary freshwater fishes, have very low dispersal facilities. In such groups almost any range disjunction is secondary, caused either by geological events (plate movements, mountain building) or by climatic-vegetational shifts such as range disjunctions caused by the Pleistocene ice caps or by post-Pleistocene habitat shifts.

Groups with poor dispersal facilities are particularly useful in determining the age of taxa. If such a group is found in South America and Australia but not on the northern continents, it is probably an old Gondwana element that goes back to the time when Australia was still connected with South America across Antarctica. If poorly dispersing groups are found in Africa and South America, they presumably go back to the time before Africa separated from South America. Most faunas, however, are a mixture of old and more recent elements (Mayr 1976:552–564) and these elements have to be carefully discriminated to establish a reliable chronology. The New Zealand kiwis and moas, for instance, are among the remnants of the old Gondwana fauna that ruled when New Zealand was still in contact with Antarctica, but most or all of the other birds of that island are descendants of more recent transoceanic immigrants from Australia or Melanesia.

The facility with which certain groups can cross water gaps is documented by the numerous colonizations of the Hawaiian Islands or of even more remote Pacific islands such as the Marquesas and the Australs. In the case of easy dispersers, it is often difficult to determine the source area of colonization and thus the nearest relative. Some biogeographers tend to underestimate the dispersal ability of most organisms. The lizard family Iguanidae is otherwise restricted to the Americas, but surprisingly there is a species in Tonga and a superspecies in Fiji. All careful students of former land connections have come to the unexpected conclusion that these lizards could have reached these outposts only by long-distance dispersal.

The question is often asked, How can a pattern of distribution help in determining the nearest relative of a higher taxon? Here the basic rule is as follows: In case of doubt, the nearest relative is in the same geographic region or continent. This is well illustrated by the taxonomy of numerous Australian songbirds. When warblers, flycatchers, shrikes, nuthatches, tree creepers, and (wood) swallows were found in Australia, it was only natural that they were placed with the corresponding groups of the holarctic region. There were, however, enough puzzling aspects of the ecology and distribution of these Australian taxa to raise doubts. Could they not all be convergent adaptive radiations of indigenous Australian stocks? That this seems to be the case was eventually shown by Sibley and Ahlquist (1983) with the help of DNA hybridization.

The nearest relative, however, is not always found in the same geographic region. The nearest relative of the llamas is not a South American animal but the old world camel. The story of the marsupials is more complex. Today they are restricted to South America and Australia (except for the opossum that successfully reinvaded North America). Fossil marsupials, however, have been found in North America, Europe, Africa, and Asia. Geologists are now able to date the age of fossils and of all former land connections after the Pennsylvanian rather accurately. As far as the marsupials are concerned, this leads to the well-substantiated conclusion that North America was the original source area. From there, one radiation invaded Europe (and from there Africa and Asia) early in the Tertiary across a north Atlantic land bridge; this radiation soon became extinct. A far more successful radiation colonized South America and from there expanded across Antarctica to the Australian region. We are fortunate that it is now possible to provide reasonably accurate dates for the biota of the various plates and plate remnants.

A primitive relic sometimes survives in peripherally isolated areas, such as the tuatara on New Zealand and the rich fauna of lemurs on Madagascar, a phenomenon used by Matthew (1915) and other zoogeographers as a basis for broad generalizations. On the whole, however, the more primitive branches of a major radiation are most likely to be found near the original source area, with the expanding colonizations evolving farther and farther away from the ancestral location.

The theory of plate tectonics has shed considerable light on the evolutionary history of many animal taxa, particularly that of poor dispersers. It has helped resolve a number of taxonomic puzzles. In 1922 Michaelsen made the unexpected discovery that the earthworm fauna of India is more closely related to that of Africa and South America than it is to that of the rest of Asia. He explained this on the basis of Wegener's theory of continental drift, according to which the Indian plate was formerly part of Gondwanaland and established contact with the Asian plate only in the middle of the Tertiary. Many other biogeographic puzzles have since then been explained in the light of plate tectonics. However, even today there are biogeographic puzzles which have not been solved by the better understanding of geographic history brought about by plate tectonics.

Many genera of insects that are now found only in the southern continents (southern Africa, South America, Australia) were once believed to be part of the Gondwana fauna. Surprisingly, some of these genera have since been found as fossils in Oligocene Baltic amber. It is now clear that the southern distributions of these genera are relicts of formerly much wider distributions (Ander 1942). Such cases warn us not to become too assertive in trying to reconstruct the connections between present distribution, the Earth's history, the fossil record, and taxonomic relationship.

THE WEIGHTING OF CHARACTERS

The more characters a taxonomist uses, the more likely it is that the information provided by different characters will be discordant. According to one set of characters, species A and B are more closely related to each other than they are to C, but according to another set, species A and C are more closely related than either is to species B (Figure 11-2). Homoplasy and mosaic evolution are responsible for such seeming contradictions. There once were great hopes of resolving these conflicts with the help of numerical methods (Chapter 11), but these hopes were only partially fulfilled. Unfortunately, the use of different phenetic methods (Presch 1979) often results in different phenograms, and the same problem is encountered with the use of different cladistic methods. Aware of these difficulties, numerical taxonomists are increasingly including weighting in their methods (Funk and Wheeler 1986). Evidently in these cases different characters provide different information.

What the taxonomist must ask is which of the discordant characters should be trusted more. How can it be determined which of two conflicting characters gives the more reliable information on relationship? The process by which this is determined is called *weighting*. There are no ironclad rules about weighting. Often the relative weight of different characters is different at different levels of the taxonomic hierarchy. The fact that some characters are more significant than others is by no means a new insight. Experienced taxonomists have always insisted that characters differ in the contribution they make to the soundness of a classification. Darwin (1859:414–417), for instance, gave some good empirical rules concerning the usefulness of certain characters. Owing to their crucial importance, weighting criteria must be discussed in detail.

Correlated Characters

Taxonomists generally agree that they rely more on correlated character complexes than on any other clue to relationship. However, in the weighting of characters a strict distinction must be made between two kinds of correlations, one of very high and the other of very low weight. Characters that are *functionally correlated* are of low weight and do not deserve to be treated as separate characters because, being members of a single functional complex, they are redundant. Characters that are not functionally correlated but are phenotypic manifestations of the same ancestral well-integrated gene complex are of high weight. The diagnostic characters of the Deuterostomia and the Chordata are examples. One might call this kind of concordance *phyletic correlation*.

When comparing the species of two well-defined higher taxa, one always finds that the species in one taxon share certain functionally uncorrelated characters that are not displayed by the species in the other taxon. We infer that the joint possession of characters by species of a higher taxon is a result of common descent and assign high taxonomic weight to such concordant, that is, phyletically correlated, characters. The character that is concordant with the greatest number of other characters has the highest weight. Each species of a higher taxon, of course, has many additional characters that do not belong to such correlated character complexes, yet most higher taxa are distinguished by the possession of correlated character complexes.

Whether an association of characters represents functional or phyletic correlation is not necessarily evident at first sight; this sometimes requires a careful functional analysis. In a few cases, there is no sharp line between two types of correlation. Many phyletically correlated character complexes may have originally started as functional complexes in which the genetic integration was retained even after the functional correlation had broken down because of a change of function of individual components.

The construction of classifications with the help of the computer allows a numerical approach to weighting. After a cladogram has been constructed, one finds that the presence or absence of certain characters is highly correlated with the branching of the cladogram but that this is not true for other characters that presumably are homoplasic. If the best correlated characters are given a higher weight in a second run, an improved cladogram will result. This procedure is repeated until no more improvement results, usually after three to five iterations. For further details on this method, see Farris (1969:374–385) (Chapter 11). When characters are weighted by direct evaluation, the following considerations must be attended to.

A Priori and a Posteriori Weighting The Aristotelians and their successors often assigned a priori weights to certain characters. Cain (1959*a*) pointed out the fallacy of this approach. Neither function, conspicuousness, nor any other known aspect of a character gives it a priori a greater weight than other characters have. Indeed, the same structural difference may have high weight in one taxon and low weight in a related taxon. Nor should the taxonomist confuse the weight of a character with its usefulness in a diagnostic key. Any rigid following of the dictates of a few arbitrarily chosen a priori characters leads inevitably to a classification that is not "natural," i.e., that has a low predictive value.

Adanson and the empirical taxonomists of the eighteenth century rejected *a priori weighting* and replaced it with an empirical process perhaps best called *a posteriori weighting*. Adanson was satisfied with merely eliminating useless and redundant characters; when the natural groups of animals were better understood, later authors attempted increasingly to assess the relative merits of each character. The value of this approach, in spite of individual errors, is substantiated by the fact that all good classifications proposed in the 100 years after 1859 resulted from such a posteriori weighting.

The acceptance of evolutionary theory after 1859 did not change this method as such. It did, however, provide scientific justification. It became evident why some characters are better indicators of natural groups than are others. Different characters contain very different amounts of information about the ancestry of their bearers. *Weighting can be defined as a method for inferring the phyletic information content of a character.* If character *a* indicates assignment of a species to genus *A* and character *b* indicates assignment to genus *B,* we must determine which of the two characters has the higher information content. It is neither necessary nor even possible to give a precise numerical value to the relative weight of each character. Qualitative statements are usually more important than quantitative ones. Knowing whether a species has a chorda is more important than a thousand measurements in assigning it to the correct phylum.

The scientific basis of a posteriori weighting is not entirely clear, but difference in weight somehow results from the complexity of the relationship between *genotype* and *phenotype*. Characters which appear to be products of a major and deeply integrated portion of the genotype have a high information content about other characters (which are also products of this genotype) and are thus taxonomically important. Other kinds of characters, such as *monogenic* and *oligogenic* characters, along with superficial similarities, convergences, and narrow adaptations, have a low information content concerning the remainder of the genotype and are thus of low value in the construction of a classification. What the descendants of a common ancestor share is not an aggregate of independent characters but a whole well-adapted harmonious genotype. Such a genotype has a considerable evolutionary inertia, and it appears that adaptively needed modifications can be superimposed on it without destroying it. Indeed, one might speculate that the characters in a phyletic line that are most intimately tied up with the basic well-integrated genotype are the most conservative.

Simpson (1962:499) described good taxonomic characters as "readily observable characters that are believed to be fairly constant within taxa but different between taxa at any pertinent level." This is a good empirical description, but it requires painstaking study of large series of specimens to demonstrate that characters are "fairly constant" and a delimitation of the taxa (based on a study of characters) before it can be shown that the characters are different between them. This description also cannot cope with the difficulties produced by the various types of similarity. Indeed, it is precisely this problem of how to determine what is a good character, a character with high weight, that has been a main source of concern among taxonomists.

Methods of a posteriori weighting have been discussed by Harrison (1960), Hennig (1950, 1966*a*), Maslin (1952), Remane (1952), Simpson (1961:82–106), Throckmorton (1969), and Wheeler (1986). No overall treatment of this subject has been published, and it is still cloaked in uncertainty. The classical approach has been to work backward from classifications that have produced what appear to be natural groupings and then to study the characters that delimit such natural groups. Tooth structure in mammals, wing venation and structure of the genitalia in insects, and the structure of the bony palate in birds are taxonomic characters that are fairly constant within such groups and are therefore given high weight. Each of these favored taxonomic characters fails occasionally or, in some groups, frequently. The number of cervical vertebrae (7) is a class character in mammals, while it is not even a generic character in birds, where it fluctuates from 23 to 25 in the genus *Cygnus* (swans). High weight is given to certain characters because generations of taxonomists have found that these characters are reliable in permitting predictions about association with other characters and in permitting the assignment of previously unknown species.

Evolutionary taxonomists may have to employ weighting at two stages in classification: in the delimitation and ranking of the recognized similarity classes and again during cladistic analysis in order to unmask cases of homoplasy. Particularly in ranking, weighting is of great importance. For instance, different classifications of the hominoids will result depending on whether one gives equal weight to brain and protein evolution or ascribes a higher taxonomic weight to the unique brain evolution (and correlated features) of *Homo*.

It has been said with good reason that the trial-and-error method of improving a classification is ponderous and uneconomical. Undertaking a successful a posteriori weighting of characters requires a thorough knowledge of the previous classifications of a given group and an ability to make value judgments. This is why numerical methods of iterative weighting (Farris 1969) are preferred by some authors. It is to be hoped that such methods will be improved to the extent that the traditional methods of a posteriori weighting will no longer be necessary (Chapter 11). In the meantime a survey may be presented of the traditional considerations in weighting. These considerations will remain particularly important in all groups with a paucity of taxonomic characters.

Characters with High Weight

A few generalizations can be made about characters with high taxonomic weight.

Complexity Complex structures have greater weight than do simple structures even when there are more of the latter. This is one of the reasons why genital armatures in arthropods are of such great taxonomic importance. They are usually highly complex and differ even in closely related species, sometimes drastically. The probability that distantly related species will develop similar structures by convergence is extremely low. Complex ornamentations, complex cusp patterns of teeth, and complex color patterns all fall under this category.

Joint Possession of Derived Characters Hennig (1950 and later) articulated a principle that had been followed by taxonomists for generations but had apparently never before been spelled out succinctly: Taxa ought to be defined on the basis of shared derived characters (synapomorphy) and not of shared ancestral ones. The strength of this principle is self-evident. Two forms may share a primitive (ancestral) character merely because they have not yet lost it, not because they are closely related. The acquisition by two taxa of the same evolutionary novelty, however, almost invariably results from close relationship, rarely from convergence.

Constancy A character that is constant "throughout large groups of species" (Darwin 1859:415) has higher weight than does a variable character. Farris (1969) suggested that low phenotypic variability be given very high weight. Since monogenic characters sometimes have low variability, this principle is probably valid only with relatively complex (polygenic) characters.

Consistency A character that is consistently present in one group and just as consistently absent in related groups obviously has higher weight than does a character that occurs sporadically in several groups and differs merely in the frequency of its occurrence.

The Darwin Principle Taxonomists have long stressed that characters that do not serve as a specific ad hoc adaptation are likely to indicate an underlying genetic similarity. As Darwin (1859:414) put it, "The less any part of the organization is concerned with special habits, the more important it becomes for classification." This is why color pattern in birds, a cusp configuration on molars in mammals, and a particular pattern of reduction or fusion of wing veins and elaboration of sclerotic structures in the genitalia of insects are of such great value in classification. Even though a particular special configuration of these structures may have an adaptive significance that has not yet been discovered, it is more probable that the genotype responsible for the structural configuration was brought together by natural selection and has adaptive significance as a whole. The principle stating that a character that is the product of the general genotype is more likely to have high taxonomic weight than is a character that represents an ad hoc specialization has wide application.

Characters Not Affected by Ecological Shifts Most higher taxa include subtaxa (species in genera, genera in families, etc.) that have made ecological shifts. Any character not affected by such a shift has a higher weight than do characters affected by it. An example is the adaptation of the mergansers (*Mergus*) for fish eating, discussed below.

Characters with Low Weight

All conditions that represent the converse of characters with high weight (see above) have low weight, other things being equal. Every specialist knows characters which are unreliable, that is, poor indicators of relationship. They have low weight in the grouping of taxa. The characters enumerated here are usually considered to belong to this category.

High Variability Although invariable characters (for instance, "possession of two eyes" in taxa belonging to the vertebrates) have low weight, highly or erratically variable characters are equally poor indicators of relationship in this category. The pattern of branching in the arteries of vertebrates may be different not only in different individuals of the same population but even in the left and right sides of the body. Differences in arterial pattern are not nearly as helpful for classification as some authors once thought they were. Wing venation has provided important characters for the classification of insects, but Sotavalta (1964) showed that there is far more variation in this character in the tiger moths (Arctiinae) than had been known and that the traditional generic arrangement of the family, based on wing venation, is thoroughly in need of revision.

Characters known to vary greatly within groups of clearly related species and to show similarly great variation in other, only distantly related groups are also unreliable. The presence or absence of bands in snails is such a character, as is large body size in some groups of mammals. The term *variable character* may mean several things. It may mean that a character is either present or absent in members of the same population or species or that that character shows many different degrees of expression in the same species or population. It also may mean that although a character is constant within a given species, it is either present or absent in the natural group that forms the higher taxon to which that species belongs. Variability in the characters of higher taxa may mean something very different from variability in populations.

Characters that are too difficult (or time-consuming) to be determined, such as number of hairs in mammals and certain physiological constants, have low weight.

Monogenic or Oligogenic Characters A simple monogenic character, such as those involved in balanced polymorphism, usually varies independently of other characters. Such a character has a low content of phyletic information. The distribution of monogenic characters (such as albinism) in the zoological system is often rather haphazard.

Regressive (Loss) Characters Any regressive character is usually of low taxonomic weight, because such losses may happen independently in more or less distantly related phyletic lines. Taxa based on loss of eyes, loss of wings or wing veins in insects, loss of toes or other appendages in mammals and birds, loss of teeth in mammals, or loss of segments in segmented animals are often unnatural. Trends toward simplification are often erratically realized in different phyletic lines. Losses may result from special environmental conditions (eyes in caves, wings on islands in stormy oceans) or may represent tendencies in a higher taxon that are realized many times independently in subordinate lower taxa. The Sanderling (*Calidris alba*), for instance, has lost the rudimentary hind toe which is still present in related sandpipers and was for this reason placed in a special genus (*Crocethia*). However, this bird is more closely related to two species of the genus *Calidris* (*minutilla* and *pusillus*) than these species are to such other species of *Calidris* as *C. acuminata* and *C. melanotos*. Michener (1949) showed how the independent loss of characters leads to spurious similarities among saturnid moths.

Neoteny (sexual reproduction in a preadult state) has also sometimes led to spurious similarities and proposals for unnatural taxa. Since the loss of a structure or feature may occur repeatedly in independent lines, taxa based on the absence of characters are often polyphyletic and unnatural.

Narrow Specializations Characters reflecting a single selection pressure, such as loss of an organ and desert coloration in birds and mammals, have lower taxonomic weight than their conspicuousness would suggest. As Darwin expressed it (1859:414), "Nothing can be more false" than to assume "that those parts of the structure which determined the habits of life, and the general place of each being in the economy of nature, would be of very high importance in classification."

Ever since Darwin, the argument about the relative merits of adaptive and so-called nonadaptive characters has continued. The weakness of adaptive characters lies in the fact that they are often phenotypically more impressive than is their cladistic weight; more important, they tend to be subject to convergence. A particular adaptive character may emerge in unrelated phyletic lines whenever it has selective advantage, implying relationship where there is none.

For example, a group of fish-eating ducks, the mergansers (*Mergus*, etc.), acquired—in connection with their fish-eating habits—a number of ad hoc adaptations: a streamlined body and head and a long thin bill with numerous horny teeth. Some ornithologists placed them in a separate family. Later on it was shown that the mergansers agree with the golden-eye group (*Bucephala*) in the essentials of courtship display, color of downy young, internal anatomy, and protein characters. Species of the two groups also frequently hybridize with each other. The taxonomic value of the fish-eating adaptations was consequently downgraded drastically, and mergansers and golden-eyes were combined in a single tribe. Characters associated with shifts in the food niche are particularly susceptible to a rapid attainment of conspicuous differences or, conversely, convergent similarities. The weakness of nonadaptive characters is that they are often old ancestral conditions (plesiomorphic characters), which have to be used with the same caution that applies to all other ancestral characters.

Structural adaptations subsequent to the invasion of a new food niche may be acquired very rapidly and generally have low taxonomic weight. In birds, for instance, particularly songbirds, the bill is highly plastic, and the bills characterizing flycatchers, finches, warblers, thrushes, and shrikes have developed many times independently. Different subspecies of the same species may have dramatically different bill types, indicating how quickly the shift from one type to another can occur. Teeth and jaw structures of cichlid fishes in African lakes also illustrate the principle of the rapidity of nutritional adaptations. Similar dramatic shifts characterize many differences that result from a single selection pressure, such as locomotory specialization and secondary sexual characters resulting from sexual selection (e.g., birds of paradise).

THE RELATIVE VALUE OF DIFFERENT KINDS OF CHARACTERS

Each kind of character adds to the total information. Very often the taxonomist is limited to preserved material, in other words, essentially to morphological characters, and to the geographic and ecological information that came with the specimens. Whenever morphological analysis produces equivocal results, as is often the case, every effort should be made to supplement the basic information with molecular, behavioral, or other additional information. Different kinds of characters are used in the discrimination of species than are used in the determination of relationships among higher taxa. It is only within the most recent past that an effort has been made to determine the relative weight of different kinds of characters at different levels of the taxonomic hierarchy.

Redundant Characters

As pointed out above, it is not necessarily true that the more characters a classification is based on, the better it is. There is a law of diminishing returns, because different components of the phenotype (different characters) may be pleiotropic or functionally correlated aspects of the same information in the genotype. *A character is redundant if it is a necessary correlate of other characters*. Large size alone may make species *A* appear more similar to species *B* than it is to a third small species *C* even though elimination of the size factor and its universal correlated effects may reveal that *A* is actually far more closely related to *C* than it is to *B*. Proportions are often far more sensitive indicators of relationship than is mere size, although the effects of allometry must be evaluated carefully (Gould 1966). A tendency toward an increase in meristic elements may actually be a single factor, even though it may affect all parts of the body and all appendages that show meristic elements. Scoring each element separately would grossly distort actual similarity.

The same peril of redundancy is present with *functional character complexes*. On the basis of an analysis of over 50 characters, Verheyen (1958) wrongly placed the diving petrels (a family of Tubinares) in the same order as the auks (actually relatives of the gulls) because nearly all the characters he used were taken from locomotory adaptations for wing diving. The degree of inferred relationship among the anthropoid apes was distorted until quite recently because the classification was based almost entirely on arboreal locomotory and vegetarian masticatory adaptations. In the skulls of birds, the bill, tongue, soft palate, bony palate, jaw muscles, and other structures are all adapted to the particular food niche a given species utilizes. Individual aspects of these structures cannot necessarily be scored as so many independent characters.

One can count and evaluate in classification only characters that are reasonably independent of each other. Exactly what independent is and how this can be determined is still controversial. Reducing the weight of characters according to their degree of correlation is only part of the answer, because characters may be correlated for two entirely different reasons: function or phylogeny.

Summary

A classification based on phyletic weighting has numerous advantages. It is the only known system that has a sound theoretical basis; it has greater predictive value than do other kinds of classification; it stimulates a character-by-character comparison of organisms believed to be phylogenetically related; and it encourages the study of additional characters and character systems to improve the soundness of the classification (hence its information content and predictive value). Finally, it leads to the discovery of interesting evolutionary problems. Thus classifications based on phyletic weighting not only have scientific advantages but are actually best able to satisfy the demands of the practice by having a greater total information content than do artificial systems.

CHARACTERS AND CATEGORICAL RANK

Characters that unequivocally designate categorical rank do not exist. We cannot say that species *A* must be placed in a separate genus because it has a generic character. The degree of morphological difference between species and genera is much greater in some families than it is in others. There are stronger behavioral differences among closely related species in some groups than there are in others. The same character that is associated with generic difference in one family may vary individually in species of another family. Nor does the total level of difference tell us much. Morphs within a single population may differ more from each other than do good sibling species, at least in terms of visible morphological characters. We must always remember that it is the taxon that gives the characters, not the reverse.

Empirically, experienced taxonomists know what kind of characters are valuable in the particular groups in which they are specialists. Any difference that functions as an isolating mechanism or facilitates competitive exclusion is primarily useful for the discrimination of species. The particular system of communication that is used in isolating mechanisms (e.g., light signals in fireflies versus calls in frogs) may characterize a higher taxon, while a specific isolating mechanism may characterize a species. Characters that indicate occupation of a particular adaptive zone or that are by-products of a well-defined, well-integrated genotype generally help in the discrimination and delimitation of higher taxa. Recent studies of molecular characters have clearly shown that some of them are more useful at the species level, others are not sensitive at that level but are full of information concerning the higher taxa, and still others are subject to local polymorphism and have no taxonomic value. It is up to the taxonomist to find out what character will be most useful at a given categorical level.

Even though broad generalizations cannot yet be made, a differential significance of characters is gradually becoming apparent within each higher taxon. Different kinds of characters appear to be most useful at the following categorical levels:

1. Recognizing subspecies (geographic variation)
2. Distinguishing closely related species, particularly sibling species
3. Grouping related species into genera
4. Determining the relationship of higher taxa, from families to phyla

CONCLUSIONS

A taxonomic character is any attribute by which a member of a taxon may differ from the members of other taxa. This attribute may relate to any feature of the dead or living organism that is amenable to comparison.

1 Taxonomic characters that are conservative (i.e., evolve slowly) are most useful in the recognition of higher taxa; those which change rapidly or concern isolating mechanisms are most useful in lower taxa.
2 Taxonomic characters that are subject to parallel evolution, especially those involving loss or reduction, should be used only with great caution.
3 Taxonomic characters are expressions of the biology of their carriers. An understanding of this biology is a prerequisite for the proper evaluation of these characters.
4 The same phenotypic character may vary in value and constancy from taxon to taxon and even within a single phyletic series. The weight given to a character depends largely on its constancy in a given group.

The entire zoological classification is based on the proper evaluation (weighting) of taxonomic characters. This operation, then, is the most important as well as one of the most difficult tasks of the taxonomist.

CHAPTER **8**

NUMERICAL PHENETICS

In the 1940s and 1950s there was growing dissatisfaction with the methods of classification. When several authors could defend entirely different classifications of the same higher taxon, it was felt that something had to be wrong. To eliminate the evident subjectivity of the prevailing methods, many of them based on a few arbitrarily chosen characters, several authors proposed the adoption of strictly objective, quantitative methods based on overall similarity that would guarantee repeatable results. Proposals for such methods were made by Michener and Sokal (1957), Sneath (1957), and Cain and Harrison (1958).

Degree of similarity was traditionally one of the criteria on which the recognition of taxa had been based. Such a strictly phenetic approach was not made obsolete by Darwin's theory of common descent: Similarity, it was said, is to be expected among the descendants of a common ancestor. Therefore, bringing the most similar species and higher taxa together should automatically produce a phylogenetic classification (Sokal 1983, 1985*a*).

To overcome subjectivity, the taxonomists cited above suggested grouping species into higher taxa with the help of the computer on the basis of their "overall similarity." Such similarity, it was claimed, can be determined by recording similarities and differences in a large number of variables (preferably well over 60). *Principles of Numerical Taxonomy* (1963) by Sokal and Sneath became the classical text of the new school of *numerical phenetics,* so named because numerical methods are also used in cladistics and other fields of taxonomy. A second, greatly revised edition by Sneath and Sokal appeared in 1973. *Numerical phenetics is the methodology of assembling individuals into taxa on the basis of an estimate of unweighted overall similarity.*

CLAIMS MADE BY PHENETICISTS

The supporters of numerical phenetics felt that their methodology had numerous virtues. For instance, application of the method requires no previous knowledge of the studied taxon and its literature, only the ability to make observations and quantifications. Because these methods purportedly produce strictly repeatable results, any beginner could make classifications as good as those of a specialist. An early numerical pheneticist made the sweeping claim that after a decade or so of the application of phenetic methods, all important questions in systematics would be answered and work could be done by technicians trained solely to collect data from specimens (Ehrlich 1964). The wisdom of the computer programmers could be relied on to provide the ideal operational methods by which these ends could be accomplished. Curiously, some 25 years later, the same claims have been made for some of the new molecular methods.

Philosophers have often contended that a classification is more natural, the more characters it is based on. Since the classifications of numerical phenetics are based on far more characters than are traditional classifications, pheneticists claimed that their classifications were more natural than those of other taxonomists (Gilmour 1940, 1961).

Leading pheneticists blamed much if not most of the subjectivity and arbitrariness of conventional classifications on the influence of phylogenetic or other theoretical considerations. They urged the acceptance of a strictly inductive, theory-free approach. To be objective, a classification must emerge automatically from an analysis of raw, randomly collected data ordered simply on the basis of their similarity. Pheneticists wholeheartedly adopted Bridgman's (1945) philosophy of *operationalism,* in which "clear and possible [operational] instructions" are substituted for theories and explanations. However, the sterility of inductivism is now almost unanimously agreed upon by philosophers of science; Hull (1968, 1980) has shown most convincingly the futility of the recommendations of operationalism. Complete objectivity is impossible, and the results of a purely operational approach that ignores causation and asks no questions about explanations are biologically meaningless.

The consistent application of the principle that taxonomy must be freed of all theoretical implications led pheneticists to reject in their work any reference to species, which were replaced by the concept of the *operational taxonomic unit* (*OTU*). However, the designation OTU refers to a very heterogeneous class of entities. Some OTUs are individuals, some are populations, and some are historical entities (sensu Wiley). This equivocation leads to considerable ambiguity in the interpretation of phenograms. By the pheneticists' criteria, phenetically different individuals ought to be treated as different OTUs; however, at that point the pheneticists quietly abandoned their own philosophy. They realized that they had to utilize biological criteria to assign different sexes or other phena to the appropriate species (OTU), and such criteria are inevitably inferential. Thus, the subjectivity that is rejected above the level of the species is tacitly condoned in the assignment of phena to OTUs.

The end product of the clustering process is a phenogram. A phenogram, however, is not a classification. Various procedures have been suggested for converting a phenogram into a classification (Rohlf and Sokal 1981). The best treatment of these procedures is given by McNeill (1979).

Pheneticists employ four criteria to determine whether or not a given classification satisfactorily represents the probable relationship of the included taxa:

1 Closeness to the original similarity matrix
2 Maximal similarity within the taxa
3 Maximal predictive value
4 Stability of classification

It would seem to be difficult to avoid subjectivity in the application of some of these criteria.

The methods recommended by pheneticists are described in Chapter 11. These methods are discussed in great detail by Sneath and Sokal (1973), in various symposium volumes (Felsenstein 1983*a*), and in the periodical literature (Sokal 1986). It would seem that on the whole phenetic methods are more often used in botanical and molecular biology than in zoological systematics. The preference of botanists for phenetics is also reflected by numerical taxonomy texts authored by botanists (Abbott, Bisby, and Rogers 1985).

ACHIEVEMENTS OF THE PHENETIC APPROACH

Numerical phenetics can take much credit for the revival of interest in macrotaxonomy in the 1950s and 1960s. Their work focused attention on classification above the species level, which had been languishing for generations, and made it abundantly clear that traditional approaches to the subject were not satisfactory. Phenetics deserves credit for having introduced the computer into taxonomy. Also, because of the rigorous requirements of computer programs, pheneticists pioneered meticulous and orderly methods of character analysis. Use of the data matrix permitted rapid and accurate comparison of the similarities and differences of various taxa. It also encouraged the search for as many differentiating characters as possible.

While phenetic methods are unsuitable for the classification of distantly related taxa, they may be quite useful in classifying large genera with numerous closely related species, all with similar evolutionary propensities and hence abundant homoplasy. In these genera each species may independently realize this propensity for different characters. As a result, one finds for each *apomorphic* character a different assembly of species. To classify these species phenetically would seem the simplest answer to the dilemma. A phenetic approach is sometimes useful when a determination of the degree of similarity is the objective. A good example is the study by Daly and Balling (1978), who used a phenetic analysis to discriminate among African and European honeybees and their hybrid forms.

One of the acknowledged weaknesses of the phenetic method is the discrepancy between the endeavor to study overall similarity and the actual use of a very limited number of characters. Sibley believed he could overcome this difficulty through his method of DNA hybridization, in which an attempt is made to establish a similarity value for all the single-coding DNA of the two compared species (Sibley and Ahlquist 1983). Critics, however, have claimed that the problem of translating DNA similarities into phylogenetic trees has not been fully solved. Until the limits of this method are better understood, the method of DNA hybridization should be used with caution.

WEAKNESSES OF NUMERICAL PHENETICS

When numerical phenetics was first proposed, it was well received and adopted by numerous taxonomists. Indeed, it produced respectable results in various groups in which no good classifications had existed previously. However, disenchantment eventually set in because the promise that grouping by overall similarity results in stable and natural classifications was not fulfilled (Mayr 1965*a*, 1969:203–211, 1982:221–226; Johnson 1968; Hull 1970; Wiley 1981:269–276; Farris 1977, 1979*b*, 1981, 1985). However, the last word on the usefulness of numerical phenetics in animal taxonomy has not yet been said, and a group of enthusiastic pheneticists is still active. For arguments in favor of phenetics, see McNeill (1979, 1982), Sokal (1985*a*, *b*; 1986), Felsenstein (1984), and several authors in recent volumes of *Systematic Zoology*. There has been a lengthy controversy in *Systematic Zoology* (1977–1983) between pheneticists (Rohlf and Sokal) and cladists (Mickevich and Farris) over the respective merits of their methods. However, their disagreements remain unresolved. The conflicting claims of pheneticists and cladists have induced the authors of a number of recent taxonomic analyses to employ both phenetic and cladistic methods.

There has been a tendency in the recent taxonomic literature to refer not to phenetic versus cladistic methods but to distance versus parsimony methods. As a result of the growing availability of molecular data, in which degree of difference can be determined far more easily and more quantitatively than can primitiveness versus derived state, the use of distance methods has greatly increased in recent years (Chapter 11).

WEAK POINTS OF PHENETIC METHODS

Certain deficiencies of the traditional phenetic methods have become apparent since their introduction.

I Practical difficulties

- **A** The method is rather laborious, and the data assembly of a complete character matrix for a large number of characters and species is forbiddingly time-consuming compared with some other methods. However, a large number of characters is required in phenetics because of the insistence on the equal weighting of all characters. As a result, the relatively few characters which provide clues to relationship are diluted by redundant characters and characters that are only "noise" in the computations.
- **B** The success of the method in traditional taxonomy depends on the availability of a large assortment of variable morphological characters. However, except in some arthropods, a rich supply of such characters is not available in most groups of animals. An absence of taxonomically useful characters has traditionally been the complaint of the leading taxonomists of many major animal taxa, for instance, the sponges. (For further comments, see Mayr 1969:205–208.)
- **C** A phenetic classification cannot be improved gradually, because any addition of new characters requires a new analysis.
- **D** Numerical phenetics has failed to make a substantial contribution to the classification of any mature group or to the classification of higher taxa (orders, classes, or phyla).

II Inability to meet the claims of objectivity and repeatability

- **A** *Subjectivity:* In spite of the equal weighting of all variables (characters), the method is unable to avoid subjectivity in the choice of the characters that are analyzed. Subjectivity is also unavoidable in the handling of variation: There is nothing in phenetic theory that would permit an assembling of phena (sensu Mayr), that is, morphs, sexes, age variants, and so forth, into meaningful OTUs (species) (see above). A failure to apply biological criteria at the species level leads to clearly absurd classifications, as was demonstrated by Boyce (1964) for primate skulls. By replacing species by OTUs, pheneticists deprive their classifications of meaningful biological content. For all practical purposes they are going back to the morphological-typological species concept of the pre-Darwinian period.

B *Lack of theoretical justification of choice among clustering methods:* The choice of the clustering method is as much a matter of personal preference as is the selection of characters by which overall similarity is determined. As a consequence, "different clustering methods yield different fits by cophenetic correlation coefficients to the same resemblance matrix" (Sokal 1985*a*:13). This imparts a great deal of subjectivity to the phenetic approach.

C *Repeatability and choice of computational method:* Pheneticists claimed originally that any two systematists working independently with the same group of organisms would achieve the same results if phenetic methods were used. This claim has not been substantiated. The choice of different algorithms almost invariably yields different results, as was first demonstrated by Minkoff (1965). Numerous examples of the influence of a computational method on the resulting classification were reported by Felsenstein (1983*a*). Presch (1979) got six different phenograms when using six different phenetic methods.

D *Repeatability and selection of characters:* The claim that if a sufficiently large number of characters is used, the selection of particular characters becomes irrelevant has also not been substantiated (Fisher and Rohlf 1969). Adding new characters, particularly ones derived from different components of the phenotype or genotype, almost invariably affects the classification (see below under mosaic evolution).

III Other unsubstantiated assumptions

A *Possible differences in the importance (weight) of different characters must be ignored in the absence of reliable methods for weighting:* Although referred to by the pheneticists as nonweighting, giving all characters the same weight is, of course, also a method of weighting. The research of the last 25 years and earlier has shown the decisive importance of weighting characters.

To be sure, all modern taxonomists reject the method of a priori weighting (on the basis of physiological importance) of earlier authors from Aristotle to Cuvier (Mayr 1982*a*). However, parallel variation of characters shows that some have a far greater information content than do others, and there is no reason why this information should not be used (Chapter 7). Curiously, Sokal and Sneath are rather inconsistent in the application of their nonweighting principle; they devote an entire section of their work to the principles governing "the proper selection of characters" (1963:66–69), ignoring the fact that any deliberate choice of characters constitutes a method of weighting. The main reason why pheneticists ignore weighting would seem to be that "no one has come up with a convincing algorithm for doing it" (Sokal 1985*a*:7).

B *The amount of unweighted phenetic similarity reflects an equal amount of underlying similarity of the genotype and can be safely used for ranking:* This assumption is thoroughly refuted by situations such as the existence of sibling species, which the pheneticist is forced to assign to a single OTU; at the other extreme are taxa, such as the birds of paradise, in which closely related and interfertile species are strikingly different (Mayr 1969:208). The various recent classifications of the prokaryotes on the basis of molecular characters illustrate how erroneous the previously accepted phenetic classifications were.

C *The use of a sufficiently large number of characters makes it unnecessary to partition them into similarities due to descent and those due to homoplasy:* Similarity, pheneticists state, is to be expected among the descendants of a common ancestor. Therefore, bringing the most similar species and higher taxa together will automatically produce a phylogenetic classification. This is sometimes the case, but there are two reasons why a phenogram, except in special circumstances, cannot provide a good estimate of phylogeny. First, parallel, convergent, and reversed characters are treated as equivalent similarities in purely phenetic methods. Only through cladistic analysis can homoplasy and reversals be detected and treated as differences.

As a result of convergent evolution, taxa that are only very distantly related rarely show great overall similarity. Paleontologists, however, have called attention to the interesting situation in which only remotely related extinct phyletic lines have passed through superficially very similar grades in very different geological periods (Arkell and Moy-Thomas 1940). A clear discrimination between true homology and homoplasy is therefore of the utmost importance in classification, but this is not part of the phenetic methodology. As a matter of fact, Sokal acknowledged "that classifications based on unweighted phenetic similarity would be affected by parallelism and convergence as well as by unequal evolutionary rates in diverging lineages and hence would not necessarily yield monophyletic taxa" (1985*a*:5).

Furthermore, cluster analysis cannot discriminate a relatively homogeneous paraphyletic group (lineages A, B, and C in Figure 8-1) from a holophyletic group.

D *Taxa can be satisfactorily ranked by drawing arbitrary "phenon" lines horizontally across a phenogram:* Actually, the numerical values of phena in different taxa cannot be compared with each other because they are based on very different sets of characters and thus do not permit the establishment of a universal scale. The ranking of taxa is therefore susceptible to a good deal of subjectivity. It is a mistake to think that phenograms are phylogenetic trees. A vestige of this misconception is indicated by the frequency with which a meaningless stem appears at the bases of phenograms.

It is fundamentally unsound to quantify similarity in a comparison of entities that are as highly heterogeneous as the character complexes of different taxa (Ghiselin 1966). The conspicuous phenetic differences caused by taxonomically unimportant ad hoc specializations prove this point convincingly.

IV Weaknesses in the theoretical foundation of phenetics

A *Strict empiricism:* Pheneticists are strict empiricists, and as such they reject a consideration of theory and theoretical concepts, including the theory of evolution. Since they allow the use of raw data only, they do not consider factors that are responsible for the

FIGURE 8-1
Mosaic evolution causing contradictory information provided by different sets of characters. Lineages B and C form a sister group to lineage A according to character set 1. Lineages A and B form a sister group to lineage C according to character set 2.

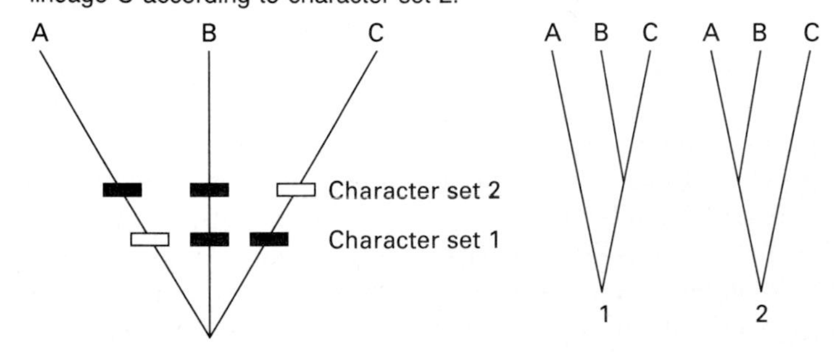

causation of a group. Common descent, of course, is such a factor in the classification of organisms. This is why organisms cannot be classified by the same methods used for books, artifacts, or any other inanimate objects (Chapter 6). However, pheneticists fail to test their taxa for monophyly. The term *natural* has an entirely different meaning when applied to the classification of evolved monophyletic taxa than it does when applied to inanimate objects classified on the basis of properties. Pheneticists, however, do not recognize that taxa are "historical groups" but "consider supraspecific taxa as classes and are therefore not especially concerned with their reality" (Sokal 1985*a*:5).

B *Failure to take mosaic evolution into account:* Different parts of the body, different stages of the life cycle, and different kinds of characters (morphological versus biochemical or behavioral) tend to have different evolutionary rates (Chapter 10). As a consequence, it will lead to different similarity estimates if either one type of character or the other is used. It is a mistake to assume that an arbitrary selection of characters will adequately reflect equivalent differences in the genotypes of different taxa. The external morphotype reveals only a very small and often highly biased portion of the genotype. How little correlation there is between phenotype and genotype was revealed by the pioneering work of Rohlf (1963, 1964), in which larvae and adults of the same 48 species of mosquitoes were compared, using over 70 characters for each age group. Even though the computational methods were the same, there was hardly any correlation between larval and adult classifications, and no light was shed on the similarities of the genotypes. Similar findings were made by Michener (1977). Sokal therefore concluded correctly "that phenetic techniques will not reach perfect congruence of classifications when these are based on different sets of characters" (1985*a*:9).

C *Conflict between a desire for complete objectivity and an awareness of evolution:* Since no distinction is made in phenetics between primitive and derived characters, pheneticists should not be able to distinguish between ancestral and derived groups. However, the terms *pregroup* (a somewhat aberrant primitive group) and *ex-group* (a specialized derived group) were proposed by pheneticists (Michener and Sokal 1957). Distance can be expressed only quantitatively, and two qualitatively very different distances may have a similar quantitative distance value.

D *Inability to serve as a heuristic device:* The strict inductionism of numerical phenetics represents a step back from the hypothetico-deductive approach of recent philosophies of science. Results achieved by induction allow for no testing, no meaningful predictions, and no interpretation of the nature of the achieved classification. Taxa clusters and the phenon hierarchy of the phenetic method lack biological meaning. They are merely a set of counts or measurements. They lack all properties of a theory and therefore cannot be tested. Furthermore, as Simpson (1964:712) said, by converting the totality of the similarity values of all individual characters into a single overall similarity value, the clustering method suffers "an enormous loss of information, mainly on the character direction and origin of differences."[1] By converting the highly revealing variation of numerous characters into a single value, the method acquires a typological aspect, which induced Simpson (1964) to conclude that phenetics has led "to retrogression in taxonomic principle... a conscious revival of pre-evolutionary, eighteenth-century principles."

OUTLOOK

The popularity of phenetic methods has greatly declined since the rise of cladistics. Indeed, it is widely believed that a method which ignores the necessity of weighting characters and the recognition of monophyletic groups cannot lead to a stable and sound classification. In combination with other methods or as a check against other approaches, phenetic methods are still much in use, but it is recommended that phenetically delimited groups always subsequently be tested for monophyly. Indeed, the method of so-called evolutionary classification consists essentially of a provisional phenetic delimitation of taxa followed by "purification" by monophyly testing when necessary.

Interestingly, the methods of numerical phenetics have found their greatest use outside the field of animal taxonomy. In the classification of objects whose groups are not the result of a causation but simply are due to the joint possession of similar characteristics, the methods of numerical phenetics are often most helpful.

THE FUTURE OF NUMERICAL PHENETICS

In recent years the decline in the popularity of phenetics seems to have been halted, indeed reversed, because of three developments. First, morphological characters are increasingly supplemented by a wealth of molecular characters that can be used for distance measures. Second, clus-

[1] In contrast to traditional methods, a phenogram does not tell what the diagnostic characters of a group of species (higher taxon) are.

tering is no longer the obligatory method in numerical phenetics, nor do distance algorithms necessarily require an assumption of an equal rate of evolutionary changes in all branches of the tree. Third, the widespread view (Chapter 10) that taxa should be relatively homogeneous has led to an increased use of phenetic methods.

An eventual synthesis of phenetics with other approaches to classification would be facilitated if pheneticists adopted three modifications of their philosophy: (1) the weighting of characters whenever the data allow for this, (2) the testing of provisionally recognized taxa for monophyly, and (3) the recognition that species have reality in nature and are not purely subjective, arbitrary inventions of the taxonomist. Since several pheneticists have already adopted one (or all three) of these modifications, there no longer seems to be a difference of principle between such a modified phenetics and the Simpson-Mayr approach.

CHAPTER 9

CLADISTICS

Even though Darwin stated emphatically in the *Origin* (1859:420, 486) that to be natural a classification must be based on genealogy, this principle was ignored by most systematists for nearly 100 years. To be sure, as noted by Hennig, some authors supported the principle and even proposed methods to implement it, yet many classifications continued to be based simply on similarity or, worse, on a few diagnostic characters. To eliminate such arbitrariness, Hennig proposed a method of classification that he called *phylogenetic systematics,* which was based exclusively on genealogy, that is, on the branching pattern of phylogeny. Hennig's approach was strongly influenced by a long-standing German tradition of considering the reconstruction of common ancestors to be the ultimate objective of *phylogeny.* Hennig's *Grundzüge einer Theorie der phylogenetischen Systematik* (1950), written in rather difficult German and poorly edited, was largely ignored until 1966, when an English translation of a revised German edition was published. Hennig also presented a succinct statement of his ideas in English in 1965. Although Hennig named his method *phylogenetic systematics,* it was based on only one of the two components of phylogeny—the branching points of lineages (*cladogenesis*)—and neglected the relative amount of subsequent divergence (*anagenesis*), the other component. To prevent confusion with the traditional concept of phylogeny, which includes both cladogenesis and divergent anagenesis, Hennig's methodology was renamed *cladistics* (or *cladism*), by which it is now more generally known (there is even a periodical called *Cladistics*). Hennig was an excellent taxonomist, and the concerns of classification were foremost in his mind. Many of his later followers, however, had little taxonomic experience and were more interested in producing better *algorithms* for the production of cladograms or in discovering patterns of character variation. Cladograms, they assumed, would automatically produce correct classifications. An enormous literature reflecting the debate about cladistics has developed in the last 25 years (Dupuis 1979). Two major texts supporting cladistics are Wiley (1981) and Eldredge and Cracraft (1980). Among the critiques are publications by Ashlock (1974, 1979), Bock (1973, 1977), Charig (1982), Felsenstein (1983*a*), Ghiselin (1984), Gosliner and Ghiselin (1984), Holmes (1980), Hull (1979), Mayr (1974*b*), and Michener (1970, 1977). Additional literature on more specific questions is cited in Chapters 6, 10, and 11 of this volume. Symposia discussing cladistics (pro and con) include Cracraft and Eldredge (1979), Funk and Brooks (1981), Joysey and Friday (1982), Felsenstein (1983*a*), Platnick and Funk (1983), and Duncan and Stuessy (1984).

The availability of Hennig's simply structured methodology encouraged many authors to revise groups with immature classifications, that is, groups that had not previously been subjected to an evolutionary analysis. This has led in the last 20 years to the unmasking of a number of previously accepted unnatural taxa, for instance, among turtles, fishes, many insect groups, and turbellarians. These cases showed what should have been evident ever since Darwin: the importance of an inference about genealogy as a prerequisite for the construction of a sound classification.

There can be little doubt that Hennig initiated a revolution in the methods of macrotaxonomy. The merits of his methodology are manifold, but two in particular should be singled out:

1 It reemphasizes the notion that because taxa are products of evolution, this fact must be carefully considered in the delimitation of taxa.

2 It employs a method that sheds light on the evolutionary history of species: a careful evaluation of taxonomic characters. An advantage of this method is that it often permits an inference about phylogeny[1] from an analysis of the characters of living forms without the use of fossil material; such an inference can subsequently be checked against fossils whenever fossils are available.

[1] Some groups of recent cladists (Platnick 1979) reject any concern with phylogeny in the construction of cladograms.

Furthermore, it suggests a legitimate and workable method for the weighting of characters. This particular method of weighting consists in partitioning characters into those derived from the nearest common ancestor (*derived* or *apomorphic characters*) and those retained from more remote ancestors (primitive, ancestral, patristic, or *plesiomorphic characters*).

This method is based on the assumption that every natural taxon (recognized as such because it possesses synapomorphic characters) can be traced back to a stem species. As simple as this basic assumption sounds, it permits various interpretations and conclusions and has given rise to much debate. Some of the objections raised against the cladistic method are discussed below.

To understand why some components of cladistics are widely accepted among taxonomists while others are rejected, one must recognize that cladistics consists of two fundamentally different processes:

1 *Cladistic analysis,* or the reconstruction of the branching sequence of phylogeny through an analysis of synapomorphic characters, a method also adopted by many noncladists

2 *Cladistic classification,* or the delimitation of taxa and their ranking in a Linnaean hierarchy based on the principle of holophyly

The two processes, the assumptions on which they are based, and the difficulties which their application encounters are discussed separately here. Indeed, they are in principle independent of each other. Since an evolutionary taxonomist, following Darwin, wants taxa to be consistent with genealogy, that taxonomist is likely also to undertake some sort of cladistic analysis. When difficulties in cladistic analysis are pointed out in the following account, they pertain to anyone attempting to infer the actual phylogeny, whether a cladist or not. The major parting of the ways between cladists and evolutionary taxonomists relates to the methods of classification.

DIFFICULTIES IN PRESENTING AN UNBIASED ACCOUNT OF CLADISTICS

There are a number of reasons why it is difficult to present an unbiased and up-to-date account of the principles and methods of cladistics. First, a considerable number of Hennig's (1966*a*) original rules were soon rejected by leading cladists. Among these rules are the following:

1 The categorical rank of a taxon is to be determined by its geological age.

2 The *deviation rule,* "which says that when a species splits, one of the two daughter species tends to deviate more strongly than the other from the common stem species'' (Hennig 1966*a*:207).

3 Each species is terminated by a split into two daughter species; i.e., all branching points are dichotomous.

4 A species begins at a branching point and ends at the next branching point in the cladogram.

The first of these rules is now universally rejected by cladists, while the other three are rejected by some cladists but accepted by others.

None of these changes affected a basic original Hennigian principle: the deliberate exclusion of any consideration of ancestral (plesiomorphic) characters and of anagenesis (''distance'') in the construction of classifications, except in regard to the usually ignored deviation rule. The only major subsequent change in cladistics occurred when Farris, employing the Wagner tree method (Chapter 11), quantified the contribution of apomorphic characters, thus paying some attention to anagenesis (Camin and Sokal 1965).

The second reason why an unbiased discussion of cladistics is difficult is the confusion produced by giving new meanings to terms that traditionally were used in a different sense. Wiley (1981:85) defended Hennig's shift of traditional terms ''because tradition has no particular place in science.'' In this claim he was entirely mistaken. To be sure, tradition has no place with respect to the acceptance of theories or concepts, but scientific terminology is another matter. Stability is perhaps the most important prerequisite of scientific terms; without it, total confusion would reign in science. No one would be able to understand which objects or concepts other scientists were talking about. The transfer of terms from one entity or process to a different one, as has been done by certain cladists, must be rejected uncompromisingly. The following transfers of terms have been particularly confusing.

1 *Phylogeny:* This term was restricted by Hennig to only one of the two components of phylogeny: branching (cladogenesis). Haeckel, who coined the term, did not define phylogeny as a branching pattern but as ''the entire science of the changes in form through which the phyla or organic lineages pass through the entire time of their discrete existence'' (1866, II:301). Thus, his main emphasis was on anagenesis.

2 *Monophyletic:* This term was coined by Haeckel in 1866 in support of Darwin's theory of common descent. He was unalterably opposed to the then-current theories of Darwin's opponents, who espoused numerous independent origins of living things. Haeckel termed such multiple-origin theories *polyphyletic*.

Haeckel applied the term *monophyletic* to all the taxa recognized in his time—for instance, Reptilia, Aves, and Mammalia—each of which he thought consisted of descendants of a nearest common ancestor. Haeckel was not concerned whether any of these taxa (e.g., the Reptilia) had given rise to divergent ex-groups (e.g., Aves and Mammalia). *Monophyletic,* then, was a term applied to taxa recognized in accordance with other criteria. As traditionally defined, any taxon was considered monophyletic whose members were all derived from the nearest common ancestor, a definition that was applied to such traditional taxa as the Reptilia, Turbellaria, and Psocidae. A taxon was considered natural if tests showed that it was monophyletic by this definition. Even though, as discussed in Chapter 10, there have been differences of opinion about the nature of the common ancestor, there was complete unanimity that *monophyletic* was a qualifying term for a taxon, not for a process of phylogeny.

This usage was universally accepted until Hennig (1950) employed the term *monophyletic* as a name for the process of descent. Instead of using the concept to test the naturalness of previously delimited taxa, Hennig used monophyly to delimit taxa. For him, a taxon is monophyletic only when it consists of the totality of the descendants of the ancestral stem species.

These are only two of the numerous terms (including relationship and homology) which the cladists have transferred to new concepts. We have come to the same conclusion as Gosliner and Ghiselin (1984:255) that "cladists have drastically altered the meaning of ... terms ... to suit their own methodological and philosophical paradigm."

Because the transfer of this term to an entirely different entity gave rise to a great deal of confusion, Ashlock (1971) proposed the term *holophyletic* for the new concept introduced by Hennig. Cladists call taxa that have given rise to descendant taxa (to ex-groups) *paraphyletic* (see below).

Cladists assert that they discover "taxa" (they mean holophyletic groups) through the study of synapomorphies. A study of the cladistic literature, however, reveals that in the better-known groups cladists begin their analysis with traditionally accepted taxa that have been delimited by other methods (Eldredge and Cracraft 1980:51) or with tentatively recognized taxa sorted by similarity (Brundin 1966:23). Indeed, most synapomorphies are discovered by searching in taxa, first based on similarity, for characters which the members of a given taxon have exclusively in common.

Relationship

In cladistic classification generally, only sister group comparisons are legitimate, not ancestor-descendant relationships. A species of a holophyletic taxon is more closely related to any other species of that taxon than it is to any species of an ancestral or sister taxon. Ancestor-descendant relationship is irrelevant in the construction of a cladogram but is of decisive importance in the delimitation and ranking of taxa (see below). Recently, some cladists have reconsidered the role of ancestor-descendant relationship. Wiley (1981:107) agreed that there is no reason to exclude this information from consideration in constructing a classification.

Another difficulty in presenting a balanced treatment of cladistics is the development of various new schools. In the last 20 years cladists have split into a number of schools that differ quite fundamentally not only in their basic philosophies but also in the degree to which they have abandoned some of Hennig's rules (Hull 1988). Wiley (1981) and some other cladists are closer in their thinking to evolutionary taxonomy; others retain most of Hennig's original rules. The pattern cladists have moved farthest from the original Hennigian paradigm (Platnick 1979, 1985; Patterson 1980). For recent critiques of these schools, see Hull (1980), Beatty (1982), Panchen (1982), and Charig (1982).

Pattern Cladistics

For a pattern cladist, the cladogram is not a reflection of the phylogeny of a group of taxa but rather the pattern of the origin of characters (Platnick 1976, 1979, 1980, 1985; Patterson 1980). These cladists, also referred to as *transformed cladists*, use much the same terminology, dendrograms, and formalism of classification that the Hennigians employ, but their philosophical attitude is quite different (Beatty 1982). Although they claim to agree with the evolutionary explanation of the diversity of nature, they deny that the pathway of evolution is knowable and also deny that their analysis has anything to do with the exploration of phylogeny. Therefore, they do not classify species or higher taxa but simply search for a "natural order," which they find in the pattern of character distribution. Like the pheneticists, they believe in a strictly inductive, operational approach that is untainted by any theory or conceptual framework. In spite of these philosophical differences, their cladograms are similar or identical to those of Hennigian cladists. For a critique of pattern cladists, see Charig (1982).

CLADISTIC ANALYSIS

Cladistic analysis endeavors to reconstruct the pathway of phylogeny. Closeness of relationship is determined exclusively by the possession of synapomorphies, which are inferred to indicate the relative recency of the common ancestor. The concept of *synapomorphy* is based on the inferred origin of characters rather than on the sharing of actual characters. Snakes are inferred to have the synapomorphy "four legs" even though they have lost their legs. This must be kept in mind when one classifies the results of reversals. It is therefore important to determine the sequence of branching events in the history of a group and to construct from it a cladogram. Cladists of Hennig's school make a number of assumptions in their analyses. They assume that branching results from a speciational event through which only two new phyletic groups originate. Each of these groups is recognizable by its *synapomorphic* characters. Each species terminates (becomes extinct) when it splits into two daughter species. The base point of a cladistic analysis is a species, and each holophyletic taxon is derived from a particular stem species.

On the basis of these assumptions, cladistic analysis should be carried out in the following manner. First, it must be established that each character to be used in the analysis is truly homologous with the other characters with which it is to be compared (Bock 1981:14–18). (For a full discussion of homology, see Chapter 6.) Since members of holophyletic groups as well as monophyletic taxa share one or several unique derived characters (synapomorphies), each character of a taxon must be evaluated to see whether it is a derived (apomorphic) or an ancestral (plesiomorphic) character (Hennig 1984:34–54). Plesiomorphic characters are also sometimes referred to as *patristic* or primitive characters, although the term *patristic* as used by Cain and Harrison (1960) does not fully coincide with the use of the word by cladists. It is inappropriate to speak of primitive taxa, for taxa referred to as primitive usually also have highly advanced specializations, such as the "duckbill" and poison spur of *Platypus*. Plesiomorphic characters are ignored in cladistic analysis because they do not help in locating branching points in a cladogram. Only apomorphic characters can do this.

Determination of Apomorphy

How is a clear-cut discrimination between derived (apomorphic) and ancestral (plesiomorphic) characters to be made? A character is clearly derived when it is restricted to a particular taxon. However, when the character is variable or is found in some but not all members of the taxon, it must be placed into a *transformation series*. This is a series of homologous characters, some of which were derived from others during evolution. For instance, if we assume that the bird feather is derived from the reptilian scale, the sequence from scale to feather is a transformation series. Only homologous characters can be placed in a transformation series. In many transformation series it is at once evident which is the ancestral and which is the derived (apomorphic) condition. In others a special investigation must be undertaken to determine the *polarity* (direction) of the series (Stevens 1980; Wiley 1981; Bishop 1982; Underwood 1982; Gosliner and Ghiselin 1984; Kluge and Strauss 1985).

Among the several kinds of evidence that permit such a determination, some are more important than others. The most important is the so-called *out-group comparison*. If one of two homologous characters that occur as variants within a single holophyletic or monophyletic group is also found in the sister group, it is the plesiomorphic character. Any character that is restricted to the monophyletic group is apomorphic (Wiley 1981:139). The assumption that the ancestral condition is represented by the character that is most widely distributed among related taxa (commonality rule) is not necessarily correct and cannot be substituted for a rigorous out-group comparison. When it is not certain which is the sister group, out-group comparisons may have to be made with several closely related taxa to avoid circular reasoning. Interpreting out-group comparisons is not always easy, particularly when there are conflicts between the inferences derived from different characters. This subject is discussed by Wiley (1981:139–176), Brooks and Wiley (1985), Watrous and Wheeler (1981), and Maddison, Donoghue, and Maddison (1984).

A second kind of evidence that helps determine the polarity of a transformation series is the sequence in which characters manifest themselves during *ontogeny* [Nelson (1978) and Kraus (1988) versus de Queiroz (1985) and Kluge (1985)]. Without getting involved in the controversy concerning *recapitulation,* it is nevertheless an empirical observation that characters acquired more recently during phylogeny often appear correspondingly later in ontogeny (Cartmill 1981; Osche 1982). Like nearly all general rules in biology, there are exceptions to this observation, particularly in regard to larval stages that have been drastically modified for the occupation of special niches.

Gosliner and Ghiselin (1984) additionally used functional criteria for the determination of the direction of evolutionary change, particularly in taxa with frequent parallelism. Some directions make sense functionally, while others do not. In such cases parsimony can be rather misleading.

In spite of the notorious incompleteness of the fossil record, the *stratigraphic sequence* in which characters first appear in fossil taxa often provides additional evidence about the polarity of a transformation series (Gingerich 1979). This type of evidence is particularly important in taxa in which character reversal is frequent (Hallam 1988).

These sources of evidence—out-group comparison, ontogenetic sequence, and stratigraphic sequence—are independent of each other, and each serves to test the validity of the others. A further and far more important test is provided by additional apomorphies. The more of them that support the same polarity in a transformation series, the more probable it is that that polarity is correct (Platnick 1977*b*), unless this is contradicted by an appropriate weighting of the characters (Chapter 7). For a critical analysis of the testing of transformation series, see Bock (1981:14–18). The phylogeny implied in a cladogram is only a hypothesis. There is always a need for further testing against biogeographic, stratigraphic, and ecological implications.

Cladistic analysis is sometimes criticized by pheneticists as speculative and thus unscientific. This criticism can be answered by two arguments. The first and more important is that the theory of common descent has been established beyond any doubt and cannot be ignored in the construction of a classification. The second is that although a particular line of descent, say, of the mammals from the therapsid reptiles, may be hypothetical, it can be inferred from a large body of evidence, particularly from an analysis of synapomorphies. This process of deriving conclusions by inference from a large body of data is not speculation but a valid and abundantly substantiated scientific method. By ignoring the evidence for the relationship of higher taxa that is provided by the theory of common descent, the pheneticists endorse a weaker, less reliable methodology.

The Cladogram

The branching pattern revealed by cladistic analysis is presented by the cladist in a diagram, called a *cladogram,* that is based on the pattern of synapomorphies. Each branching point in a cladogram represents a speciational event, which potentially gives rise to a new holophyletic taxon. No branch of a cladogram (Figure 9-1) can be based on plesiomorphic characters. Because paraphyletic taxa are not holophyletic, they are not recognized by the cladist and therefore are not indicated in a cladogram. Cladists often adopt two conventions in the construction of a cladogram: *Sister groups* are placed at the same distance from the ancestral species, and the sister taxon that has retained more primitive (plesiomorphic) characters is placed on the left branch of the dichotomy.

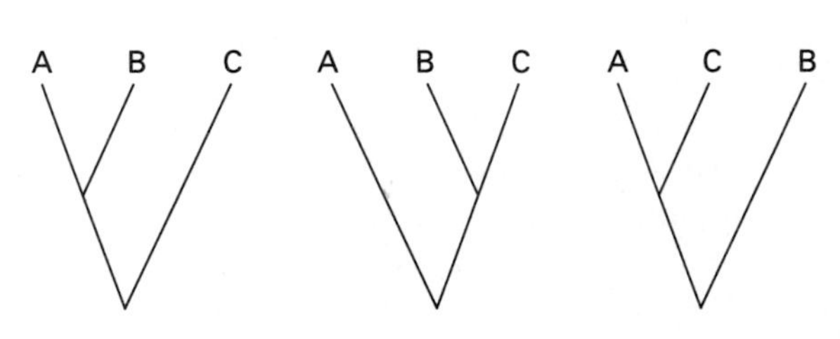

FIGURE 9-1
Dichotomous cladogram. The three nonredundant ways in which three taxa (A, B, and C) may be related on a dichotomous cladogram.

Cladograms are particularly informative in the comparison of groups that may have experienced coevolution, such as parasites and their hosts, symbionts, commensals, and pollinators. Cladograms are valuable in biogeography (Ashlock 1974), although Mayr questions the validity of the frequent claim that they permit discrimination between primary (dispersal) and secondary (vicariance) isolation.

Hennig and other early cladists considered the cladogram to represent the phylogeny of the included taxa. More recently, pattern cladists have recognized that a cladogram is actually a diagram of the distribution of synapomorphic characters. Thus its information content is limited (Figure 9-2). To refer to a cladogram as a phylogenetic tree as traditionally defined is misleading because

1 A cladogram represents taxonomic characters, not taxa.
2 A cladogram, as originally conceived by Hennig, ignores all anagenetic changes that are not synapomorphies.

When a cladistic analysis allows for the construction of several possible cladograms, that one is considered the best which is "shortest," that is, has the smallest number of changes in character states (branching points). This is called the principle of *parsimony* (Chapter 11). The best tree is the tree, as Farris (1983) formulated it, that requires "the fewest

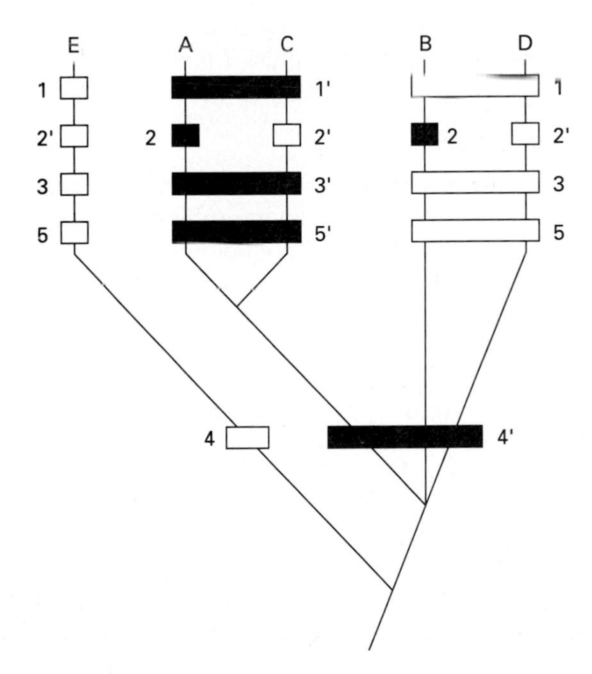

FIGURE 9-2
Cladogram. Black bars are inferred apomorphies; open bars are inferred plesiomorphies. (*From Wiley 1981.*)

ad hoc hypotheses of homoplasy." According to still another formulation, that tree is most parsimonious which minimizes the total number of evolutionary character changes needed to explain the variation in a given set of data. As will be discussed below, parallel evolution, particularly the independent loss of characters (character reversal), may cause the shortest tree to misrepresent the actual phylogeny. For discussions of the parsimony principle, see Panchen (1982), Friday (1982), Felsenstein (1978, 1979, 1983*c*), Sober (1983, 1988), Maddison, Donoghue, and Maddison (1984), Farris (1983), and Swofford and Maddison (1987).

When reference is made to the parsimony approach, what is usually meant is the construction of a minimum-length tree (Edwards and Cavalli-Sforza 1964; Kluge and Farris 1969), in other words, the minimum number of inferred evolutionary steps (Camin and Sokal 1965). Others have expressed it as the minimum number of mutational steps (Fitch and Margoliash 1967). Finally, some authors have referred to Estabrook's compatibility method as a parsimony method (Chapter 11).

Considering that the actual phylogeny is not known and that homoplasy is frequent, a parsimony analysis may produce several equally parsimonious trees. Special methods are required in such a case for determining the optimal length of a given tree (Swofford and Maddison 1987). As a result of homoplasy, the shortest tree may reflect phylogeny less well than a somewhat longer tree does. A number of methods have been proposed, though few of them are used routinely, to approach the most correct tree. They all involve some method of weighting conflicting characters and eliminating discovered homoplasies (Chapters 7 and 11).

DIFFICULTIES IN CLADISTIC ANALYSIS

Since a particular group of organisms obviously can have only a single true phylogenetic tree, a cladogram inferred from apomorphies is only an estimate of the probable phylogeny. Harper (1976) pointed out how many different cladograms can usually be based on the same set of characters as a result of the difficulties we have enumerated. There is no substance to the formerly common claim that cladistic analysis leads unequivocally to a single cladogram which in turn can be translated into only a single, and therefore best, classification. Actually, the determination of the branching points of a cladogram encounters various difficulties which require more detailed discussion.

Homoplasy

Similarity in a character found in two different taxa may be due not to homologous apomorphy but to so-called homoplasy. A homoplasy is a similarity in a character that two taxa have acquired independently. The taxonomist encounters three forms of homoplasy.

Convergence Convergence is usually a superficial similarity of two distantly related taxa; the similar characters are not homologous (Chapters 6 and 10). The wings of birds and those of insects are an example, but so are the wings of birds and those of bats *as wings* (they are homologous as anterior extremities). Convergence is rarely a problem in cladistic analysis, because it is usually restricted to a single functional system and close relationship is contradicted by numerous other characters.

Parallelism Parallelism—the independent acquisition of what seems to be the same character in related taxa—by contrast, is very common. It occurs in almost all higher taxa and is a continuing source of difficulties (Chapters 6 and 10). Parallel similarities are not homologous. The classification of the cichlid fishes of the African lakes, for instance, had to be drastically revised when it was discovered that the same jaw articulation (of which there are four types) had evidently been acquired independently by different lineages. Parallelism is rampant among passerine birds, in which similar morphological types evolved independently on most of the major continents. Examples include flycatchers, warblers, finches, titmice, shrikes, nuthatches, tree creepers, and nectar feeders. In some cases, it was possible only through the use of nonobvious characters (particularly molecular ones) to establish the true relationship and prove the existence of parallelism. Many modern classifications are probably still based in part on undiscovered parallelisms, particularly those involving groups with a scarcity of morphological characters.

When taxa are more distantly related, it is often uncertain whether a homoplasy is due to parallelism or to convergence. There is actually no good criterion for the demarcation of parallelism against convergence. The usual definition of parallelism—possession of similar features in two or more lineages which share a common ancestry—is vulnerable, because how far back should one go? Birds and bats share common ancestry, but their wings surely represent a convergence, as may all the songbird "parallelisms" listed above. A far more serious problem is posed by the propensity of genotypes to produce a certain phenotype, such as stalked eyes in certain acalypterate dipterans, which is manifested in only some of the possessors of a genotype. Parallelism in this case can be defined as homologous similarity, since the common ancestor evidently had the genetic propensity even if it was not expressed phenotypically. Gosliner and Ghiselin provided a good analysis of the various aspects of parallelism. They concluded (1984:258) that "parallelism means that taxa began with the same initial conditions, and independently underwent the same changes'' (see the discussion of parallelophyly in Chapter 10).

Reversals The loss during phylogeny of previously acquired characters is apparently far more common than was formerly realized. A lost apomorphy mimics a plesiomorphy. Paleontologists have provided abundant evidence for the frequency of evolutionary reversals, resulting in the return to a secondarily simplified structure. Bretsky (1979:149) remarked that ''evolutionary reversals may be extremely common.'' Branching sequence is strongly affected by parallel evolution and character reversal (Panchen 1982; Charig 1982; Willmann 1983). A strictly automatic application of the shortest-tree principle may lead to serious errors in such cases.

Evaluation of Homoplasy When there is discordance among characters because of homoplasy, characters should be weighted and ranked according to their quality. The following criteria are helpful in determining the character to which the highest weight should be assigned (Bock 1977:890).

1 It is a complex feature rather than a simple one, because the probability of evolving independently two or more times is considerably lower for a complex than it is for a simple feature.

2 It is a new feature rather than the reduction or loss of a feature, because the probability of a new feature evolving independently two or more times is low, while reduction or loss of a feature may occur several times independently.

3 It is a feature that has a wide range of biological roles rather than a single one, because the probability of a feature serving a wide range of roles evolving independently two or more times is less than that of a feature with a single biological role.

4 It is a feature that has an ontogenetic development that depends on a complex pattern of embryological modifications and interconnections with other developmental features rather than one that has a simple ontogeny.

Although none of these criteria is absolute or applicable in every case, a careful analysis of each case of conflicting apomorphies is likely to result in a more reliable cladogram than is an unweighted procedure. For further suggestions on the weighting of characters, see Chapter 7. The fossil record and the pattern of geographic distribution are also sometimes useful in the testing of parsimony decisions.

It must always be remembered that there is no criterion by which one can decide automatically which similarities are due to synapomorphy and which are due to homoplasy. Nor can reversals be automatically distinguished from plesiomorphies. What must be emphasized, because it is so often ignored, is the rampant frequency of homoplasy in morphological characters in many groups. For instance, it was impossible to reach consensus on the classification of the three subfamilies of Anguidae (legless lizards), because even the most parsimonious tree required the recognition of 12 homoplasies with 15 characters (Good 1987). Other recent analyses showed that more than 50 percent of all postulated synapomorphies were cases of homoplasy. The construction of monophyletic taxa in such cases must include the most careful comparative morphological and functional analysis of all characters, resulting in their proper weighting.

Choice of Characters

Since two taxa may differ by an almost unlimited number of characters, the working taxonomist is forced to make a very restricted selection. The ultimate classification may depend entirely on this selection. For instance, Rosen et al. (1981), on the basis of the characters they selected, came to the conclusion that the Dipnoi (lungfishes) are the sister group of the tetrapods, while Holmes (1985), partly on the basis of different characters, concluded that the traditional classification according to which some group of rhipidistians is the sister group of the tetrapods is far better documented. The difficulty lay in the correct identification of the synapomorphies. Similarly, the claim that birds and mammals are sister groups was based on a strict cladistic analysis (Gardiner 1982) that used a very different set of characters than did the traditional classification, in which the birds are derived from archosaurian ancestors and the mammals from the therapsids, two very different groups of reptiles; this classification was confirmed by Kemp (1988) and Gauthier, Kluge, and Rowe (1988). Nor has strictly cladistic analysis been able to establish the phylogeny of the lice, Mallophaga and Anoplura (Lyal 1985 versus Kim et al. 1987). These cases show that a cladistic analysis does not guarantee a correct cladogram unless it is based on valid apomorphies.

Polytomy

Orthodox cladistic theory asserts that in every speciation event an ancestral species splits into two daughter species, each of which gives rise to a new holophyletic taxon. All branching points in a cladogram therefore ought to be dichotomies. However, sometimes not enough characters are available in cladistic analysis to resolve branching points into dichotomies. Furthermore, speciation far more often entails the budding off of an isolated population (peripatric speciation) while the parental species continues to exist more or less unchanged. Such an unchanging parental species can bud off numerous other daughter species in the course of millions of years, sometimes several simultaneously. Evidence is not available to resolve such cases into dichotomies. Many cladograms therefore contain trichotomies or even higher polytomies. A classification based on polytomies cannot be unequivocally translated back into a cladogram without a special explanation. Modern cladists are fully aware of this problem (Wiley 1981:48–51). However, it is not very helpful to designate as incertae sedis only taxa emerging from polytomies, because some doubt or other attaches to the placement of almost all taxa; they all are incertae sedis to some degree.

Neglect of Variation

There is always the danger that the variability of a higher taxon will be underestimated if only a limited number of species are available for analysis. This may lead to erroneous judgments, such as the neglect of apomorphic characters, when it is not expressed in the limited number of analyzed species.

The outstanding importance of cladistic analysis is that it may lead to the unmasking of convergent polyphyly (Chapter 10).

To summarize, cladistic analysis led to a considerable improvement of classifications, particularly in groups that previously had been classified by means of a few key characters. It would be going too far, however, to claim that the cladistic method led to a revolutionary reclassification of all or even most groups of animals. The facts do not substantiate this claim. For instance, nearly all currently recognized higher taxa of living mammals were delimited as a result of comparative anatomic studies. The recent suggestion of a radical reclassification of birds (Sibley, Ahlquist, and Monroe 1988) also cannot be credited to cladistic analysis but resulted from the use of new characters (DNA hybridization). Several other major reclassifications, such as those of invertebrates, protists, and prokaryotes (in part admittedly controversial), were based on molecular distance methods rather than cladistic analysis.

CLADISTIC CLASSIFICATION

The traditional form of a biological classification is the arrangement of taxa in a Linnaean hierarchy of categories. After a cladist has constructed a cladogram, how are the included taxa arranged in such a hierarchy? By what criteria are taxa ranked?

Even though the various schools of cladists differ in philosophy, there is little difference among them in the concept of classification. They all more or less subscribe to a basic concept that can perhaps be articulated in the following definition: *A cladistic classification consists of a nested hierarchy of increasingly more inclusive holophyletic taxa; this hierarchy corresponds to a hierarchy of increasingly more inclusive synapomorphies.*

The cladist is convinced that this type of classification is the only one which reliably and unambiguously reflects the pathway of evolution. Therefore, a cladistic classification, among all possible ones, should have the greatest predictivity and the greatest information content. The fact that there is often a subjective element in their classifications has not been denied by pheneticists and evolutionary taxonomists. By contrast, many cladists have asserted that the cladogram-parsimony method of classification is free of subjectivity and arbitrariness. The validity of this claim is investigated in the following sections.

Back Translation of a Classification into a Cladogram

The objectivity of a cladistic classification, cladists claim, is proved by the unambiguous manner in which it can be translated back into a cladogram. Various aspects of cladistic analysis refute this claim. Any branching point that is not dichotomous allows several translations, and the choice of one over the others must be subjective. Since the infallible recognition of homoplasies is impossible, any classification resulting from an analysis of homoplasy is in part arbitrary. It is therefore erroneous to assume that a cladogram can be translated into only a single classification and, conversely, that a cladistic classification can be converted back into only a single cladogram. With a limited number of characters available, the delimitation of taxa is often uncertain, their sequence in a bushlike cladogram is usually rather arbitrary, and their ranking is also arbitrary if the taxa arising at several branching points are bracketed together at a single categorical level, as in the method of sequencing. See Chapter 11 for other reasons why those who use the method of cladistic analysis are so often unable to construct a cladogram that unambiguously reflects phylogeny.

Falsifiability

Some cladists have asserted that a cladistic classification has the virtue of being falsifiable in good Popperian fashion, an attribute not shared by any other type of classification. This claim is not valid, as has been shown by Meacham and Duncan (1987) and others. Popper's statement on falsification refers to theories based on universal laws. Since classifications are not such theories, falsification is an inappropriate consideration. Classifications by themselves are merely conventions, i.e., groupings of sets of organisms that are not by themselves falsifiable (Ruse 1979). Falsifiability is achieved only if the classification is linked to an externally falsifiable construct such as a cladogram.

The claim of Popperian correctness of cladistic theory has been widely challenged. Its inappropriateness for recasting phylogenetic inferences within a framework of hypothesis falsification has been pointed out by Kitts (1977), Hull (1979), Ruse (1979), and Cartmill (1981). Two of the objections concern the relation between cladograms and classification:

1 Any constructed cladogram is only an inference on the true phylogeny (with several other cladograms as legitimate alternatives).

2 One can usually derive several alternative classifications from a cladogram, particularly because of the absence of ranking criteria.

Finally, Meacham and Duncan (1987:87) demonstrated profound difficulties with regard to the falsifiability of ancestor-descendant relationships.

It is evident that the claim of objectivity for the cladistic method can at best be made only for cladograms, and, as we have shown, the numerous ambiguities of cladistic analysis (parallelisms, reversals, unresolved polytomies) make it impossible for cladograms to be called objective.

DIFFICULTIES IN CLADISTIC CLASSIFICATION

A far more serious problem exists for cladistic classification. Although it is perfectly clear in theory what a cladistic classification should be (a nested hierarchy of holophyletic taxa), there is considerable uncertainty about how to convert a cladogram into such a hierarchy. After Hennig's failure to develop a sound system of ranking, most cladists have avoided a discussion of the principles and practical methods of classification. Among the exceptions are Nelson (1974*a, b*) and Wiley (1979, 1981). However, their treatments deal primarily with formal questions, such as the categorical designation of fossil taxa and the placement of the stem species of holophyletic taxa. Wiley perceptively recognized that many, if not most, taxa (from the species up) originate by budding rather than by splitting and suggested how this can be handled in a cladistic classification. However, he nowhere admitted that the rather formidable difficulties in classification that confront cladists are largely the consequence of the adoption of Hennig's arbitrary principles.

Cladists have criticized taxonomists of the Simpson-Mayr school for being subjective. Actually, when it comes to converting cladograms into classifications, cladists are far more subjective than traditional taxonomists are. They have adopted a series of arbitrary conventions that are in complete conflict with the otherwise universally adopted classifying principles enumerated in Chapter 6. Indeed, a cladistic classification is merely a phylogenetic model that lacks all the properties one expects in a true classification. If all a cladist wants is a phylogeny, so be it. Traditionally, however, a taxonomist expects more than that from a classification. Taxonomists who still feel that way will turn to the so-called method of evolutionary classification (Chapter 10). However, it may be useful to point out which cladistic rules or principles lead to the unsatisfactory nature of cladistic classifications.

The Principle of Holophyly

The traditional method of zoological classification has been to delimit taxa on the basis of their shared characteristics and then to test those provisional taxa for monophyly. As Wagner (1980:175) said quite correctly: "The [phylogenetic] tree is based upon taxonomy, not the reverse (taxonomy being defined here as phenetic classification, clustering by relative amounts of resemblance and difference)." The advantage of this procedural sequence is that it leads the construction of a classification composed of taxa that are relatively homogeneous yet are consistent with genealogy.

By contrast, the cladist delimits taxa by the principle of holophyly. For the cladist, a taxon consists of the stem species and all its descendants. The origin of even a single apomorphic character may require the removal of a species from the assembly of its sister species and its placement with its most divergent descendants.

To be sure, it is legitimate, indeed necessary, to apply the principle of holophyly in order to locate the branching points of a cladogram, but the functions of a cladogram are very different from those of a classification. There is nothing in any theory of classification that would require one to rely on the principle of holophyly. Indeed, the artificiality and heterogeneity of holophyletic taxa have been the source of most adverse criticism of cladistics. A holophyletic taxon is recognized by the cladist solely on the basis of the shared possession among its species of one or more synapomorphic characters. Each of these holophyletic groups is in principle recognized as a separate taxon.

A holophyletic taxon that corresponds to a traditional monophyletic taxon, such as the class of birds, is reasonably homogeneous. In many other cases, however, holophyletic taxa are very heterogeneous, either because in a lengthy phyletic lineage the early stem groups are very different from later ones and particularly from the ultimate crown group or because some relatively undifferentiated stem group gave rise to a highly divergent ex-group, such as the family of turbellarians that gave rise to the parasitic trematodes and cestodes.

A cladist is forced by the principle of holophyly to classify all *stem groups* with the *crown group* of the same lineage. This practice may result in very heterogeneous assemblages, because the lower stem groups of a phyletic lineage are likely to share far more characteristics with the lowest stem groups of sister lineages than they do with the crown group of their own lineage (Chapter 10). The cladistic theory of relationship forces the cladist to consider all members of a holophyletic taxon to be more closely related to each other than any of them is to a sister group or ancestral taxon. To translate this into terms of human relationships, it would mean that Charlemagne (A.D. 742–814) would have to be considered more closely related to his now living descendants than to his own mother or his brothers and sisters.

Holophyletic taxa that are delimited by combining relatively undifferentiated stem groups with their highly derived ex-groups are usually very heterogeneous (Chapter 10). Their recognition results in classifications that are in conflict with the widely held concept that to be of practical value, a taxon should be reasonably homogeneous.

Another disadvantage of the recognition of holophyletic taxa is the wide separation of stem groups in lengthy phyletic lines from their nearest relatives on sister lineages. The removal of stem groups from the assemblage of their nearest relatives results in the creation of paraphyletic taxa.

A paraphyletic group is a monophyletic taxon (as traditionally defined) which does not include all the derived ex-groups. Paraphyletic taxa are not included in a cladistic classification by an arbitrary fiat, because they are not holophyletic. Because the reptiles have given rise to two ex-groups—birds and mammals—and are therefore called paraphyletic, they are not recognized in a cladistic classification. This taboo against paraphyly is responsible for most of the departures of cladistic from traditional classification (Michener 1977). It requires the destruction of most traditional taxa and leads to the recognition of heterogeneous taxa joined together merely by a few synapomorphies. Paraphyletic groups such as the Agnatha, Rhipidistia, Reptilia, and Insectivora are far more readily recognizable than are the taxa into which cladists break them and indeed than are most holophyletic groups. (For a discussion of paraphyly, see Chapter 10.)

Dichotomy and Budding

Hennig developed the cladistic method on the basis of the assumption that new species originate by the splitting of a stem species into two daughter species, hence on the principle of dichotomy: The parental species disappears with the birth of the two daughter species. He saw the origin of new higher taxa in an equivalent manner as a dichotomous process. Much research of the last 50 years has indicated, however, that budding is a far more frequent way of originating new taxa than is splitting. When a new species originates by means of peripatric speciation, this has no effect on the parental species from which the neospecies has budded off. However, by Hennig's criteria, a species which has given rise to a new species by budding thereby becomes paraphyletic and has to be removed from the classification, even though it has not been affected by the budding event. The same argument pertains to higher taxa, most of which evidently originated by the budding off of an enterprising new species that was successful in a new niche or adaptive zone. The parental taxon continued to flourish unchanged in its traditional niche, but it has become paraphyletic by cladistic definition and must be excluded from the classification. It is now fully evident that the proposal to disqualify paraphyletic groups from recognition in classifications is not only impractical and destructive but scientifically untenable.

The Theory of Ranking

Hennig's theory of classification was on the whole based on the assumption of equal rates of evolution. In the process of budding, however, the rates of divergence are usually drastically different. The parental species may be virtually static while the neospecies may be undergoing precipitous evolutionary change. To be sure, in his deviation rule Hennig acknowledged the occurrence of unequal rates in sister lineages, but this insight is not reflected in cladistic classifications. Most of his followers ignore the deviation rule altogether.

As to ranking, Hennig laid down two rules: (1) The products of each dichotomy are to be given the next lower categorical rank from that of the parental taxon, and (2) both sister taxa are to be given the same rank. As far as rule 1 is concerned, every critic of cladistics has singled out this provision for special criticism. Following this rule would have led to an impossibly large number of categorical levels, since the number of branching points in phylogeny is exceedingly large. For instance, in his classification of the class Insecta, Boudreaux (1979) required 13 taxonomic ranks between class and order. In McKenna's (1975) cladistic classification of mammals, a large number of unfamiliar rank designations were used. Even in the most favorable case of a completely symmetrical dichotomous system (Figure 9-3), with each species giving rise to two species of the next lower category, a minimum of 20 categorical levels would be required for 1 million species. In an asymmetrical hierarchy the number of required categories might go into the hundreds of thousands. The only practical solution is to abandon the demand to have a shift in

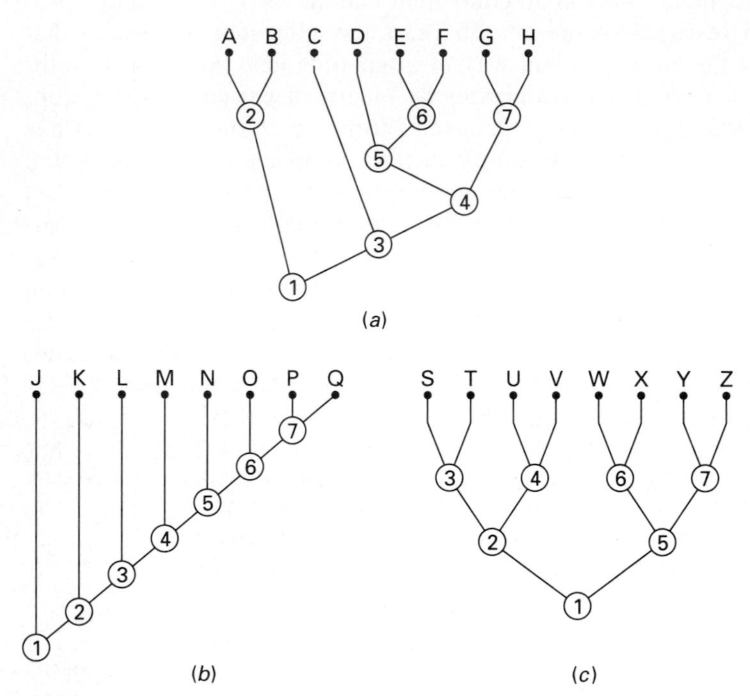

FIGURE 9-3
Dichotomous cladograms. While n taxa require $n - 1$ nodes on a dichotomous cladogram, the number of hierarchical levels needed depends on the configuration of the dendrogram. (*a*) A typical intermediate cladogram requiring five hierarchical levels. (*b*) A completely asymmetrical "comb" tree requiring seven hierarchical levels. (*c*) A completely symmetrical "bush" tree requiring three hierarchical levels.

categorical ranking with each dichotomy, and this was indeed done by later cladists (see below in the section on sequencing).

Rule 2 deals with the ranking of sister groups, which are the taxa to which a speciational event has given rise. "A sister group . . . is hypothesized to be the closest genealogical relative of a given taxon exclusive of the ancestral species of both taxa. Two sister groups are hypothesized to share an ancestral species not shared by any other taxon" (Wiley 1981:7). For Hennig (1966*a*:156), it was a "fundamental principle" of phylogenetic classification that sister groups have to have the same rank.

However, it was soon found that some sister groups have only a few *autapomorphic* characters (or none), while others evolve rapidly and acquire numerous autapomorphies. Indeed, sister groups, such as birds and crocodilians, are sometimes so different that cladists, in violation of Hennig's rule, have not hesitated to give them different ranks. The method of sequencing also ignores rule 2.

A New Synapomorphy and Change of Categorical Rank

According to cladistic theory, the origin of a new character coincides with a speciation event. However, the stem species of a new holophyletic taxon produced in this manner may have only the first one of the numerous apomorphies which this higher taxon acquires during its subsequent history. For instance, the particular species of archosaurian reptiles which gave rise to the class Aves might have had only a single apomorphic character that is now characteristic of birds, say, featherlike scales. It is therefore quite misleading to place at the branching point in the cladogram all the apomorphic characters eventually acquired by the new higher taxon. Cronquist (1987:35) rightly rejected this: "It should not be supposed that all of the apomorphies on a given internode were achieved at one time, in a great leap forward. On the contrary, the assumption must be made that they were achieved separately and individually in a series of speciation events." There is nothing natural about a classification in which the stem species of a new holophyletic taxon is automatically raised to the categorical rank which the taxon eventually achieves. No origin of a new species ever consists of a categorical saltation to the rank of a new order or class, as was asserted by Goldschmidt (1952).

Whenever an adaptive radiation occurs within a phyletic lineage, it results in the production of heterogeneous holophyletic taxa. The Archosauria of the cladists, for instance, consist of various reptiles and birds; their Choanata consist of the rhipidistian fishes and the terrestrial tetrapods; their Doliopharyngiophora consist of some (not all) turbellarians and the parasitic trematodes and cestodes. There is very little one can say about such heterogeneous holophyletic taxa except that they share a few synapomorphies, sometimes only one. It would be much better to present the information about their branching relationships in a cladogram than to destroy traditional taxa such as the Monera, Protista, Turbellaria, Amphibia, and Reptilia, which have served for up to 250 years as keys to information storage and retrieval.

These problems are encountered not only at the level of the higher taxa but also at the level of family and genus. In case after case, a holophyletic taxon, although impeccably formed in accordance with cladistic principles, is so heterogeneous that it seems useless for all practical purposes.

Restriction to Diagnostic Characters

It has traditionally been one of the basic rules of classifying (Chapter 6) that in the establishment of classes one should make use of "as many characters as possible" (Mayr 1982*a*:181, 193). By contrast, cladists use only synapomorphies for the delimitation and ranking of taxa. It is agreed that plesiomorphic characters cannot be used for the construction of cladograms, yet there is no reason why they should not be used in ranking. Indeed, ancestral characters are sometimes among the most evident characters of a taxon, particularly if the sister groups have developed numerous autapomorphies. Traditionally, and quite rightly, it has been one of the most elementary rules of classifying that classes should be based on the totality of characters. Evidently there is a major difference between the philosophy of those who construct their classifications strictly from a few diagnostic characters (synapomorphies) and that of those who delimit and rank taxa by evaluating the totality of characters. It is useful to discuss the conceptual basis of this difference in more detail.

There has been long-standing uncertainty about the meaning of taxa as recognized in a classification. During the reign of downward classification, as practiced by Linnaeus and his predecessors (Mayr 1982*a*:147–179), the major objective of classifying was to facilitate identification. Clear-cut diagnostic characters were the earmarks of a good classification; taxa had to be monothetic (Chapter 6). This philosophy led to the adoption of many highly artificial classifications. When upward classification, that is, classification by grouping, was more generally adopted, a consensus developed that one should get away from "single-character classifications." This movement culminated in numerical phenetics, with its aim of basing classifications entirely on "overall similarity."

Rebounding from the extreme of phenetics, the cladists unwittingly returned to the ideal of pre-Linnaean taxonomy: facile diagnosis. They delimit their taxa in a way that permits the construction of a hierarchy "that allows the most useful, succinct and informative diagnoses that completely describe the data" (Farris 1979*b*:511). Synapomorphies are stressed for the sake of their diagnostic value. Cladists proceed as if a monothetic diagnosis were sufficient to characterize a taxon. Many branching points in a cladogram are based on changes in a single character or at most a very few characters. For this reason cladistic classifications may have the vulnerability of any single-character classification. Since paraphyletic taxa cannot be described in this manner, cladists believe that they cannot be made part of a classification.

Farris is not alone in this stress on diagnostic characters. Nelson and Platnick (1981:304) confirmed that "systematics in general consists of the search for defining characters of groups." Indeed, many of their classifications are nothing but diagnostic keys. What is forgotten in this approach is that the diagnostic characters are very few in comparison to the number of other valid characters of a taxon and that a far more useful, and hence natural, classification results from delimiting and ranking taxa on the basis of *all* their valid characters, as is the objective of evolutionary classification (Chapter 10).

It would seem that cladists are rather inconsistent in their neglect (for ranking) of the information content of plesiomorphic characters. For cladists, loss of information has been one of the principal objections to the compatibility method (Chapter 11), which is therefore "no longer scientific. . . . Scientists should not discard data, for if they do, there is no longer any connection between observation and theory" (Kluge 1984:28). And yet the amount of information discarded by the compatibility method is a mere fraction of that discarded by cladists when they rank taxa. By their own argument, then, delimiting and ranking taxa entirely on the basis of synapomorphies would be "no longer scientific." Michener (1970:19) likewise criticized the cladistic approach to classification, stating that cladistic classifications "imply nothing about the changes in the number of other [nonsynapomorphic] characters and little about phenetic differences between forms."

Neglect of Autapomorphies

Considering what high diagnostic capacity *autapomorphic characters* have, the extent to which they are neglected in cladistic classifications is surprising. Autapomorphic characters are best defined as *characters that evolve in only one of two sister groups*. For instance, all the characteristics of birds that evolved after they separated from their archosaurian sister groups are avian autapomorphies. Characters that are synapomorphies within a group (for instance, birds) are autapomorphies in comparison with the characters of a sister taxon. The importance of these characters lies in the fact that they permit an objective description and measurement of the amount of anagenesis. The characters by which a species or group differs from its sister species or group are autapomorphies.

Since Hennigians neglect the use of anagenetic information in the construction of a classification, they minimize the usefulness of autapomorphic characters. To be sure, these characters are not useful for cladistic analysis, but this is not true in regard to classification. In order to be able to ignore such characters in their construction of classifications, cladists offer a definition of autapomorphy that is thoroughly misleading, calling it "a character evolved from its plesiomorphic homologue in a single species" (Wiley 1981:123), as if this definition were not equally true of any apomorphic character. Since cladists do not have a theory of ranking, they are unable to make use of autapomorphic characters, even though Hennig did refer to autapomorphies. The failure to exploit the information content of the autapomorphic characters of sister groups is another reason for the arbitrariness of many cladistic classifications. It represents a further invalidation of the claims of cladists that their classifications are more objective than those of other taxonomists.

Absence of a Theory of Ranking

One of the elementary rules of every theory of hierarchical classification is that hierarchical rank is determined by amount of difference (Chapter 6). Cladists have criticized both pheneticists and evolutionary taxonomists by saying that the amount of difference is a subjective criterion that is responsible for the arbitrariness of the classifications of their opponents. Cladists also translate their cladograms into the Linnaean hierarchy, yet we have failed to find in their writings any clear-cut theory of ranking, even though Hennig himself admitted "that determining the absolute rank of systematic categories [taxa] is a very serious and important problem" (1966*a*:154). He then rejected making degree of morphological difference a decisive criterion and reaffirmed the primacy of branching points by demanding that sister groups be given the same categorical rank (1966*a*:156). He emphatically rejected the suggestion "that the different velocities of phylogenetic development have to be taken into account in determining the absolute rank" of taxa (1966*a*:157). Ultimately Hennig decided, although with evident misgivings, that geological age is the most meaningful ranking criterion for the higher categories.

His followers have abandoned both of Hennig's basic criteria (equal rank of sister groups and geological age) not only because they led to too great an imbalance between explosively evolving and slow groups (including living fossils) but also because strict coordination of sister groups led to a veritable explosion of categorical levels. To prevent this, most cladists have adopted a convention called *sequencing* (Nelson 1974). According to this convention, descendant groups in asymmetrical trees may be given the same categorical rank and listed in the order of their branching sequence (see Wiley 1981:207 and Figure 6-3). This is an application of the traditional method of listing members of a group of taxa in primitive to derived order, and it greatly reduces the number of categorical levels needed in a classification. Unfortunately, this convention is difficult to apply to "bush trees," that is, symmetrical trees, in which the rate of branching is about the same in most or all parts of the cladogram. It is worth mentioning that if several taxa in a cladogram are placed at the same categorical level, the resulting classification cannot be unambiguously translated back into a cladogram.

Sequencing was the first methodological departure from Hennig's insistence on "branching points only" as the basis of classification. By permitting the assignment of the same categorical level to the products of several consecutive branching events—provided that they had not diverged too strongly—cladists introduced a consideration of anagenesis. The place where one should shift from sequencing to a change of categorical rank is signaled by the occurrence of a major gap between nodes in a cladogram. For example, if in the comb cladogram (Figure 9-3*b*) a large anagenetic gap were found between nodes 3 and 4, then species M through Q could form a higher taxon.

Anagenesis was also given recognition in Kluge and Farris's (1969) method, which they claimed is a computerization of W. H. Wagner's "groundplan/divergence method." This method was used in an attempt to quantify "distance... measured by the number of different characters that become specialized." (For further developments of this method, see Chapter 11.) However, this quantification was restricted to synapomorphies, as Farris (1977) emphasized in his description of the method of *special similarity*. Again, all other aspects of anagenesis, particularly all ancestral characters, were excluded from consideration.

To repeat what was said at the beginning of this analysis, there seems to be no statement anywhere in the cladistic literature that provides clear norms for the ranking of taxa. It would seem that the ranking criteria of cladists are no less subjective than those of evolutionary taxonomists. Indeed, as will be shown in Chapter 10, evolutionary taxonomists have developed an extensive set of guidelines, while one can search the cladistic literature in vain for a similar set of norms.

Neglect of Ancestor-Descendant Relationship

Curiously, ancestor-descendant relationships, one important source of information on the relative amount of differences among taxa, are almost completely ignored by cladists when they construct classifications. Some of the consequences of ignoring this relationship were described by Meacham and Duncan (1987).

In a cladistic classification no higher taxon can give rise to another higher taxon.

The cladograms of most cladists do not recognize ancestral species "because they cannot be identified with certainty." This is not a valid argument, because a lack of certainty is equally true for synapomorphies and sister groups (Holmes *in litt.*). Indeed, as Bock (1977), Bretsky (1979), and Hull (1980) have shown, ancestor-descendant relationships can often be tested more easily than can sister group relationships. Apomorphy distributions are usually sufficient to establish ancestor-descendant relationships, and the occasional occurrence of reversal (often recognizable by means of character weighting) is no more serious a problem than it is in other operations of cladistics.

Placement of the Ancestral (Stem) Species

Since the stem species gives rise to two holophyletic sister taxa, it cannot be included in either taxon without violating the principle of holophyly. However, it can form a holophyletic taxon together with the two holophyletic taxa to which it has given rise. Wiley (1981:233) recognized the problem and departed from traditional cladistic practice by suggesting that a stem species be placed in a paraphyletic monotypic higher taxon. This completely stultifies the whole classificatory procedure. Unless there is a good fossil record, it is impossible to determine whether the stem species already had the synapomorphies of the two daughter taxa. Deciding which fossil is closest to the ancestral species is also difficult, but it can sometimes be determined with reasonable probability by combining morphological analysis, stratigraphic occurrence, and biogeography (Hull 1980; Prothero and Lazarus 1980; Wiley 1981:106–107).

Inability to Deal with Fossil Taxa

Most cladists exclude fossils from their classifications or at best list them as *plesions*. However, since extinct taxa are as much a part of the total genealogy of the animal kingdom as living species are, the responsibility for placing them in the classification cannot be evaded (Chapter 6). A classification that omits extinct taxa is simply incomplete and therefore quite misleading in many respects; it is not a general reference system.

A particular problem is posed by periods of mass extinction. Many flourishing groups had a few survivors after such an extinction, and the survivors quickly gave rise to new higher taxa. All extinct taxa of this type are considered paraphyletic. The cladistic dogma demands that the paraphyly of the ancestral taxa be ignored or broken up by tracing the ancestry of the survivors. On that basis, all fossil taxa except those terminal ones which became extinct have given rise to descendant taxa and are thus paraphyletic and must be excluded from a cladistic classification. By not attempting to delimit homogeneous phyletic groups whose chronological boundaries often coincide with divisions of the geological time scale, cladistic classifications ignore a major and economically important objective of paleontology. Every discussion of fossil taxa by a cladist indicates how many problems are created by the rigidity of cladistic principles (Chapters 6 and 10).

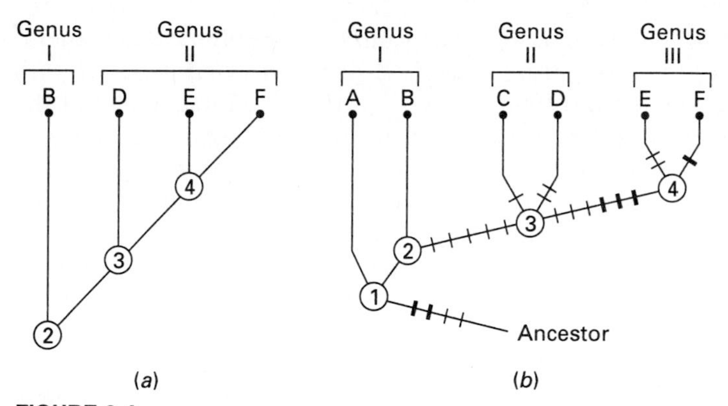

FIGURE 9-4
Evolutionary systematic phylograms of the *Bergidea* group before (*a*) and after (*b*) revision by Ashlock (1985), illustrating how adding two species (A and C) and proposing one new genus (III) affects classification. Width of tick marks indicates relative strength of characters. For the effect on cladistic classification, see text.

Instability

Cladists are well aware of the profound changes that result from the application of their procedures to traditional classifications. After a period of instability during which most groups are properly analyzed, they feel that classifications will settle down and again become stable. This view, however, is probably incorrect.

Cladistic classifications are inherently unstable for two reasons. First, when different cladists work in the same group, minor disagreements which produce only minor differences in their interpretations of cladogenesis may cause one cladist to portray a given traditional group as holophyletic and another to show it as paraphyletic. Thus, even minor disagreements can produce major differences in two cladistic classifications of the same group. Examples of such disagreements can be found in classifications of the jawless fishes (Agnatha) (Halstead 1982) and the book, biting, and sucking lice (Psocoptera, Mallophaga, and Anoplura) (Hennig 1981 with comments by Schlee therein; Boudreaux 1979[2]).

The second reason why holophyletic classifications are unstable is that the introduction of new taxa can change holophyletic groups into paraphyletic ones. Such a change is illustrated in Figure 9-4, where the two phylograms portray a small group of insects before and after revision

[2] Boudreaux's cladistic classification of the insect orders is interesting in that he makes no changes in the composition of any insect order other than the Mallophaga and Anoplura, yet it is inconceivable that all the other orders of insects are holophyletic. Indeed, many orders, for example, Mecoptera and Diptera, may well be paraphyletic.

by Ashlock (1985). The first shows a group of two holophyletic genera, genus I being monotypic, from Chile and Argentina, and genus II with three species, all from the Juan Fernández Islands. In the course of the revision, species A and D were added to the group, and the large gap between nodes 3 and 4 was used to justify the delimitation of genus III. The addition of species A to genus I makes that genus paraphyletic, and the trifurcation at node 3 makes genus II potentially paraphyletic.

A well-known cladist suggested that the solution to this problem is to place species A in one genus and all the other species in a second genus. The effect of following this suggestion would be to place two very similar species in separate genera, to add one continental species to an insular genus to which it bears little resemblance, to require the naming of a new genus for species A, and to synonymize genus II with genus I. This disruption of the existing classification and the creation of a morphologically and biogeographically heterogeneous group demonstrate what the discovery of a single species can do when the rules of cladistic classification are followed strictly.

THE CONSEQUENCES OF THE ADOPTION OF CLADISTIC PRINCIPLES FOR THE CLASSIFICATION OF ANIMALS AND PLANTS

Followers of Hennig have praised cladistics as the only method that can produce an objectively sound classification. Our analysis has shown, however, that this claim has no validity. Cladistic analysis is as good a method of analyzing the phylogeny of a group as any other, and in many cases it is apparently the best method. However, this cannot be said of the conversion of a cladogram into a classification. Hennig's set of arbitrary rules for achieving this conversion lead to the delimitation and ranking of taxa in conflict with what traditionally has been considered a sound classification. Sound and stable classification is what biologists traditionally have demanded from taxonomists.

Cladistic classifications are characterized by their emphasis on difference and their neglect of major groupings and major gaps among taxa. As a result, most cladistic classifications are characterized by extreme splitting. To give an example, in a recent reclassification of the avian family Capitonidae (barbets), this group was broken up into eight higher taxa (subfamilies and tribes), three with a single genus and five with two genera each (Prum 1988). A pragmatic taxonomist might well ask whether it is necessary to express every phylogenetic conclusion nomenclaturally; even worse, this procedure forces the name of a specialized ex-group (Ramphastidae, Toucans) on the diverse family of barbets that occurs on three continents.

Ultimately, the deficiencies of cladistic classifications are due to the failure of cladists to appreciate the difference between the functions of a model of descent and those of a classification. Hull rightly pointed out that "biological classification cannot be made to reflect very much about phylogeny without frustrating other functions of scientific classification" (1979:437). There is no need to demand that a classification simultaneously serve as a diagram of descent. The genealogical information can easily be supplied by publishing a cladogram together with an evolutionary classification.

Those of Hennig's arbitrary rules discussed above which are particularly responsible for preventing the construction of sensible classifications are listed here once more: (1) delimitation of highly heterogeneous taxa in accordance with the principle of holophyly, (2) rejection of the use of ancestral characters even though they are by far the majority of the characters in most taxa and often characterize taxa far better than a few diagnostic synapomorphies do, (3) return to the pre-Linnaean ideal of downward classification, with its reliance on few or single characters, and (4) rejection of the recognition of polythetically characterized taxa, with its implicit denial of the frequency of parallel evolution.

These arbitrary rules, if consistently applied, would play havoc with existing classifications. In some branches of zoology they have already produced great instability as well as uncertainty about how to handle certain situations. Most devastating would be the destruction of a large number, possibly the majority, of the well-known classical taxa of the taxonomic literature. If cladistic classification were adopted, "group after group [such as prokaryotes, amphibians, and reptiles] would have to be broken up or dissolved, and replaced by taxa that may reflect the sequence of phyletic branching but have little to do with overall similarities that can be grasped by biologists in general" (Cronquist 1987:36).

The most deplorable effect of the destruction of a high percentage of existing classifications would be the simultaneous destruction of an important information retrieval system in which the traditional classifications have served as the key for most of the history of taxonomy (Halstead 1982; Cronquist 1987). Little compensating benefit would seem to result from the breaking up of traditional groupings that have endured for up to 200 years. Instability is actually built into the cladistic system, because when new evidence requires a change of the branching sequence of the cladogram, all taxa arising above that point in the cladogram may have to be reclassified and given new ranks.

Cladistic classification leads to the total destruction of the classification of fossil organisms. All currently recognized taxa of fossils (except terminal ones) gave rise to descendant taxa and are thus paraphyletic by cladistic definition. Not being able to handle ancestor-descendant relationships, cladists arbitrarily declare paraphyletic taxa to be inadmissible in a classification and replace them with heterogeneous holophyletic lineages.

An inability of cladists, after the failure of Hennig's two ranking criteria (geological age and equal rank of sister taxa) to develop a new set of objective criteria for ranking, has resulted in a purely subjective approach to the ranking of taxa. Indeed, by basing ranking quite arbitrarily on the few (per taxon) synapomorphic characters of monothetically delimited taxa, the cladists' system of classification is far more subjective than that of their opponents. These deficiencies of cladistic classification have been pointed out by Mayr (1974*a*) and many subsequent critics but have been openly ignored by the cladists. Usually they try to deflect attention from this weakness of their methodology by defending cladistic analysis even though this is not the part of cladistics that is being criticized.

Owing to these numerous, indeed seemingly fatal, shortcomings of cladistic classification, some authors have attempted to develop a theory of classification which combines the virtues of the traditional theory of classification (Chapter 6) with a due regard for the theory of common descent. This approach, frequently referred to as the theory of evolutionary classification, is dealt with in Chapter 10.

APPENDIX: Weighting and Anagenetic Analysis[3]

The main objection to cladistics raised by classical taxonomists has been that it is difficult if not impossible to convert a cladogram that shows the relationship of holophyletic taxa into a balanced Linnaean hierarchy that consists of relatively homogeneous taxa.

A number of methods have been proposed to achieve this end. For instance, one can divide the study group at internodes where larger phylogenetic gaps are indicated by the presence of larger numbers of changes in character states. Cladogram-drawing computer programs that vary the length of internode lines to reflect the number of character states on an internode are useful in this procedure. Figure 9-5 shows the result of using the cladistic program PAUP on the demonstration data used in Chapter 11. (The longest internode on this dendrogram is 3–4, which divides the demonstration taxa into two groups, a paraphyletic ACF and a holophyletic BEGDH.) The phenetic analysis identified the same two groups. Counting character states on the internode actually calculates the Manhattan distance between pairs of nodes or between node and terminal taxon. Because it preserves the branching pattern of a cladogram and also provides a visual measure of a relative amount of evolution, the resulting dendrogram is a form of phylogram.

Character states, however, are not of equal weight. In the demonstration ex-

[3] This discussion was contributed by P. Ashlock.

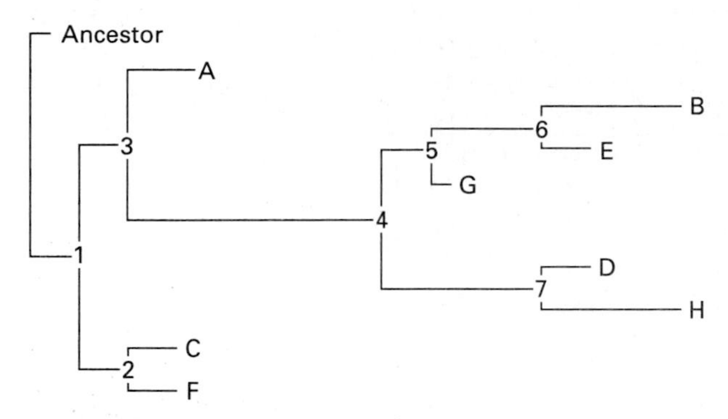

FIGURE 9-5
Cladogram of the demonstration insects (Figure 11-1) produced by the PAUP computer program. The length of a horizontal line reflects the number of characters on that line. Lines are proportional to the number of characters on a line.

ample based on a data set discussed in detail in Chapter 11, characters were weighted by guessing their quality for cladistic analysis, and many computer programs give additional weight to enhance the program's ability to find the most parsimonious tree. In the final tree, however, each character state is still given a weight of 1; i.e., the presence of a feature is given the same value as the color of a feature is. It would be desirable to alter the weight of the character states in some way that would reflect their quality, but this cannot be done in a systematic manner with any current method before the tree is completed.

The following system of weighting, which uses the completed tree as a point of departure, was worked out by P. D. Ashlock and D. J. Brothers and was used by Brothers in 1975 in his study of aculeate Hymenoptera. The version described here is simplified.

The general principle is to use the distribution of character states on the cladogram to assess the weight given to each character state. Parallel character states (which appear more than once on the cladogram) and reversals are given less weight than are unique character states. On the other hand, the more taxa that show a given character state—a measure of the "success" of that character state—the more weight it is given.

The Ashlock and Brothers system is not a precise measure of anagenesis, but weighting is done on the basis of specified principles, and it is an improvement over the equal weighting of character states. Some especially important character states may be given weight if the need for this can be convincingly justified. The simple calculations needed for anagenetic analysis are listed in Table 9-1.

Character states that appear two or more times on the cladogram are obviously less significant than are those which evolved only once. (The number of times each parallel character state appears on the demonstration cladogram is indicated by a number immediately preceding the character–character state mnemonic.) An appropriate method of adjusting the value of parallel character states

TABLE 9-1
ANAGENETIC CALCULATION

Node	AGV	STV	ASTV*	TGV	%TGV*
1–2	1.50	2	1.41	2.1	6.0
2–C	1.00	1	1.00	1.0	3.0
2–F	0.50	1	1.00	0.5	1.5
1–3	1.00	6	2.44	2.4	7.0
3–A	1.00	1	1.00	1.0	3.0
3–4	6.00	5	2.24	13.4	40.0
4–5	1.00	3	1.73	1.7	5.0
5–G	0.00	1	1.00	0.0	0.0
5–6	2.00	2	1.41	2.8	8.0
6–E	1.00	1	1.00	1.0	3.0
6–B	1.83	1	1.00	1.8	5.5
4–7	2.50	2	1.41	3.2	9.5
7–D	1.00	1	1.00	1.0	3.0
7–H	1.83	1	1.00	1.8	5.5
				$\Sigma = 33.7$	

Note: AGV = anagenetic gap value; STV = subtended taxa value; ASTV = adjusted subtended taxa value, a root of STV; TGV = taxonomic gap value, AGV × ASTV; %TGV = percent taxonomic gap value, TGV/sum of TGV × 100.
*Rounded.

is to give each one a value equal to the reciprocal of the number of times it appears in the cladogram. A character state that appears twice, then, has a value of 0.5. In the cladogram in Figure 9-6, character states 2–A3Sw$_1$ and 1–TbSp$_1$ are given a weighted value of 0.5.

The reversed state of a character (a derived character state that resembles a more primitive character state) also deserves less weight, even though, as often happens, the reversed character state appears only once in the confines of the study group (e.g., ColC$_0$). In a larger study, of course, ColC$_0$ would appear as a derived feature in some ancestral group. To account for this eventuality, the first appearance of a reversed character state can be recorded symbolically by placing it at the bottom of the study-group tree in parentheses to indicate its symbolic nature. This internode (base–1), which leads to the common ancestor of the study group, need not enter the calculations directly, but placing a reversed character state here as a bookkeeping measure ensures that the reversed character state will be given diminished weight. Here, ColC$_0$ appears twice (once on the stem) and can be given a value of 0.5. Similarly, character state CrVn$_0$ (reversed twice from CrVn$_1$) is also entered in parentheses at the base of the tree and is given a value of 0.33 each time it appears on the tree.

Individual character state values are summed to provide an anagenetic gap value (AGV in Table 9-1) for each internode. In the example, internode 1–2, with two character states, one a parallel character state appearing twice on the tree, has a total AGV of 1.5. Internode 2–C has a single character state that appears nowhere else on the tree and is scored 1.0. Internode 3–4 has six unique character

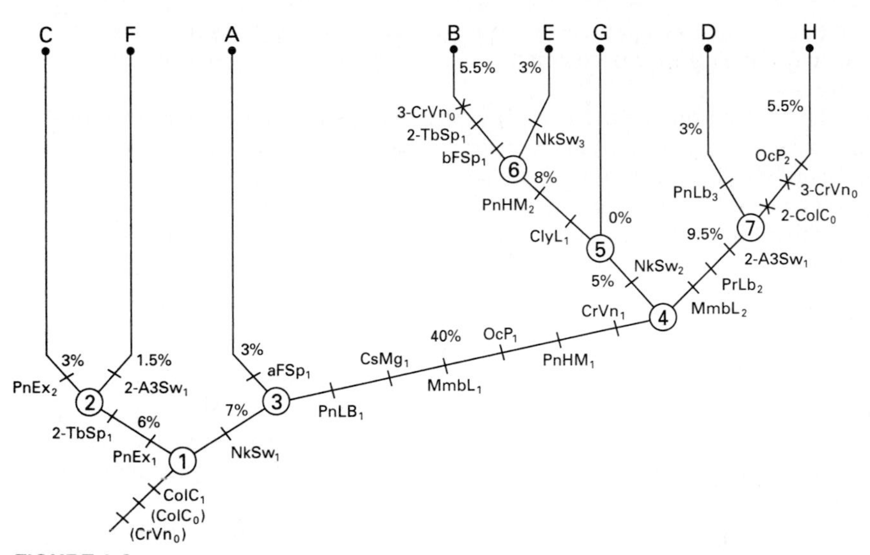

FIGURE 9-6
Phylogram based on the cladogram in Figure 9-5. Percentages are derived from the proportion of the total of the weighted characters that appear on each internode. See Figure 9-7 for the method used in drawing the tree.

states for a score of 6.0, while internode 7–H has one unique and two reversed character states, valued at 0.5 and 0.33, respectively, for a total AGV of 1.83. Values for all internodes have been entered in Table 9-1, which will be used for all further calculations.

The more taxa that have a given derived character state, the greater the survival utility that character state may be presumed to have. Conversely, a character state that is found in a single taxon or species has been minimally tested, and its value for survival in the future is uncertain. It follows, then, that the more taxa that possess a given character state, the more weight that character state should receive. Each internode on a tree subtends a number of taxa, and these numbers have been entered on Table 9-1 under the heading STV (subtended taxa value) for each internode. Here the study taxa are single species, but terminal taxa in other studies may represent genera with many species each. Experiments have shown that using the raw number of taxa places too much weight on this factor. A more useful procedure is to use the square root of the number of subtended taxa in small or moderate-sized studies. In studies that involve very large numbers of taxa, higher root values should be used. Brothers (1975) used cube roots because his base taxa contained up to 20,000 species. One should use a root value that produces a low-to-high ratio that roughly approximates the low-to-high ratio of the anagenetic gap values. The new values are called adjusted subtended taxa values (ASTV) and are rounded to two places (column 4 of Table 9-1).

The anagenetic gap values are multiplied by the adjusted subtended taxa values to get a taxonomic gap value (TGV in column 5 of Table 9-1). Finally, these values construed as a percentage of the sum of the taxonomic gap values (i.e., the total change on the tree), rounded to one decimal place, are calculated (%TGV in Table 9-1).

The admonition that the size of the phylogenetic gap should be proportional to the size of the group—meant to ensure that small or monotypic groups are highly distinctive while larger groups may be less distinctive—is observed automatically in this method by being incorporated in the TGV or %TGV calculations. To divide the group A through H into two taxa, one need only observe that the largest gap is 40 percent at internode 3–4.

Figure 9-6 shows the cladogram in Figure 9-5 altered to a phylogram. Instructions for the modifications are shown in Figure 9-7. This method works only in small studies; in a complex cladogram, lines often overlap. Since the tree is not meant to be exact, problems can sometimes be avoided by slightly offsetting the lines, or the taxa can be revolved around a particular node. Note that in Figure 9-6 taxa B, E, and G have been revolved around node 5 so that taxon B does not overlap taxon D. Another solution would be to break the phylogram into suitable pieces.

The final phylogram (Figure 9-6) provides a clear summary of the data: The group is divided into two higher taxa by the large gap on internode 3–4. The %TGVs are entered by each internode so that close values can be interpreted

FIGURE 9-7
Phylogram construction. A ratio is calculated for the paper width between the first taxon (C) and the last taxon (H), say, 200 mm, and the sum of the %TGV values for the outside internodes of the phylogram (i.e., 2–C, 1–2, 1–3, 3–4, 4–7, 7–H, which = 71; 200 mm/71 = 2.8 mm; round off to 3; 3 × 71 = 213 mm, the final distance between C and H). Each %TGV is multiplied by the ratio (3) to provide a length in millimeters for the proportional values requested in the figure.

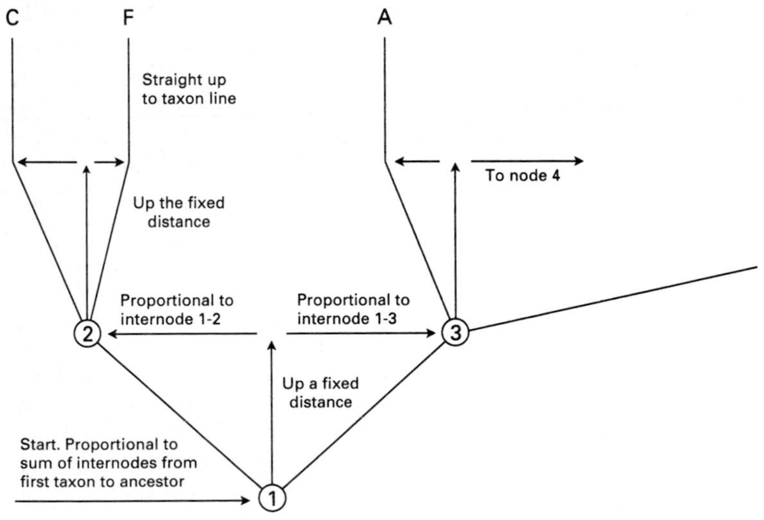

correctly. Finally, since all character states have been entered by character mnemonic codes, following the progress of evolutionary change is easy.

We know of only two attempts so far to provide for a numerical method which simultaneously attempts to convey both similarity and branching pattern. One is Estabrook's (1986) method of convex phenetics; the other is a method proposed by Hall (1988). Only further empirical applications and mathematical-statistical analysis will permit us to determine the efficiency and validity of the new methods.

Attempts to construct phylograms on the basis of molecular character states face great difficulties. The fact that much molecular evolution is neutral or nearly neutral makes the molecular clock possible, and neutral distances can be used to establish the chronology of the branching points. As important as this is, it must not be forgotten that changes in neutral (or "junk") DNA may be simply evolutionary noise, useless for the determination of evolutionary divergence. What the taxonomist wants to classify are the results of genuine evolutionary change, and this is reflected in nonneutral DNA changes. No method has been suggested that can partition DNA changes into those which are significant in evolution (and thus for the taxonomist) and those which are not. For the time being it is necessary to infer the evolutionarily significant changes by their effect on the phenotype, more specifically, on the morphology of the compared taxa.

CHAPTER **10**

EVOLUTIONARY CLASSIFICATION

Traditionally, classifications have been expected to group items that seem to belong together because of their similarity or identical causation (Chapter 6). After Darwin's discovery that the diversity of the organic world can be explained by common descent, phylogeny was accepted as one of the indispensable bases of biological classification. In his brilliant analysis of the principles of classification in the thirteenth chapter of the *Origin of Species* (1859:411–458), Darwin wrote that "the *arrangement* of the groups within each class, in due subordination and relation to other groups, must be strictly genealogical in order to be natural."

From 1859 on evolutionary taxonomists have relied on Darwin's guiding principles for the construction of classifications. Mayr (1942:280) stated that "the 'natural system' of the modern taxonomist is based on phylogeny and...the higher taxa are monophyletic. In striving for this ideal, shortcomings are due to insufficient knowledge, and the only intrinsic difficulty of a phylogenetic system consists in the impossibility of representing a 'phylogenetic tree' in linear sequence." Since common descent is likely to produce similarity, a genealogical approach usually produces classifications not too different from those based on weighted similarity. However, a number of evolutionary processes (discussed below) may result in a discordance between genealogy and the kind of similarity traditionally relied on by taxonomists. For instance, sister groups may diverge at very different rates. The possibility of conflict between genealogical splitting and evolutionary divergence was known to Darwin, and he recommended that it be taken into consideration in the ranking of taxa, stating that "the *amount* of difference in the several branches or groups, though allied in the same degree in blood to their common progenitor, may differ greatly, being due to the different degrees of modification which they have undergone, and this is expressed by the forms being ranked under different genera, families, sections or orders" (1859:420).

For these reasons, followers of Darwin base their classifications on two criteria: genealogy (common descent) and, as Darwin said, "the different degrees of modification which [descendants] have undergone," that is, degrees of difference. Such taxonomists are usually referred to as followers of *classical* or *evolutionary classification*. Their methodology has been explained by Cain and Harrison (1960), Simpson (1961), Mayr (1969), and Bock (1977). For detailed critiques, see Wiley (1981:240–269) and Farris (1977). It has been asserted (Cracraft 1983) that there is a difference between the two schools in the consideration of the ontological status of higher taxa, but this is not correct. Both the holophyletic taxa of the cladist and the monophyletic taxa of the evolutionary taxonomist are *historical groups* (Wiley 1980) and thus are equally natural (see below).

The evolutionary school has also been called the Simpson-Mayr school because its principles were developed most fully in the writings of those authors. This label applies only to the macrotaxonomy practiced by Simpson and Mayr (they differ considerably in their ideas on species and speciation). In spite of significant differences within this school with respect to the concept of monophyly, the heterogeneity within evolutionary classification is less than that between the different schools of cladistics.

The difficulty in the evolutionary approach lies in determining how to combine criteria as different as similarity and descent in the construction of a classification. The cladist first reconstructs inferred descent in a cladogram and then attempts to assign the branches, parts of branches, or nodes of the cladogram to taxa. That this procedural sequence is not satisfactory was shown in Chapter 9 in the discussion of cladistic classification. It is quite difficult to convert a cladogram into a practical classification without violating several universally recognized principles of classification (but see the Appendix at the end of Chapter 9).

Traditionally (but see below) the evolutionary taxonomist therefore reversed the sequence, first sorting the taxa into a provisional classification that reflected the appearance of the taxa, i.e., basically their overall similarity. Such provisional taxa can be delimited by inspection, as were such traditional groups as birds, beetles, and butterflies, or can be clustered by numerical methods whenever visual clustering encounters difficulties. In a second step such provisional taxa are tested for monophyly, that is, for their status as descendants of the nearest common ancestor. The most suitable test in most cases is a cladistic analysis.

By using this procedural sequence, the evolutionary taxonomist is able to satisfy both demands of a sound phylogenetic classification: (1) the delimitation of taxa formed by the universal principles of classification (Chapter 6) and (2) the monophyly of those taxa. Such a taxonomist is also able, whenever necessary, to make use of the methods of the two other schools of taxonomy: those of numerical phenetics to perform a tentative sorting of the pattern of variation in nature and those of cladistic analysis to test the monophyly of the phenetically delimited provisional taxa.

Practitioners of the evolutionary method have always been dissatisfied with the necessity of dealing with similarity and monophyly sequentially. For this reason there have been a few recent proposals of algorithms that would combine a phenetic approach with a cladistic approach (Estabrook 1986; Hall 1988) (see the Appendix at the end of Chapter 9). These endeavors are still at a somewhat experimental stage, but the recognition of the merit of a combined approach gives hope that a sound methodology may soon be available.

Evolutionary classification differs quite fundamentally from phenetics by (1) considering species not to be human constructs but phenomena of nature and (2) considering taxa to be products of evolution and thus the result of a manifest cause that cannot be ignored in the delimitation and ranking of taxa. Evolutionary systematists therefore undertake a phylogenetic analysis in which the monophyly (origin by common descent) of all accepted taxa is tested. By accepting the need for cladistic analysis, evolutionary taxonomists also accept the thesis that synapomorphies provide valid evidence for the naturalness (common descent) of a taxon and that a careful determination must be made of which characters are ancestral and which are derived (Chapter 9).

There are minor differences of emphasis between evolutionary taxonomists and cladists in their application of cladistic analysis. For instance, evolutionary systematists test synapomorphies primarily to recognize and reject polyphyletic taxa, not to create taxa. When Hennig (1966*a*) proposed his cladistic theory of classification in 1950, most major taxa of animals had already been described and reasonably well characterized, and the similarity of the included species unquestionably played a role in their original recognition. It is these traditional taxa which the evolutionary taxonomist uses as the basis for cladistic analysis. An evolutionary taxonomist tests the naturalness of a study group by asking whether all the members of that taxon are descendants of the nearest common ancestor.

DIFFERENCES BETWEEN CLADISTIC AND EVOLUTIONARY CLASSIFICATION

The most important difference between a cladist and an evolutionary taxonomist is a philosophical one. For a cladist, the genealogy automatically supplies the classification. The evolutionary taxonomist not only considers the reconstruction of phylogeny a logically distinct task from the making of a classification but also believes that a classification does not have to reflect every detail of genealogy as long as all the taxa are monophyletic (Hull 1979; Felsenstein 1983*a*). This philosophical difference results in a number of operational differences between the two schools (Charig 1982:417–437). Evolutionary classifications differ from cladistic ones in the following ways.

1 Taxa are not delimited on the basis of holophyly but through an evaluation of resemblance and difference, provided that they are monophyletic.

2 Ancestral (plesiomorphic) characters are given appropriate consideration because they often contribute strongly to the aspect, and hence the classificatory status (Chapter 6), of a taxon.

3 The term *monophyletic* is retained in its traditional conception of a qualifying adjective for a taxon; monophyly is not used as a method for delimiting taxa. Taxa that are based on convergent polyphyly are, however, rejected.

4 Sister groups are not necessarily given the same rank but are ranked on the basis of degree of difference (see below).

5 Accepted in the evolutionary classification are monophyletic taxa that have given rise to one or several ex-groups, as the reptiles have to birds and mammals, even though these taxa are rejected by cladists as paraphyletic.

6 Evolutionary classification also makes use of a broad range of evidence derived from biogeography, ecology, and paleontology instead of basing and ranking taxa entirely on the outcome of an analysis designed to delimit holophyletic taxa.

Adherents of evolutionary classification find that a classification constructed by their methodology is a better general reference system than is a purely cladistic classification. Evolutionary classifications delimit and rank taxa not only on the basis of a few diagnostic (synapomorphic) characters but on the basis of the totality of characters, including ancestral ones. This provides a much broader basis for comparisons.

Phyletic lineages may change so drastically during their evolution that early and late members of a taxon delimited holophyletically may be far more different from each other than the early members are from their sister groups. Jefferies very perceptively pointed out that in many fossil lineages two stages can be distinguished, "stem groups" and a "crown group" (1979:449). Cladists and evolutionary taxonomists often differ in their treatment of stem groups (Figure 10-1). When delimiting a holophyletic taxon, cladists combine stem groups with the crown group to which they have given rise and with which they share synapomorphies (or shunt them aside as plesions). Evolutionary taxonomists, by contrast, may unite stem groups with their sister groups if they share numerous ancestral features and differ only by a few autapomorphies. This may result in the recognition of a parental taxon that consists of the stem groups of several lineages, while the specialized crown groups are ranked as separate taxa. Some cladists have argued that stem groups are invariably more closely related to the derived crown group than they are to their sister groups (Farris 1979*b*), but this is based on a strictly genealogical

FIGURE 10-1
Crown groups I and K and stem groups A through H. A cladist might recognize two holophyletic taxa, A + C + E + G + I and B + D + F + H + K; an evolutionary taxonomist might recognize three monophyletic taxa, C + A + B + D, E + G + I, and F + H + K. If stem groups A, B, C, and D are combined in a single taxon, no two stem groups differ by more than six characters. When EGI and FHK each are combined in a taxon, no two groups differ by more than four characters. If stem groups and the crown group of holophyletic lineage ACEGI are combined in a taxon, A differs from I by nine characters; the same is true of the difference between B and K in the holophyletic taxon BDFHK. The numbers (after each a) refer to the apomorphic characters in this tree.

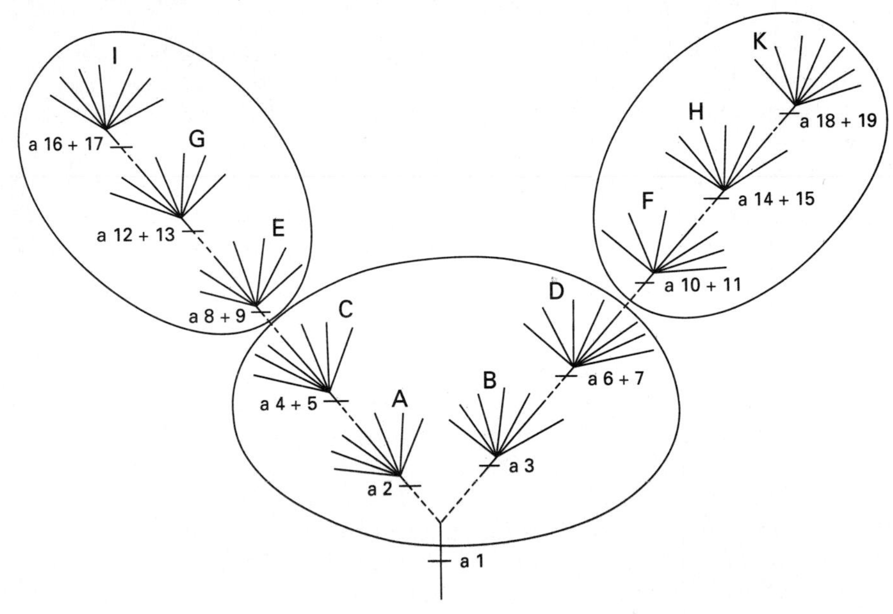

definition of the term *related*. (See Chapter 9 for a more detailed discussion of holophyly.)

The early species, genera, and sometimes families (stem groups) of a new phyletic lineage are usually far closer to equivalent taxa of sister lineages than they are to the eventual crown group of the new lineage. Indeed, the "Adam" of a new phyletic lineage is almost invariably nearly indistinguishable from the other members of the ancestral taxon. The first ancestor of the birds, for instance, long before *Archaeopteryx*, was presumably a rather conventional dinosaur or thecodont but one with featherlike scales. Unlike the cladist, the evolutionary taxonomist will not take this stem taxon out of the ancestral taxon, with which it agrees in almost every regard. *Eohippus* provides a similar illustration. If *Eohippus* is as much in the ancestry of the tapirs as that of the horses, as some paleontologists claim, how can it be included in the family Equidae? If this relationship were confirmed, it would be less confusing to recognize a primitive group of Perissodactyla which gave rise to the horse family, the tapirs, and several extinct families. The cladist might shunt this ancestral group aside as a plesion.

Since this consideration is the reason for some of the major differences between cladistic and evolutionary classifications, it is helpful to illustrate it with another example. Somewhere in the Reptilia the synapsid lineage originated which eventually produced taxa with the typical mammalian attributes (warm-bloodedness, hair, middle-ear bones, milk production). Where in this lineage should one place the beginning of the mammals? Should one combine the mammals with their reptilian ancestors into a single holophyletic taxon? The synapsid reptiles that gave rise to the mammals were a highly diverse and very long-lived group, appearing in the Pennsylvanian and surviving until around the end of the Triassic, a span of some 120 million years. The earliest forms are about the most primitive reptiles known, whereas the skeletons of the latest members of this lineage are almost indistinguishable from those of mammals. The cladist forms a holophyletic taxon, including everything from the lowest of these primitive reptiles up to the mammals. This requires breaking up the reptiles as a taxon and removing the primitive synapsids from their reptilian sister groups, with which they have infinitely more characters in common than they do with their ultimate descendants, the mammals. It is reasonably certain that the earliest synapsids had none of the mammalian characteristics of epidermal covering (hair), teeth, jaw articulation, endothermy, and milk glands of their later derivatives.

In a case like this—and there are literally scores—the evolutionary taxonomist terminates the synapsid reptiles at the level where the lineage begins to share more characters with the recent mammals than it does with the stem synapsids. The evolutionary taxonomist also leaves the stem synapsids with the other reptiles with which they share so many patristic characters. Admittedly, drawing such a line of "balance" is arbitrary, particularly since it cannot be determined from fossil remains when the most diagnostic mammalian characteristic—milk produced by mammary glands—originated. Traditionally, the change in jaw articulation has been considered a convenient dividing line between mammal-like reptiles and mammals. The acceptance of such a demarcation between an ancestral taxon (therapsid reptiles) and a descendant taxon (mammals) has the advantage of delimiting two reasonably homogeneous taxa that are far more meaningful physiologically, ecologically, and behaviorally than is a single holophyletic taxon that consists of an assemblage of synapsid reptiles and genuine mammals. Making the first synapomorphic character in a lineage, rather than the ensemble of characters, the arbiter of relationship results in single-character classifications with all their vulnerability. Figure 10-2 illustrates diagrammatically the different delimitation of taxa by cladists and evolutionary taxonomists.

Relatively homogeneous taxa delimited on the basis of a balanced consideration of both ancestral and derived characters are more practical than are taxa delimited by synapomorphies alone. The cladist calls this method subjective and arbitrary, and it is in comparison with the appli-

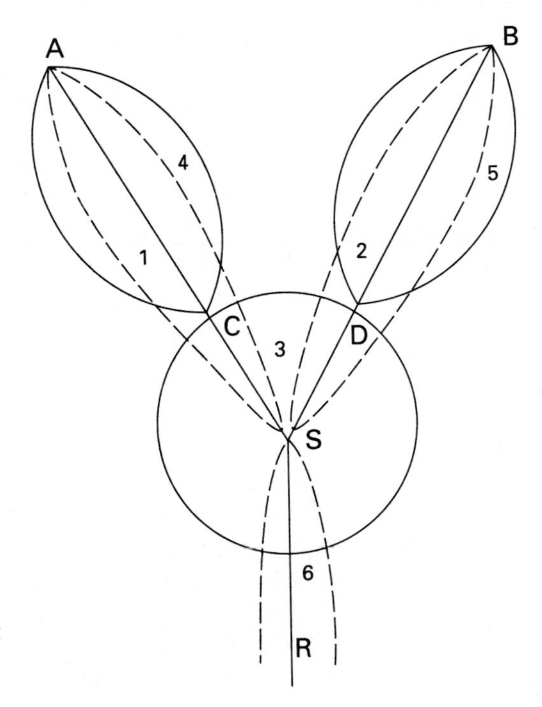

FIGURE 10-2
Translation of a cladogram into an evolutionary (solid lines) or a cladistic (broken lines) classification. The cladist recognizes taxa 1(SA), 2(SB), and 6(SR); the evolutionary taxonomist recognizes taxa 3(SCD), 4(CA), and 5(DB).

cation of the principle of holophyly. However, the cladist ignores how arbitrary and subjective it is to decide that holophyly should serve as the almost exclusive basis of classification.

The Treatment of Sister Groups

Sister groups are given the same categorical rank in Hennigian cladistics. This may lead to rather unbalanced classifications when sister groups have highly unequal rates of evolution. One sister group may still be very like the parental taxon while the other has acquired numerous autapomorphic characters. (See Chapter 9 for a discussion of sister groups and autapomorphy.)

The importance of autapomorphy is well illustrated by a comparison of birds with their reptilian sister group (Figure 10-3). Birds originated from the branch of the reptiles, the Archosauria, which also gave rise to pterodactyls, dinosaurs, and crocodilians. Although birds and crocodilians share a number of synapomorphies that originated in the archosaurian lineage after it had branched off from the other reptilian lines,

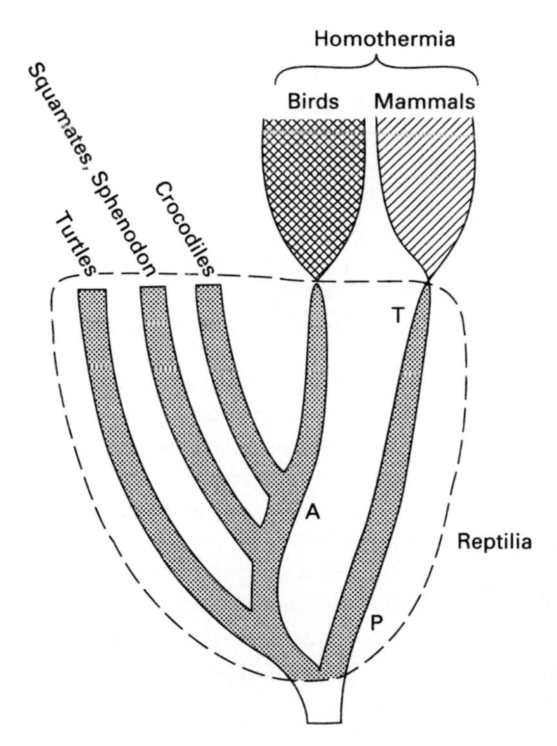

FIGURE 10-3
A phylogeny of the Recent classes of Amniota, with the monophyletic (though paraphyletic) Reptilia, the two holophyletic ex-groups Birds and Mammals, and the invalid convergently polyphyletic taxon Homothermia. A = Archosauria, P = Pelycosauria, T = Therapsida. (*After Carroll 1988.*)

crocodilians are on the whole still very similar to other reptiles because of their joint possession of numerous ancestral characters. Birds, by contrast, in connection with their shift to aerial living, have acquired a vast array of new autapomorphic characters.

The evolutionary taxonomist feels that whenever a *clade* (phyletic lineage) enters a new adaptive zone that leads to a drastic reorganization of its members, greater taxonomic weight must be assigned to the resulting transformation (acquisition of new autapomorphies) than to the sister group status. This is effected by assigning higher categorical rank to a sister taxon with numerous autapomorphies.

Important differences between cladistics and evolutionary classifications are described in Table 10-1.

TABLE 10-1
DIFFERENCES BETWEEN CLADISTIC AND EVOLUTIONARY CLASSIFICATION

	Cladistics	**Evolutionary classification**
	Analysis	
Aim	Construction of a cladogram	Construction of a phylogram
Parsimony	Important criterion of best cladogram	Testing for homoplasy by various criteria
Major role of synapomorphies	To recognize branching points	To discover and reject convergent polyphyly
	Classification	
Delimitation of taxa	By holophyly	By monophyly and possession of greatest number of shared characters
Minimal number of characters needed	A single synapomorphy	As many characters as possible
Ranking of sister groups	Same categorical rank	Different rank if they differ by sufficient autapomorphies
Kinds of characters used	Only diagnostic synapomorphies	All homologous characters, including ancestral ones
Basis of classification	Branching only	Branching and evolutionary divergence
Stem groups	Always with crown group	Sometimes with sister group
Diagnosis of taxa	Only monothetic admissible	Polythetic also admissible
Autapomorphies	Generally ignored	Important in ranking of taxa
Categorical ranking of taxa	No method available, hence arbitrary	By degree of difference

THE LINNAEAN HIERARCHY

Ever since Linnaeus, degrees of relationship have been expressed by combining increasingly more distantly related taxa into taxa of increasingly higher categorical rank (Chapter 6). The endings of the names of higher taxa indicate their categorical rank, and the hierarchy of categories facilitates information retrieval by permitting an easy survey of taxa. But what method should one use to determine categorical rank? As was shown in Chapter 9, cladistics does not have a theory of ranking, at least not since the principles proposed by Hennig proved to be impractical. Degree of difference is the criterion used in classical taxonomy (and by Simpson and Mayr) to determine ranking. This inevitably introduces a certain amount of subjectivity, but no other criterion has proved superior.

In spite of all the arbitrariness it introduces, the Linnaean hierarchy is extremely useful in classification and is therefore retained by nearly all taxonomists.

HOW TO CONSTRUCT AN EVOLUTIONARY CLASSIFICATION

An evolutionary classification, as has been stated, utilizes two semiindependent sources of information: branching points and degrees of difference among taxa. To attain the most informative ranking, not only synapomorphies but, whenever useful, autapomorphic characters and even ancestral ones (*symplesiomorphies*) are used. The cladist dogma that plesiomorphic characters cannot be used in classification is without justification. It is correct to say that they cannot be used in cladistic analysis to locate branching points, yet they may be of great importance in classification when they represent the dominant characters of a taxon. Therefore, in evolutionary classification all kinds of homologous characters are used in ranking.

The two criteria of evolutionary classification—monophyly and totality of shared characters—cannot be applied simultaneously. They must be applied consecutively, and this raises the question of which should be applied first. In view of the unsolved difficulties of converting a cladogram into a practical classification (Chapter 9), adopting the traditional procedure—first recognizing provisional taxa by the totality of their shared characteristics and then testing the provisional taxa for monophyly—is recommended. In the first step one may be able to adopt traditionally delimited taxa, but in poorly known groups one may have to delimit taxa de novo with the help of some phenetic procedures. These provisional taxa are then thoroughly tested for monophyly (presence of synapomorphies, explanation of homoplasies). By establishing the ancestor-descendant relationships of these taxa, one can construct a phylogram.

Autapomorphic characters are given appropriate weight in evolutionary classifications, since they are often among the most important characters in a taxon. Stem groups are combined with stem groups of sister lineages rather than with the crown group of their own holophyletic lineage whenever this leads to the recognition of more homogeneous taxa. Therefore, a phylogram of higher taxa does not necessarily agree entirely with a cladogram based on the same organisms. For instance, a phylogram can show one taxon arising from another, something a cladogram cannot do. Relevant cladograms should be published together with evolutionary classifications to reveal any differences.

The curious objection has been made that evolutionary classification cannot quantify relationship, but neither can cladistics. The sharing of a nearest common ancestor (and synapomorphies) is not a quantification. Actually, the procedure of evolutionary classification can be quantified by methods outlined in the Appendix at the end of Chapter 9 and in Chapter 11.

Mosaic Evolution and Rate of Change

Partitioning characters into apomorphic (derived) and ancestral ones is of course only one method of weighting characters (Chapter 7). Because they are rather conservative, ancestral characters are sometimes more important in classification than are rapidly changing characters. As indispensable as derived characters are for the determination of branching points, the conservative ancestral characters may actually tell us more about the total genotype. It is now clear that different portions of the genotype may evolve at very different rates, a process termed mosaic evolution (Chapter 6). As a result of this inequality, different conclusions may be reached about the rank of a taxon depending on the selection of the characters on which the decision is based (Mayr 1969:228). The relationship of human beings and chimpanzees is an apt illustration (Chapter 6). Mosaic evolution is responsible for the fact that different classifications are generated whenever new sets of characters are used (Rohlf, Colless, and Hart 1983).

In many large genera, both animal and plant, an occasional species seems to have experienced a veritable morphological revolution, so that it differs enough to have been described as a separate genus even though it branches off very close to one of the more typical species of the parental genus. A striking instance was described for an offshoot of the plant genus *Clarkia* (Sytsma and Gottlieb 1986). There will probably always be disagreement about whether such a species should be classified by branching point or by phenotype.

Monophyly

Even though cladists and evolutionary taxonomists agree that taxa should consist of groups of relatives united by common descent, they have not reached a consensus on how to translate this basic concept into a methodology. Traditionally, a taxon was considered monophyletic if it consisted of descendants of a common ancestor (Chapter 9), but it has remained controversial how to define a common ancestor. From Haeckel to 1950 a taxon was called monophyletic if it was derived from a single ancestral taxon, but more fine-grained analysis in the last 40 years has resulted in several rather different definitions of monophyly. The argument over which of them is the best has resulted in a large literature. Most important are articles by Tuomikoski (1967), Ashlock (1971, 1972, 1979, 1984), Farris (1974, further explained by Platnick 1977*b*), Holmes (1980), Duncan, Phillips, and Wagner (1980), Charig (1982), and Meacham and Duncan (1987).

THE TRADITIONAL CONCEPT OF A COMMON ANCESTOR

It is important for the sake of historical accuracy to realize that the concepts of monophyly and the common ancestor were introduced into taxonomy when (Darwin notwithstanding) students of phylogeny were still dominated in their thinking by idealistic morphology. Even after they had adopted evolution for the construction of phylogenies, morphologists still continued to derive one "type" from another. The writings of comparative anatomists from Gegenbaur and Huxley to Remane and Romer clearly demonstrate archetypal thinking, which is well illustrated by Schindewolf's (1969) presentation of phylogeny as a progression of archetypes. Even Simpson never succeeded in emancipating himself entirely from this thinking, hence the absence in his entire opus of a discussion of the multiplication of species and the role of species in the origin of new higher taxa.

It was in this period of archetypal thinking that almost all the traditional classifications of higher taxa were proposed. Therefore, monophyly simply meant that one morphological archetype was exclusively derived from one ancestral one, not from several. Even when authors referred to the common ancestor in the singular, they had the ancestral type in mind, and only in the rarest cases a single individual or a single stem species. This simplistic, comparative anatomy-based approach eventually had to be given up for two reasons. One was that the acceptance of gradualistic Darwinian thinking and modern speciation theory made it inevitable to conclude that every higher taxon had started with a founder species. The other was that so many fossil representatives of each anatomic "type" were found that one could no longer evade questions such as, Which particular group of reptiles gave rise to the mammals? A seeming need for greater precision finally culminated in the demand that the old archetypal definition of monophyly be replaced by a definition stating specifically that the common ancestor was a single ancestral species. As logical as this solution seems at first sight, it generated a number of difficulties.

VARIOUS CONCEPTS OF MONOPHYLY

Holmes (1980), in a thorough analysis of the concept and term *monophyly,* classified the numerous concepts in the literature into four classes. Actually, there are slightly different versions within some of these classes. Furthermore, this classification does not place sufficient emphasis on the crucial differences between the various concepts. These concepts all agree in rejecting convergent polyphyly. Their differences can be brought out better if three instead of four classes of monophyly concepts are distinguished. They differ in the upward and downward delimitation of a monophyletic lineage, in other words on where to draw the line against ancestors and against descendants.

Holophyly

The simplest and most unambiguous but in some ways also the most unrealistic delimitation is Hennig's (1950, 1966*a*), which includes in a monophyletic taxon not only the stem species but *all* its descendants. This concept of monophyly, along with the correlated concept of the monophyletic taxon, differs to so great an extent from the traditional concept of monophyly that Ashlock (1971) renamed Hennig's concept *holophyly*. Every holophyletic taxon is also monophyletic, traditionally defined, but many if not most classical monophyletic taxa are not holophyletic because they do not include some of the ex-groups to which they have given rise. The reasons why evolutionary taxonomists reject basing the recognition of taxa on holophyly are described in Chapter 9.

Evolutionary Taxonomy

The definition of monophyly most widely accepted among evolutionary taxonomists is as follows: *A group is monophyletic if all the included spe-*

cies and their ancestors are derived from the most recent common ancestral species, which is also included in this taxon. Ordinarily, only taxa that qualify under this definition are recognized. This definition does not require that ex-groups be included in a monophyletic taxon. Hence, the Reptilia, which gave rise to the Mammalia and Aves, are a monophyletic taxon for the evolutionary taxonomist, while the cladist calls such a taxon paraphyletic (Chapter 9).

The definition of monophyly must include the specification "the most recent common ancestral species"; the wording "descent from a single common ancestral species" is not sufficient, because if one goes back far enough, any two species, say, a beetle and a fish, have a common ancestral species.

Even though the concept that each phyletic group is ultimately derived from a single common ancestral species is implicit in Darwin's gradualistic theory of common descent, it was rarely stressed in the phylogenetic literature during the first 50 to 80 years after Darwin. Instead, most authors considered it a sufficient qualification for monophyly if a taxon had descended from the nearest ancestral "taxon." Even though this definition lacked precision, it avoided some of the difficulties caused by parallel evolution.

A determination of monophyly ordinarily proceeds under the fiction that new taxa originate in single lineages and that the origin of a new apomorphy is a unique event. All the available evidence, however, indicates that this does not reflect reality. This realization has led to the proposal of a third kind of definition of monophyly.

Monophyly Allowing for Parallelophyly

Organisms have a genotype and a phenotype. Difficulties in the definition of monophyly arise when the genotype has hidden propensities that are not expressed in the phenotype. Several descendant lineages of an ancestor with such a genotype may independently evolve a similar or identical phenotype in a parallel manner. If recent analyses of homoplasy have taught us anything, it is how often species and higher taxa have unexpected concealed propensities that are expressed in some of the descendants but not in others. Well-known examples are the irregular occurrence of stalked eyes among the acalypteran flies (Hennig 1966*a*:117) and certain spermatheca shapes in *Drosophila*. Throckmorton (1965:228) stated that "parallelism is the rule rather than the exception for individual characters in *Drosophila*." Simpson (1961:121) claimed that "parallelism is a widespread phenomenon in evolution, and it is not uncommon to find that some generally recognized taxon arose by parallel evolution through two or more lineages from different ancestral taxa.'' The same finding has been made by various botanists, e.g., Cronquist (1968:16–17).

The similarities acquired by the members of such taxa are clearly due to their common ancestry. They can be postulated to have been derived from the genotype of the nearest common ancestor. Mayr (1965*a*:79, 1968*b*:548) has repeatedly stressed the need to consider genotypic similarity in delimiting taxa and determining common ancestors. There has been much soul-searching among students of phylogeny as to whether a group or taxon in which a diagnostic character originated by such parallelophyly qualifies as monophyletic (see Figure 10-6).

Even though a definition of monophyly based on the inclusion of the stem species appears at first sight to be the most logical solution, it faces a number of difficulties. It is in conflict with established practice because taxa were traditionally derived from ancestral higher taxa, not from specified species (Simpson 1961; Cronquist 1987). It is impractical because, owing to the imperfection of the fossil record, we will never know what particular species was the stem species of a flourishing higher taxon (Harper 1976). Indeed, we do not know of a single case where a fossil species has been unequivocally established as the stem species of a new higher taxon. This indicates that the stem species solution is an instance of ''the fallacy of misplaced concreteness'' (Whitehead 1948:75, 85). In the case of birds, we do not even know from what family of reptiles they arose, not to mention from what genus or species. The situation is a little more favorable in the mammals, with their superb fossil record. Their nearest cynodont relatives have been rather well identified, yet one cannot pinpoint which species or genus was the immediate ancestor of the first mammal.

Is it acceptable science to tie a monophyletic taxon to an ancestral species that is unknown and forever unknowable? In these circumstances it is not surprising that a taxonomist as experienced as Simpson felt that a definition of monophyly tied to a single stem species can ''readily be shown to be undesirable in principle and usually inapplicable in practice'' (1961:123).

When Simpson said this, he thought that the class Mammalia ''almost certainly arose from several different taxa within the order Therapsida'' (1961:121), thus exemplifying *parallelophyly*. Subsequent research cast doubt on this interpretation and induced Simpson ''no longer to accept [his] former definition in its too precise form'' (1971:192). However, Simpson still considered the thesis of a (narrowly defined) monophyletic origin of the mammals as ''simplistic'' (1971:192–193). He also thought that the horse genus *Merychippus* had arisen from more than one species of *Parahippus* that shared the same propensity for developing a new tooth structure (Figure 10-4). To be able to include such cases of seeming

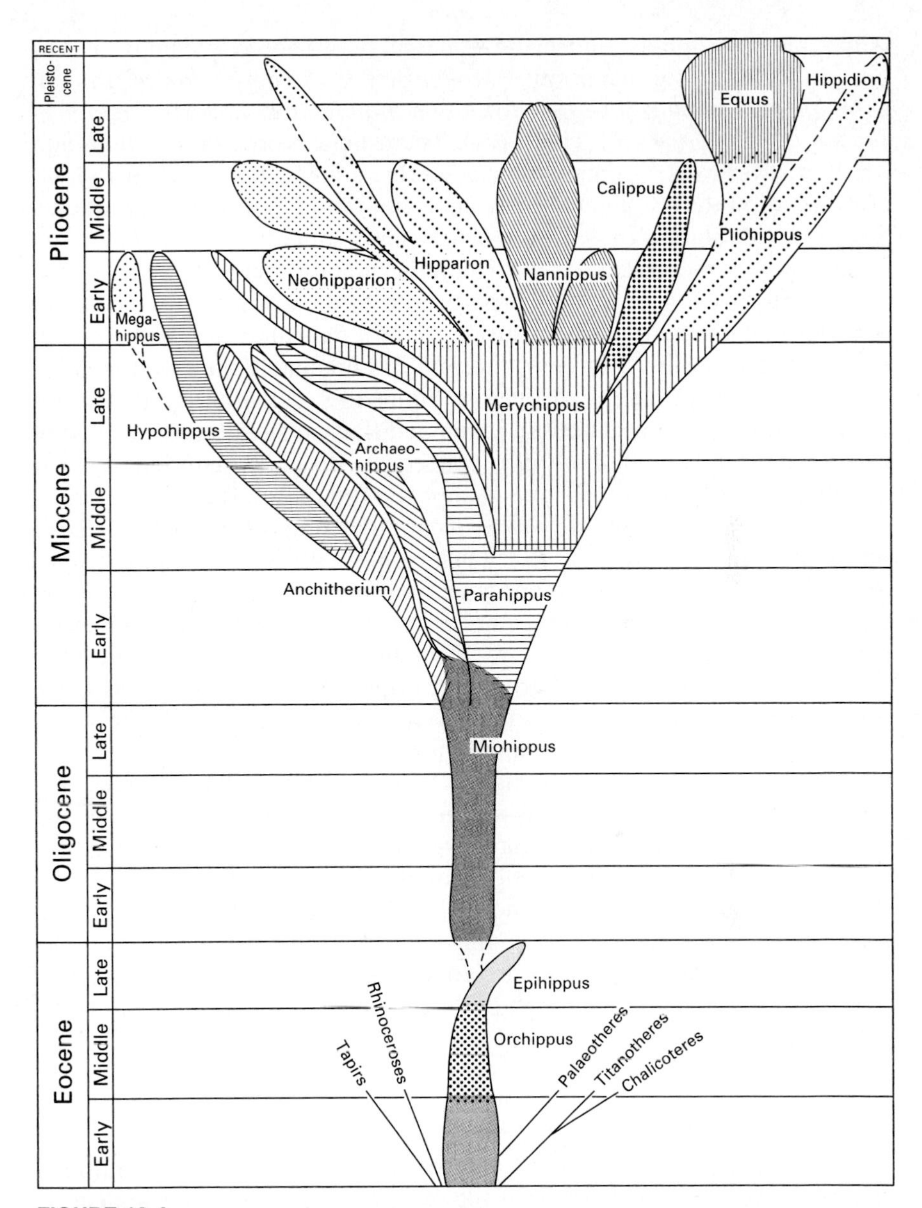

FIGURE 10-4
A tentative phylogeny of the horse family (Equidae). Each of the lower stem genera (*Miohippus, Parahippus,* and *Merychippus*) gave repeated rise by parallelophyly to the next higher level of taxa. (*After Stirton.*)

genotypic monophyly under the concept of monophyly Simpson proposed the following definition: "*Monophyly is the derivation of a taxon through one or more lineages from one immediately ancestral taxon of the same or lower rank*" (1961:124). The crucial words in this definition are "or lower rank." Simpson included this qualification in order to limit consideration to the "immediately ancestral taxon." Attempts to ridicule Simpson's definition by saying it would allow one to consider birds and mammals a monophyletic group Homotherma are therefore invalid. The ancestral group of lower rank from which birds are derived consists of dinosaurs or thecodont archosaurians, while mammals are derived from cynodont therapsids, two very different "immediately ancestral taxa."

Another objection raised against Simpson's definition is that it includes reference to an ancestral taxon and that no such reference should be included in the definition of a concept as broad as monophyly. One should refer only to groups. We cannot see the reason for a replacement of taxa by groups. After all, a classification arranges taxa in a Linnaean hierarchy, and we agree that taxa must be monophyletic. But how are we to determine that taxa are monophyletic if our method allows us to determine only that certain groups are monophyletic? What advantages do groups have over taxa? To be sure, the delimitation and ranking of taxa is subjective, but so is that of groups in a hierarchy of groups. Also, the delimitation of taxa may have to be modified as our information becomes more complete, but so does the delimitation of groups.

The fact that the taxa of higher or lower rank which we test for monophyly are provisional has nothing to do with the definition of the concept of monophyly. Therefore, we cannot see any reason why taxa should be disqualified as objects of monophyly tests. Do those who make this claim confuse defining monophyly with testing for monophyly? To escape such confusion, a taxonomist must always remember that the same principle that applies to the concepts of species and homology is true for the concept of monophyly: The definition is independent of its testing in individual cases.

POLYPHYLY

One of the reasons why many authors rejected Simpson's definition is that they thought it would permit the recognition of polyphyletic taxa. Is this a valid objection? To answer this question we must first determine what is meant by the term *polyphyletic*. In the older literature, when monophyly was defined in terms of descent from an antecedent type or higher taxon, the term always referred to similarity resulting from convergence. Polyphyly meant *convergent polyphyly* (Figure 10-5), that is, derivation from unrelated ancestors. When definitions of monophyly

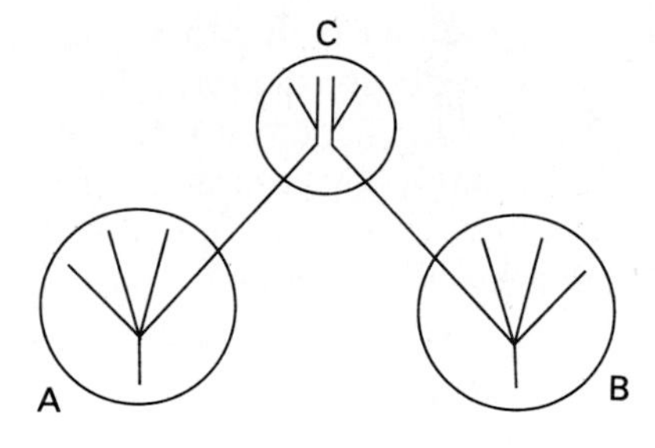

FIGURE 10-5
Convergent polyphyly. Ex-groups of taxa A and B have by convergence become so simllar that they were combined in the polyphyletic taxon C.

which stressed descent from a single ancestral species were adopted, it was discovered that there was another kind of polyphyly, caused by the parallel acquisition of the same apomorphies in several independent lineages derived from the same ancestral species. When a new taxon is delimited on the basis of these independently acquired apomorphies, it has to be called polyphyletic. However, this polyphyly, which may be called parallel polyphyly (*parallelophyly*), is something entirely different from convergent polyphyly (Figure 10-6). The existence of these two kinds of polyphyly has been discussed by a number of authors. Charig (1982:408–409) referred to them as proximate and remote polyphyly.

In the case of parallel polyphyly (parallelophyly), the common propensities or potentials of members of the ancestral taxon (due to their common origin) may lead to parallel but independent evolutionary change in several sublineages, resulting in the derived taxon. In such a case all the members of the derived taxon are descended from the same nearest ancestral taxon in which the propensity originated. Such a taxon is monophyletic in the sense of Simpson and Mayr because it is a single nearest ancestral taxon that gave rise to the new taxon. Even Hennig apparently does not consider such monophyly as polyphyly (1966*a*:146),

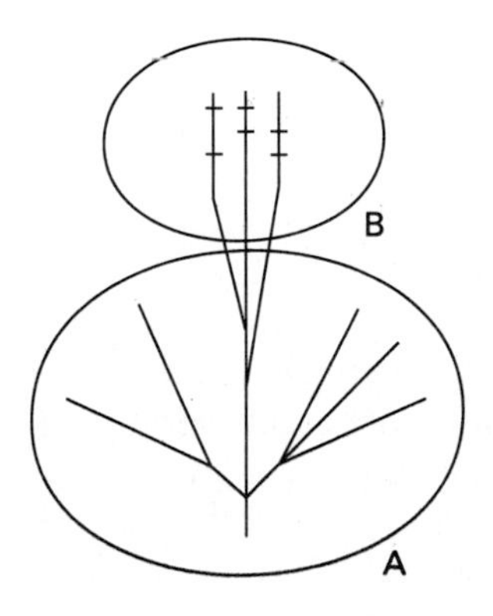

FIGURE 10-6
Parallelophyly. Three clades of taxon A have in a parallel manner evolved into taxon B, owing to the possession of the same genotypic potential.

since he considers a polyphyletic group to be one that is due to *convergent* similarity. Tuomikoski also does not hesitate to use the term *parallel apomorphy* when it is due to "the agreement in capacity to develop parallel similarity" (1967:141). Convergent polyphyly, by contrast, exists when a "phenetic" taxon actually consists of two or more groups of species that evolved independently by convergence, with each group having derived from a different nearest ancestral taxon. Huxley's Homotherma [Gardiner's (1982) Haemothermia], a taxon proposed for the warm-blooded vertebrates (birds and mammals), illustrates convergent polyphyly excellently.

Early in the history of taxonomy species were often grouped together because of superficial similarity. The Coelenterata and the Echinodermata, for instance, were combined by Cuvier and Agassiz into Radiata because both groups were radially symmetrical. However, the Radiata are polyphyletic, because coelenterates and echinoderms are unrelated rather than descended from a nearest common ancestor. As traditionally defined, a taxon is polyphyletic when it is composed of members of two or more unrelated phyletic lineages. *Unrelated* means that their nearest common ancestor has also given rise to lineages placed in other taxa that are not ancestral to the polyphyletic taxon.

The Muscicapidae of the standard ornithological classifications were polyphyletic under this definition, because this "family" included Eurasian as well as Australian flycatchers whose common ancestor gave rise to numerous other phyletic lineages that are classified as separate families. Both Simpson and Mayr reject the acceptance of taxa based on convergent polyphyly. To date, no unequivocal case of a taxon is known to us that originated by parallelophyly. Such evolution is, however, suggested by the fossil record of the horse genera *Miohippus, Parahippus,* and *Merychippus*. If these cases were confirmed, then a definition of monophyly based on a stem species would be inappropriate. Some version of Simpson's definition (with the proper emphasis on taxa "of lower rank") would then seem appropriate.

Parallelophyly does not invalidate the concept of monophyly because a genotypic propensity for a character, even if it is expressed phenotypically only in later descendants, would have originated in a single species. However, if such a species did not yet have the phenotype of later species in its lineage, it would have to be included in the ancestral taxon. In view of these uncertainties, it would seem advisable not to be too rigid and dogmatic about the definition of the term *monophyletic*. Also, cases of very remote parallelophyly may grade into convergent polyphyly (Charig 1982).

The major difference among the three concepts of monophyly lies in how they are used in classification. The cladist delimits taxa on the basis of holophyly, while the adherents of concepts 2 and 3 use monophyly to test the correct delimitation of taxa. Whether one includes a reference to the stem species in the definition of monophyly would seem to be a rather academic matter, since such a species, as well as its actual characteristics, is unknown and presumably forever unknowable. Hence, in actual practice, the nearest known common ancestor is usually a higher taxon.

The Status of Paraphyletic Taxa

Monophyletic taxa that have given rise to one or several ex-groups are called paraphyletic by cladists. For instance, cladists call the Reptilia paraphyletic because they gave rise to birds and mammals. Cladists do not recognize paraphyletic taxa and exclude them from their cladograms. Traditional taxonomists, however, consider paraphyletic taxa to be legitimate members of classifications. Paraphyletic taxa, properly delimited, are commonly far more homogeneous than are the holophyletic taxa into which cladists merge their parts (Figure 10-7). They usually have precisely the characteristics one would like taxa to have in a practical classification (Charig 1982:429–432).

One of the reasons why cladists reject paraphyletic taxa is that they cannot diagnose them monothetically. For evolutionary taxonomists this is no objection, because they characterize and delimit taxa using all their characters, not just a few diagnostic ones. The reptiles, which are amniotes with ectothermy, can be well characterized polythetically even though they do not have a single exclusive diagnostic character (Gans and Pough 1982). The main reason for the cladists' rejection of paraphyly is that it is in conflict with the primacy of holophyly, which is their concept of relationship.

All taxa show a mixture of ancestral (plesiomorphic) and derived (apomorphic) characters. Even though only derived characters are decisive in cladistic analysis, we know of no good reason why ancestral characters should be excluded from consideration in the construction of a classification. Diagnosis, after all, is only one of the objectives of a classification.

The earliest reptilian ancestors of the mammals share far more characters, even though they are ancestral ones, with several reptilian sister groups than they do with their remote descendants. In other words, they are far more closely related to their sister groups than to their ultimately derived descendants, the hairy, warm-blooded, milk-producing mammals. This is ignored by cladists because of their restrictive definition of relatedness.

Because most traditional higher taxa are ex-groups of ancestral taxa, the cladistic method forces its followers to consider most traditional taxa

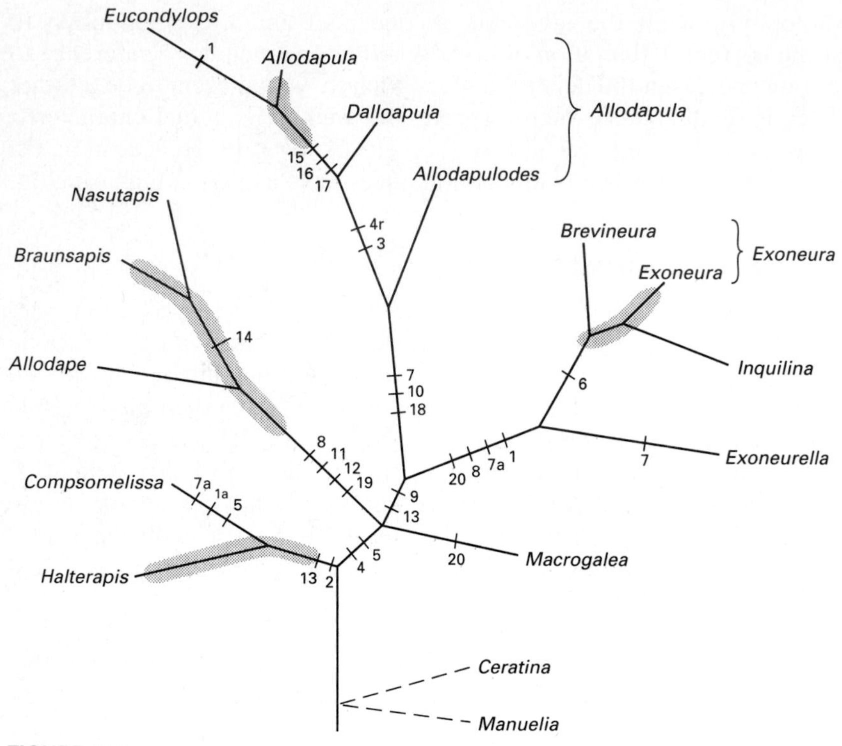

FIGURE 10-7
Phylogram of the 16 genera of allodapine bees. Four genera (stippled) have produced derived genera and are thus "paraphyletic." (*From Michener 1977.*)

to be paraphyletic. This includes all fossil taxa except a few terminal ones. For instance, all major ancestral taxa of the mammals, Synapsida, Therapsida, Theriodontia, and Cynodontia are paraphyletic. In accordance with their principles cladists should ignore such taxa. However, they cannot suppress the fact of their existence, and so they delimit them differently or dismiss them as plesions. Unfortunately, this procedure causes them to obscure the genealogical position of these taxa, an item of information that should be of the greatest importance to a cladist. Rejecting these traditional taxa or giving them nontraditional new delimitations violates the concept of a stable classification as a key to information retrieval.

In well-known groups, analysis leading to an evolutionary classification usually supports the existing classification, while the cladistic recognition of holophyletic taxa often results in the establishment and naming of an entirely new set of taxa. For further discussions of why traditionally delimited monophyletic groups are so often superior to holophyletic groups, see Charig (1982:429–432) and Haszprunar (1986). We cannot see that anything is gained by breaking up traditional taxa by arbitrarily adopting the principle of holophyly as the basis of classification.

How questionable it is to reject paraphyletic taxa is shown by the fact that no cladist would refuse to recognize the Reptilia as a natural taxon if birds and mammals did not exist. Since neither birds nor mammals have any retroactive effect on the characteristics of their ancestors, there is simply no biological reason for not recognizing the Reptilia. Most high-ranking paraphyletic taxa, such as reptiles and turbellarians, can be recognized at once, even by a nonexpert, while it may require a great deal of special expertise to recognize the holophyletic taxa into which they are broken up by cladists.

The recognition of paraphyletic taxa does not preclude an occasional informal reference to holophyletic groups in order to illustrate certain phylogenetic trends or groupings. The term *archosaurians,* which refers to birds and crocodilians together with their extinct ancestors and relatives, can indeed be used in evolutionary discussions. However, to include all the therapsid or even synapsid Reptilia in an expanded version of Mammalia, as is done by some authors (e.g., Ax 1984), seems more misleading than helpful.

The ontological status of higher taxa, i.e., an understanding of what they actually are, is important for determining whether the recognition of paraphyletic taxa is legitimate. That higher taxa are not classes is evident, but are they individuals or something else? We agree with Wiley (1981:75) that they are not individuals, for they lack the internal cohesion of individuals; rather, they are *historical groups*. If historical continuity were used in order to claim that taxa are individuals, then one would recognize only a single individual for all forms of life because one could not find a place to make a cut between an ancestral taxon and a descendant taxon. If such cuts were allowed, then cuts between a paraphyletic taxon and its descendant ex-group would be as legitimate as those between any taxon and the ancestral taxon that gave rise to it.

Phylogenetic Trees

Phylogenetic trees differ from cladograms in a number of ways. The taxa are delimited by shared, carefully weighted characteristics, not by holophyly, but are tested for monophyly. Polythetically characterized taxa are allowed. When there are several possible branching diagrams, the one that is most probable on the basis of all available evidence, not merely on the basis of parsimony, is preferred. Gaps between taxa are properly considered in the ranking of taxa. Rank in the categorical hierarchy, other things equal, is determined by degree of difference. Ancestor-descendant relationships are carefully weighed. The lengths of the branches of the tree (between internodes) and the angles at the branching points are chosen to indicate the amount of evolutionary divergence (anagenesis) (Chapter 11).

ADVANTAGES OF EVOLUTIONARY CLASSIFICATION

In summary, an evolutionary classification or a truly phylogenetic classification employs both cladogenesis and anagenesis to arrive at taxa that are genealogically circumscribed and reasonably homogeneous. As products of natural processes, these are natural groups, not created in the sense that phenetic groups are created. The analysis on which the classification is based provides a historical explanation of the existence of the taxa and a justification of their formal recognition. The dendrogram on which characters are deployed, with the accompanying discussion, portrays many hypotheses that others may contest. Unlike too many older phylogenetic classifications, which tended to be authoritarian, the systematist's reasons for his or her conclusions are available for all to judge.

Because the goal of evolutionary systematics is to find homogeneous monophyletic groups, the resultant taxa, unlike those of strictly holophyletic classifications, are made up of reasonably similar organisms. Such groups are easier to remember and recognize (identify) than are many of the groups proposed by cladists. One can say more about homogeneous taxa than about holophyletic taxa, whose members need share only a single synapomorphic character.

Because divisions between taxa are based on gaps where many characters have separated lineages and much evolution has taken place, a maximum of characters is available to delimit taxa. While classifications of poorly studied groups usually change radically after any method of analysis is thoroughly applied, well-studied groups usually remain stable with the evolutionary approach, even when cladistic analysis may suggest minor changes or when new species or characters are added. Similar changes or additions can cause major upheavals in cladistic classifications.

A major use of classifications is to serve as a key to the vast information storage system of the zoological literature. That is, they serve as indexes to what is known about each taxon, including the analysis that justifies them. We see no advantage in adopting a cladistic classification that burdens the memory, yields taxa that are hard to delimit, and otherwise confounds the information retrieval process. All these difficulties are the price paid for the single advantage of being able to approximate the cladogram from the formal classification. The price is far too high for this slight and questionable advantage.

PRACTICAL CONSIDERATIONS IN THE CONSTRUCTION OF A CLASSIFICATION

Considering the diverse uses to which classifications are put, it is not surprising that taxonomists have always wanted to produce the "best" classification (Chapter 6). It must be stable yet flexible; must reflect the evolutionary history of the included taxa; must be convenient for the storage and retrieval of information and hence must consist of relatively homogeneous taxa; must be easy to use not only by the specialist but by all zoologists; and should serve as a reliable foundation for comparative studies. It is our conviction that an evolutionary classification in which not only branching but also evolutionary divergences and both ancestral and derived characters are duly considered comes closest to fulfilling this set of demands (Chapter 6). Furthermore, evolutionary taxonomists, with their more flexible guidelines, are in a better position than are cladists to make use of all available information on biogeography and life history in the construction of a truly helpful classification.

In the process of classification one deals with living beings, not with preserved remnants of morphotypes. Thus biological taxonomists make use of everything they can learn about behavior, ecology, life history, distribution, physiology, or any other attribute of the living organism in making decisions about relationship. Those who delegate the construction of classifications to the computer tend to assume that the output of the computer is the last word. Actually, uncertainties remain even with the best computer programs. It is recommended that numerical classifications be considered as tentative; all less than secure relationships should be tested by means of other biological criteria and should be subjected to a number of reviews that will be outlined in the following sections.

Even when one follows the principles of evolutionary classification, it is usually possible to base a number of different classifications on the same phylogenetic tree. The same phylogeny can sometimes be translated into several different classifications because three operations of the taxonomist cannot be carried out without an element of arbitrariness: (1) the delimitation of groups that we formally recognize as taxa (i.e., their size); (2) the rank in the hierarchy assigned to a given taxon (e.g., tribe, subfamily, family); and (3) the position of a taxon in the sequence of taxa.

Figure 10-8 illustrates the rather arbitrariness with which some authors converted the information provided by phylogenetic trees into classifica-

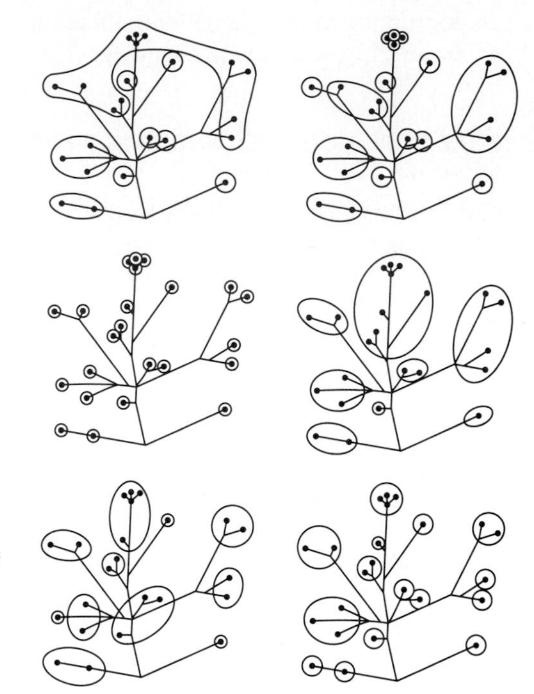

FIGURE 10-8
Subjective placement of 23 genera of ferns in families in classifications by six different authors. Each point represents a genus and each circle (around one or several points) a family. (*After Wagner 1969.*)

tion. It illustrates instances of extreme splitting, unbalanced treatments, and disregard for monophyly in some of these alternative schemes. Such subjectivity can be avoided only if one adopts precise standards.

In choosing between various options, we must depend on our judgment as to which of them will produce the most practical result. The conscientious taxonomist is concerned with the optimal size of taxa (neither too large nor too small), with gaps between taxa, with a practical (not excessive) number of categorical ranks, with the use of similar ranking criteria in different major taxa, and with other inevitably subjective considerations. Taxonomists believe that if such collateral criteria are duly taken into consideration, a more useful classification will be achieved than that which will result from an attempt to translate automatically the results of a numerical analysis into a Linnaean hierarchy.

Translating a multidimensional phylogenetic tree into a linear sequence permits many alternative arrangements. A taxonomist will be guided in his or her choice by the fact that the major function of a classification is to be useful. For instance, a classification that attempts to express every shade of relationship (by splitting) makes information retrieval exceedingly difficult and thus defeats its own purpose; this explains the adoption of the method of phyletic sequencing by post-Hennigian cladists. The taxonomist must always remember that the key to a filing system does not contain all the information contained in the file.

CRITERIA FOR DELIMITATION AND RANKING

Since evolutionary taxonomists do not accept the rule of cladistics that every holophyletic lineage is a separate taxon, they must develop a different set of principles by which to convert a phylogenetic tree into a classification. They base their recognition of taxa and their ranking on a balanced consideration of seven criteria:

1 Distinctness (size of gap)
2 Degree of difference
3 Evolutionary role (uniqueness of adaptive zone)
4 Grade characteristics
5 Size of taxon
6 Equivalence of ranking in related taxa (balance)
7 Stability

All seven factors must be weighed before each decision is made, although their respective importance differs from case to case. Whenever a taxonomist encounters a group of species that seems to merit recognition as a new higher taxon, that taxonomist must ask questions concerning these seven points, such as the following: Is the new taxon sufficiently different to merit separation and naming? Is it of a size that is convenient for information retrieval? The taxon should be formally recognized only after these questions have been answered in the affirmative. For a critique of some of these criteria, see Wiley 1981:240–269.

Distinctness

The greater the gap between two clusters of species, the greater the justification for recognizing both as separate taxa. The size of the gap is measured not merely in terms of phenetic distance but, more importantly, in terms of the biological significance of the difference (see "Evolutionary Role" below).

The gaps between taxa are a result of evolution. Speciation, extinction, adaptive radiation, unequal rates of evolution, and other evolutionary phenomena are responsible for the existence as well as the unequal size of the gaps that separate higher taxa. Not only the gaps but also the clusters of species separated by gaps are realities of nature. When a cluster of species is large and heterogeneous and is not subdivided by clear-cut gaps, the determination of whether it is profitable to separate it into several taxa requires careful evaluation (Michener 1957, 1963:153). Few aspects of the evolution of higher taxa are more useful for the delimitation of taxa in a classification than are gaps, yet cladists have pronounced the use of gaps in the construction of a classification as "defective" because "it ignores the continuity of evolution" (Farris 1979*b*:485). This documents the cladists' exclusive focus on vertical evolution. Continuity of evolution in the vertical dimension does not necessarily produce continuity in the horizontal dimension.

Gaps among higher taxa are due to either one or both of two causes: extinction and divergent evolution. Taxonomists whose classifications make use of gaps are thus able to include these two important evolutionary processes in their classifications, something cladists fail to do. Furthermore, due attention to gaps permits the construction of classifications that are more practical than those in which gaps are ignored. There is, of course, no conflict between the recognition of gaps between known or living taxa and the acceptance of complete evolutionary continuity within each phyletic lineage.

Degree of Difference

When speaking of the degree of difference between two clusters of species, the taxonomist usually thinks of the "distance" between the means of the two groups of species. This value, however, must be considered in conjunction with two other data: the scatter of the clusters and the gap between them. The denser and more uniform a cluster of species is, the better justified its recognition as a separate taxon usually is (Mayr 1969:235).

The more characters that have been used to determine the degree of difference between two taxa, the more reliable this measure is likely to be. The validity of taxa based on a single character or a few characters is often refuted by further analysis. The search for characters must be equally intensive throughout the studied group; all subgroups must be thoroughly investigated for characters.

Evolutionary Role

Systematic zoologists have increasingly given weight to ecological considerations in the ranking of taxa. Almost any prosperous higher taxon is descended from a founder species that succeeded in shifting to a promising new adaptive zone. Consequently, it is legitimate to consider the ecological role of a taxon when one is determining its categorical rank. Well-delimited higher taxa almost invariably have a definite ecological meaning. Cats, dogs, horses, and woodpeckers fill well-defined ecological niches or adaptive zones in nature. A consideration of its adaptive and evolutionary role is thus an important element in the categorical ranking of a higher taxon.

Ecological considerations are also important in ranking taxa that belong to a single phyletic lineage. Evidence is mounting that the polytypic species *Australopithecus africanus-afarensis* (sensu lato) is the immediate ancestor of *Homo,* yet the human-ape *africanus* with a mean brain size of 450 ml filled such an entirely different niche from *sapiens* with a brain of 800 to 1500 ml that generic separation would seem to be abundantly justified. The more different the adaptive zone occupied by two taxa, the more pronounced the gap between them.

Grade Characteristics

Huxley (1958) proposed that groups of animals "similar in general level of organization as distinct from groups of common genetic origin" (Simpson 1961:125) be called *grades*. As Simpson (1961:126) confirmed, "It is a frequent phenomenon in evolution that whole groups of animals, with numerous separate lineages, tend to progress with considerable parallelism through a sequence of adaptive zones or of increasingly effective organizational levels."

In general the word *grade* designates a shared morphological level and is most often used to characterize the morphology of a group of species that occupy the same adaptive zone. However, like the word *cline,* the word *grade* does not necessarily designate a taxon. The word *woodpecker* in its ecological sense designates both an adaptive grade and a well-circumscribed taxon (Picidae). The word *tree creeper* also designates an adaptive grade, but this grade has been reached by several unrelated groups of birds (Certhiidae, Climacteriidae, Dendrocolaptidae, and several other more isolated avian genera) and therefore does not represent a taxon. The latter situation, in which several unrelated taxa have reached the same grade, has been designated by Simpson (1961:128) as *grade by convergence*. Because such grades, as Simpson rightly implied, are polyphyletic, they are irrelevant for reconstructions of genealogy. They are classes, as Ghiselin (1980) stated correctly, not historical groups (taxa), and "the concept of a grade has little meaning, unless it is applied to related animals only" (Simpson 1961:128).

Evolution from the origin of life to the more advanced animals progressed in stages. The prokaryotes share a common Bauplan characterized by the absence of a nucleus and (symbiontic) cellular organelles. In spite of the enormous diversity of prokaryotes (various kinds of archaebacteria, "eocytes," eubacteria, etc.), they represent a definite evolutionary grade, which, in spite of its heterogeneity, it is convenient to recognize under a common name (Monera). After the origin of the eukaryotes some 1.5 billion years ago, a vast diversity of one-celled organisms originated, which subsequently gave rise to various lineages of fungi, plants, and animals (Corliss 1981). Not only are these lineages diverse, many of them combine characteristics of animals and plants. Indeed, in the aggregate they are almost as different from each other as they are from fungi, plants, and animals together. However, they have one thing in common: They are one-celled. Traditionally they have been recognized as the kingdom Protista. No one questions that different protists gave rise to the different multicellular ex-groups (metazoans, metaphytes, fungi) that arose from the protists, but in many respects these stem group protists are closer to sister groups among the protists than they are to the derived multicellular crown groups. We cannot see that anything is gained by breaking up traditional taxa by arbitrarily adopting the principle of holophyly as the basis of a reclassification.

Even though grades are not taxa, one can distinguish among taxa some with and some without gradelike characteristics. The class Aves clearly has gradelike characteristics. A classification based exclusively on branching pattern cannot make use of the gradelike characteristics of certain taxa. An evolutionary classification may attempt to do so by means of appropriate ranking.

Size of Taxon

The number of species included in a taxon constitutes its size. Because a classification functions like the key to a filing system, its subdivisions—the taxa—should meet the need for greatest possible efficiency; that is, they should ideally all be of approximately equal size to facilitate information retrieval. As a taxonomist once said, all one can say about the genus is that it should be neither too large nor too small.

The fact that the two processes of phylogeny—branching and divergence—are thought to be relatively independent of each other is responsible for the fact that an ideal equality in the size of taxa has not been achieved. Continuous speciation (branching without visible divergence) results in the production of large taxa, while divergence without branching leads to the evolution of monotypic taxa or, in an extreme case, to a series of superimposed monotypic higher taxa. An African species, the aardvark (*Orycteropus afer*), is the sole representative of a monotypic genus, family, and order (Tubulidentata).

The independence of branching and divergence results in a frequency distribution of genera and species usually referred to as a *hollow curve*

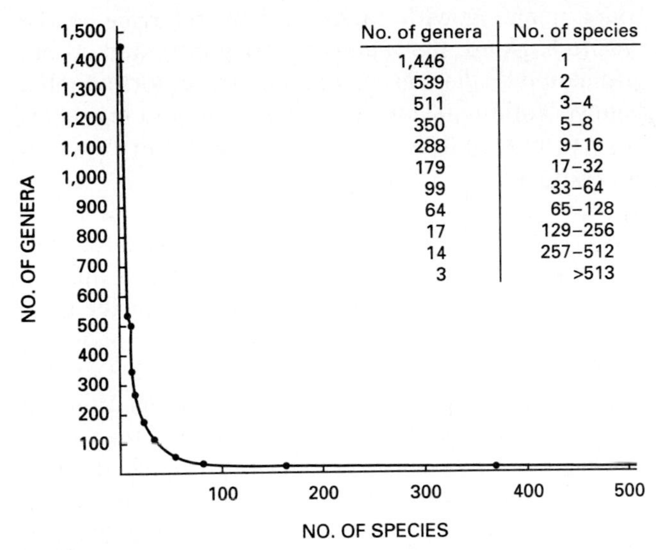

No. of genera	No. of species
1,446	1
539	2
511	3–4
350	5–8
288	9–16
179	17–32
99	33–64
64	65–128
17	129–256
14	257–512
3	>513

FIGURE 10-9
Example of a hollow curve: The number of species in 3510 genera of weevils (Curculionidae). (*After Kissinger 1963.*)

(Figure 10-9). Clearly, the processes of evolution often thwart the systematist who tries to follow the recommendation that taxa should be, so far as possible, of approximately equal size. Still, excessively large genera and an excessive number of monotypic genera reduce the usefulness of a classification in terms of information retrieval. This is why it is recommended that the size of the gap justifying the separation of a higher taxon should be inversely correlated to the size of that taxon.

Most higher taxa of organisms contain some 20 to 30 percent monotypic genera. The okapi, the ostrich, the giant panda, and many other well-known animals are monotypic and often are the only representatives of an entire family or order.

By contrast, several well-defined natural genera contain over 1000 or even over 2000 species. Among well-known large taxa are the North American freshwater fish genus *Notropis,* with about 120 species, several genera of weevils (Curculionidae) and the ant genus *Pheidole* with more than 1000 species each, and the genus *Drosophila* with more than 1500 species.

It is not advisable to break up well-defined large groups such as the genus *Drosophila*. Here it is more useful to introduce subdivisions such as the subgenus. The liberal use of the informal category *species group* is an even more useful device as an aid to memory. *Drosophila* specialists have no difficulty communicating with each other by referring to the *repleta* group, the *willistoni* group, the *melanogaster* group, and so on. Many subgeneric names could be dispensed with if more authors adopted the use of species groups. Nothing is gained by chipping off a few aberrant species from a very large group and making them the types of monotypic genera or subgenera.

Splitters and Lumpers Except in the case of dense, well-isolated species clusters, there is unfortunately no reliable method for determining unequivocably the proper size of a taxon. This is the reason for the eternal argument between *splitters* and *lumpers,* two types of taxonomists found among specialists in almost every group of organisms. Splitters try to express every shade of difference and every degree of relationship through the formal recognition (naming) of separate taxa and their elaborate categorical ranking. Lumpers try to produce a classification in which the emphasis is placed on relationship and which avoids burdening the memory with too fine a division of taxa. We believe in broad taxa and therefore side with the lumpers.

Splitting is particularly deleterious at the level of the genus. The generic name is part of the scientific name of an organism and indicates affinity more than does the name of any of the other higher categories. It is of particular importance for the work of the general zoologist that genera be broadly conceived. Specialists must remember that they are not writing only for a few cospecialists. For a more detailed treatment of the problem of splitting and lumping, see Mayr 1969:238–241.

Equivalence

Categorical rank in related taxa should be equivalent. It is incumbent upon the systematist dealing with the ranking of a taxon at one categorical level to consider implications for all the related taxa at that level and for the categories immediately above and below it. The degree of distinction of a species group expressed by ranking it as a genus should ideally be the same in mammals, insects, mollusks, and every other group of animals. Unfortunately, for historical reasons, this ideal has not yet been reached. The orders of birds, for instance, are probably no more distinct than are the families of insects. Recent molecular studies have revealed that the problem is aggravated by mosaic evolution. Species of *Rana* (frogs) and *Bufo* (toads) are morphologically very similar to each other but show molecular differences that would indicate generic or even higher rank among mammals or birds.

Old taxa, that is, taxa which had their major period of diversification rather early in the history of life, have had time to evolve extremely large differences in their molecular sequences. If one adopted for the ranking of taxa the yardstick used for the classification of vertebrates, one might have to divide the protists into a large number of phyla and even superphyla. Indeed, one might be tempted to abolish the Protista altogether and break them up into the rather distantly related groups of which they are composed. It seems to us that this would not be a very constructive solution. The protists, one-celled eukaryotes, represent a characteristic grade in the evolution of organic diversity, and retaining them as a kingdom certainly facilitates information storage and retrieval far better than disintegrating them would.

What is altogether inadmissible, however, is the raising of a single taxon, say, a family, to the rank of order and the concomitant raising of all the subdivisions within this taxon without regard to the consequences for other families in this taxonomic group. In taxa that occur on several continents, one may find that specialists on different continents have very different criteria for recognizing taxa and for categorical ranking. Such a lack of balance can be very misleading for the general zoologist. In short, the recognition of taxa and their ranking must be undertaken in a manner that does not upset the equivalence of categories in related taxa.

Stability

The usefulness of a classification as a communication system stands in direct relation to its stability, which is therefore one of the basic prerequisites of such a system (Chapter 6). The names for the higher taxa serve as convenient labels for the purpose of information retrieval. Terms such as Coleoptera and Papilionidae must mean the same thing to all zoologists in order to have maximum usefulness. During revisionary work on higher taxa a determined effort must be made to disturb the stability of the currently prevailing classification and nomenclature as little as possible and to maintain, if not improve, its information retrieval qualities. Widely accepted classifications should never be abandoned except for very good cause, because any change of classification impedes information retrieval. The principles of evolutionary classification are far more compatible with stability than are those of cladistics (Chapter 9).

CHAPTER **11**

NUMERICAL METHODS OF PHYLOGENY INFERENCE

Traditionally authors followed a rather arbitrary procedure to infer a phylogeny. An author simply decided that owing to the possession of certain character states, taxon a was closest to taxon b. The cheetah, the author would say, is of all cats the one closest to the dogs because of its long legs and its habit of hunting by running instead of by stealth. Alternatively, the author might say that the three forest anthropoids—orangutan, gorilla, and chimpanzee—are most like one another and as a group are decidedly different from the hominids and the gibbons. It was not very difficult to assemble a number of character states with which to back up such judgments. This subjective approach, however, too often led to glaringly wrong phylogenies. Even before Darwin, certain farsighted systematists searched for more objective and reliable methods.

To find such methods has been the aim of numerical phylogenetics. Many numerical methods have been proposed in the last 30 years, but no consensus has been reached as to what methods are the best for inferring phylogeny (Farris 1983; Felsenstein 1982, 1988*b*; Sober 1988). Disagreement occurs at two levels: philosophical and practical.

The philosophy of the systematist may dictate a preference for certain methods. Should we choose the simplest hypothesis that explains the data at hand and thus use parsimony as our guide, or should we choose the hypothesis which confers the highest probability of observing the data under a particular model of evolution? Which model is appropriate? Should we weight evidence provided by different traits of the organisms?

No matter what answers are given to these questions, appropriate numerical methods can be applied, and computer programs are available for nearly all of them. It is becoming increasingly probable that different numerical methods and computer programs should be used for different tasks. There are indications that the relationship of well-delimited higher taxa (with little homoplasy but much autapomorphy) should be analyzed by programs different from those used to establish the relationship of species in large genera where parallel homoplasy may overwhelm the limited development of restricted apomorphies. Sokal (1985*b*) has asked questions about such differential suitability of algorithms, although his findings seem somewhat counterintuitive.

Historically, numerical methods were classified into phenetic and cladistic ones. However, since some so-called phenetic methods can be converted into cladistic ones and vice versa, it is more informative to distinguish between *distance* methods, in which the similarity or distance of taxa from each other is determined, and *character data* methods, in which dendrograms are constructed that consist of taxa that share derived characters. Some molecular methods produce only distance data. More prominent examples come from DNA-DNA hybridization and microcomplement fixation. One can also convert character data into distance data by calculating measures of distance between taxa on the basis of the character data. Character and distance data must be analyzed differently. Character-based methods, which begin with a taxon-by-character matrix of character states, include parsimony, compatibility, and maximum-likelihood methods. Distance methods begin with a taxon-by-taxon matrix and include methods originally developed by pheneticists for classificatory purposes as well as newer methods specifically developed for phylogeny inference, e.g., the Wagner distance method (Farris 1972).

All schools of numerical phylogeny inference have as a goal the construction of a treelike diagram, or dendrogram. Two sorts of dendrograms are of interest to us. A *cladogram* endeavors to represent the branching pattern of the phylogeny. A *phylogram* not only reflects the branching pattern of the phylogeny but also attempts to portray the amount of anagenetic divergence among the taxa by the length of the branches and the angles between them. Inference of these branch lengths is generally a step in the construction of cladograms. This chapter explains how such dendrograms are constructed.

These dendrograms are a form of graph. The term *graph* as used in numerical taxonomy indicates a series of points, which represent the taxa under study, connected by lines. Dendrograms contain no loops and are thus minimally connected graphs; they are known in graph theory as trees. Trees in the sense of graph theory may be either undirected or directed. Figure 9-5 shows a directed tree "rooted" toward its ancestor. Any starting point could have been chosen. However, if the starting point is correctly chosen, directed trees approximate evolutionary pathways. Methods of calculating directed trees are an important component of some techniques of computer cladistic analysis. For a discussion of graph theory, see Penny (1982).

The computer is employed whenever it can shorten the labor of computation. Computer taxonomy is an exceedingly active field. More than a score of textbooks, symposium volumes, and readers have been published in the last 25 years, and not only *Systematic Zoology* but also *Systematic Botany, Taxon, Cladistics,* and other periodicals have published hundreds of papers on the theory underlying computer methods. Finally, computer program packages, some inexpensive and able to run on a desktop personal computer, are available to those who want to analyze their taxonomic data (see the Appendix at the end of this chapter). As these programs vary in their hardware requirements, it is wise to know something about the desired programs before investing in a computer. Because new programs are often released and old ones are periodically upgraded, it is impossible to discuss them here in detail. The specialist literature or colleagues must be consulted for further details and to discover which programs are at the moment considered the best and where they can be obtained.

In this chapter several numerical methods for inferring cladograms and phylograms are described, starting in each case with a method suitable for hand analysis of small groups of taxa. This is followed by a sample of numerical methods that are possible but tedious without a computer. Reading and comprehending the description of a procedure can be taxing unless the procedure is actually tried (it is like reading a computer manual without a computer on which to work things out). It is helpful, then, to obtain a small data set and try out one or two of the methods.

As discussed more fully in Chapter 7, the word *character* has been used in numerical taxonomy in two different ways, often in the same paper. In reference to attributes in a description or a key, this term is correctly used for a distinguishing feature, for example, blue eyes rather than brown ones. However, in numerical taxonomic analysis, the term is incorrectly used in a collective sense (e.g., eye color), and such "characters" are divided into character states. To avoid this confusion, Ashlock proposed the word *signifer* for the collective, i.e., the feature that varies (eye color), and the term *character* or *signifer state* for the blue or brown color of the eye. We realize that the character–character state terminology has been too widely adopted to be easily dislodged. Therefore, any endeavor to restore the traditional meaning of the word *character* would cause considerable confusion. Hence, although with considerable reluctance, we use *character state* for what traditionally has been called a character.

THE DATA SET

The same data are used for all the examples in this chapter so that the various methods can be compared. The data are based on a set of eight imaginary insects, A through H, shown in Figure 11-1. Figure 11-2 illustrates a closely related form and shows the location of the various characters.

Table 11-1, which defines specific character states, lists characters by mnemonic names that are used to recognize them after they have been entered on a cladogram. The defined character states of each character

FIGURE 11-1
Imaginary insects A through H. Characters from these insects serve as demonstration data in this chapter. Note that the bodies are elongated and the heads are exserted compared with those of the closely related insect shown in Figure 11-2.

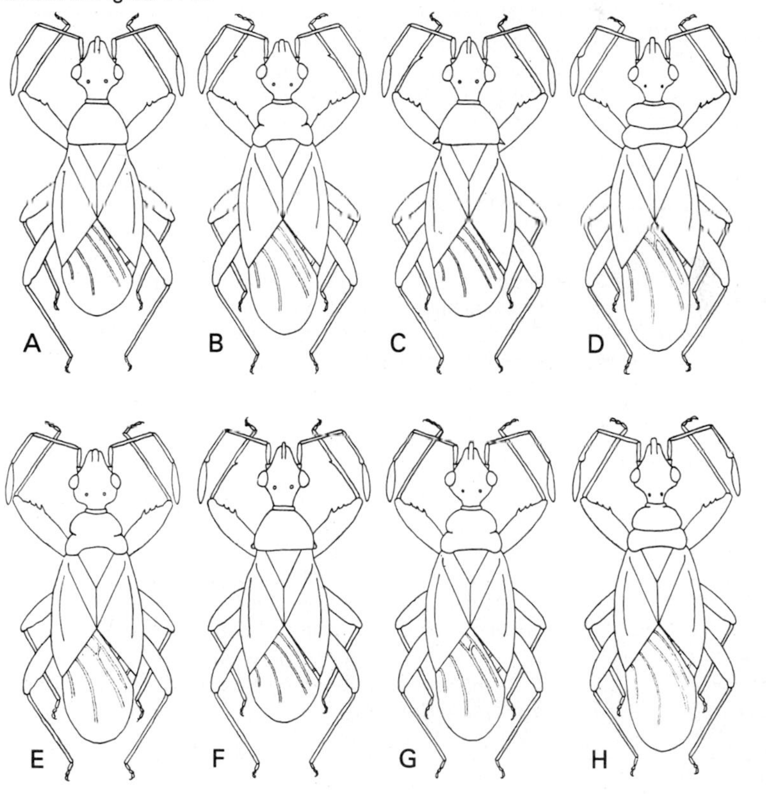

TABLE 11-1
CHARACTER MNEMONICS, CODES, AND STATE DEFINITIONS

Character			
No.	**Mnemonic**	**Code**	**State definition**
1	PnLb	0	Pronotum medially without a transverse groove
		1	Pronotum medially with an incomplete transverse groove producing lobes
		2	Pronotum medially with a complete transverse groove producing lobes
		3	Pronotum with anterior and posterior lobes greatly swollen
2	CsMg	0	Costal wing margin curved in at base
		1	Costal wing margin curved out at base
3	NkSw	0	Neck curved in behind eyes
		1	Neck curved out behind eyes for about length of an eye
		2	Neck curved out behind eyes for more than length of eye
		3	Neck quadrately swollen behind eye
4	ClyL	0	Clypeus projecting beyond rest of head
		1	End of clypeus even with rest of head
5	ColC	0	Collar incomplete
		1	Collar complete
6	PrEx	0	Pronotal expansion absent from posterior angle
		1	Pronotal expansion rounded apically
		2	Pronotal expansion acute apically
7	CrVn	0	Cross vein absent from wing membrane
		1	Cross vein present in wing membrane
8	MmbL	0	Wing membrane shorter than head and pronotum together
		1	Wing membrane longer than head and pronotum together
		2	Wing membrane longer than basal part (corium) of wing
9	PnHM	0	Pronotal hind margin straight
		1	Pronotal hind margin lightly curved
		2	Pronotal hind margin deeply curved
10	OcP	0	Ocelli on an imaginary line drawn between posterior eye margins
		1	Ocelli less than one eye length behind imaginary line on posterior eye margins
		2	Ocelli one eye length behind imaginary line on posterior eye margins
11	A3Sw	0	Antennal segment 3 not swollen
		1	Antennal segment 3 swollen apically
12	aFSp	0	Forefemur with two spines apical to large spine
		1	Forefemur with one spine apical to large spine
13	bFSp	0	Spine near base of forefemur absent
		1	Spine near base of forefemur present
14	TbSp	0	Foretibia without a spine
		1	Foretibia with a spine

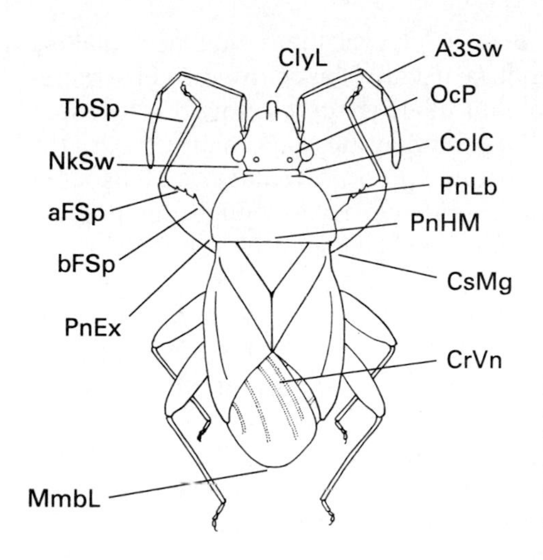

FIGURE 11-2
Character locations for imaginary insects. This insect serves as an out-group for cladistic analysis of the demonstration insects. Character codes are defined in Table 11-1.

are coded; that is, they are each assigned a number from 0 to 3. Table 11-2, the data matrix, lists the character and state codes for each of the eight imaginary insects.

Like any systematic analysis, numerical analysis begins with a search for differentiating character states. Organizing the character states into a data matrix requires a much more rigorous description of each character and character state than is usually found in the list of character states that is customary in traditional descriptions of taxa. Such lists all too of

TABLE 11-2
DATA MATRIX

		A	B	C	D	E	F	G	H
1	PnLb	0	1	0	3	1	0	1	2
2	CsMg	0	1	0	1	1	0	1	1
3	NkSw	1	2	0	1	3	0	2	1
4	ClyL	0	1	0	0	1	0	0	0
5	ColC	1	1	1	1	1	1	1	0
6	PnEx	0	0	2	0	0	1	0	0
7	CrVn	0	0	0	1	1	0	1	0
8	MmbL	0	1	0	2	1	0	1	2
9	PnHm	0	2	0	1	2	0	1	1
10	OcP	0	1	0	1	1	0	1	2
11	A3Sw	0	0	0	1	0	1	0	1
12	aFSp	1	0	0	0	0	0	0	0
13	bFSp	0	1	0	0	0	0	0	0
14	TbSp	0	1	1	0	0	1	0	0

ten skip some features of some taxa or change their sequence, making comparison difficult. The data matrix usually has a row of cells representing each character and a column of cells representing each taxon; ideally, every cell must be filled. Assembling the data matrix forces the systematist to provide completely parallel descriptive data for all taxa under study, which is good systematic practice. Even though this process was not invented by pheneticists, their insistence on its routine use has been one of their major contributions.

Skill in delimiting characters and their component character states contributes much to a systematic analysis. The principles of such delimitation are exactly the same in every analysis and are briefly discussed below.

The simplest characters to code are those with two character states: present versus absent, red versus white, and so on. These characters may be coded 0 and 1; other symbols such as − and + may be used if the computer program permits. However, characters often have more than two component character states. A structure may be in front of, next to, or behind another structure, or an area may have four, five, six, or seven teeth, spines, spots, hairs, or fin rays. Such integral variables as actual measurements are often directly listed as character states in a data matrix, but measurement ranges may also be used. A structure that ranges in length from 10 to 18 mm might be coded as three character states: 10–12 mm (coded 0), 13–15 mm (coded 1), and 16–18 mm (coded 2). Such a character could be correctly coded as two character states only if the data were bimodal. If the mean and mode are both near 14 mm, more character states are needed.

When the code value 2 is given to a particular character state, adjacent code values (3 and 1) should be given to the character states that are most similar to character state 2. This is easy when a character falls into a linear array, but some characters do not. For example, a group of regular polygons from triangle to hexagon can be coded triangle 0, square 1, pentagon 2, hexagon 3, but a rectangle added to this series cannot logically be put at either end or between any two members. It is clearly most similar to the square, but the triangle and the pentagon, with three and five sides, respectively, must be placed on either side of the square. One way to handle such situations is to use a branched array. The rectangle may be coded 2′, indicating a branch from the square, 1. A longer rectangle then would be coded 3′, while a rhomboid would be coded 2″, indicating still another branch from the square. Some computer programs allow one to specify that the character evolves in a branched array.

The use of hand methods and some computer programs is simplified when branched characters are recast into linear ones (Sokal and Sneath 1963; Farris, Kluge, and Eckardt 1970). This recasting sometimes requires a bit of thought. In the polygon example, the original set of regular polygons formed a linear series in which the character was the number of sides. The addition of the rectangle added a new element, length of side. Separating out the new element creates two linear series, or two characters, the first concerned with the number of sides and the second concerned with the relative length of sides. Thus, reducing the complexity in a character can be done by a judicious increase in the number of characters employed. The number-of-sides character can be coded as before, and the second character—relative length of sides—can have all regular polygons coded as 0 (sides equal) and the rectangle coded as 1 (sides unequal). Recoding of branched characters into linear ones is becoming less relevant, as the latest programs do this automatically.

When the final form of each character has been decided upon, the data matrix may be drawn and the cells must be filled with appropriate character state codes for each character and each taxon. In a hand analysis, it is helpful to place taxa in a preliminary order of similarity before drawing up the matrix.

DISTANCE METHODS

Measures of distance between taxa can arise from the observational techniques used (for example, melting point temperatures in DNA-DNA hybridization studies) or can be obtained by transforming character data into distance data. These data are gathered into a matrix of distances. Finally, relationships among taxa are determined from the distance matrix by means of a process of *cluster analysis*.

Many methods for converting character data to distance data and for the analysis of distance data were developed by pheneticists. These computerized methods (algorithms) were usually not developed with the goal of reconstructing phylogeny but to produce a classification. They are described in the major texts on numerical phenetics (Sneath and Sokal 1973; Rohlf and Sokal 1981). Other descriptions of these methods are given by Jardine and Sibson (1971), Duncan and Baum (1981), Dunn and Everitt (1982), Abbott, Bisby, and Rogers (1985), and Sokal (1986). Among symposium volumes on the subject are those of Heywood and NcNeill (1964), Cole (1969), and Felsenstein (1983*a*). Atchley and Bryant (1975) and Bryant and Atchley (1975) have reprinted many significant journal articles. Numerous analyses have been published in *Systematic Zoology* (particularly from 1962 to 1970), *Taxon*, and *Systematic Botany*.

In recent years, distance methods have been developed specifically with phylogeny inference in mind; they are described below under "The Wagner Distance Method" and "Other Distance Methods."

Distance or Similarity Matrix

If the distance measure is obtained by transforming character data, as it is in our example, such a measure is calculated from the greatest number of available characters to avoid subjectivity. The processing of 50 to 100 or more characters is a time-consuming activity. Part of it, the preparation of a data matrix, cannot be avoided; however, as shown below, other steps in the procedure, such as the calculation of similarity coefficients and the clustering of taxa, can be greatly facilitated by computer programs.

Distance measures can be calculated by comparing character states of several taxa to determine the overall similarity or difference between each possible pair of taxa. The taxa to be classified are thoroughly searched for differences, and the differences are coded and assembled into a data matrix (Table 11-2). Then a coefficient of similarity (any of various distance or correlation measures) between each possible pair of taxa is calculated by comparing all character states of each taxon with those of every other taxon (or OTU). The coefficients are assembled into a *similarity matrix* (Table 11-3).

The point has been made that the matrix discussed here could equally well be called a dissimilarity or distance matrix. Both the columns and the rows in a similarity matrix are for taxa. The cells contain a measure of similarity or distance between each possible pair of taxa, exactly like the tables on many road maps that give the distances between cities. A matrix for n number of taxa requires calculation or acquisition of $(n^2 - n)/2$ similarity measures. A matrix for 10 taxa, then, requires 45 calculations, and a matrix for the 8 taxa of the data set requires 28 calculations.

Manhattan distance is the simplest similarity coefficient to calculate and, fortunately, one of the most useful. It received its name because the

TABLE 11-3
SIMILARITY MATRIX: MANHATTAN DISTANCE

	Taxa							
	A	**B**	**C**	**D**	**E**	**F**	**G**	**H**
A	X	11	5	11	11	5	8	11
B	11	X	12	10	4	12	5	10
C	5	12	X	14	14	2	11	14
D	11	10	14	X	8	12	5	4
E	11	4	14	8	X	14	3	10
F	5	12	2	12	14	X	11	12
G	8	5	11	5	3	11	X	7
H	11	10	14	4	10	12	7	X

distance corresponds to the distance around corners of city blocks. It is calculated by finding the sum of the absolute differences between the character states of each character for each possible pair of taxa. For example, in Table 11-2 the Manhattan distance between taxa A and B is 11, between C and F it is 2, and between D and H it is 4. These values are assembled into a taxon-by-taxon similarity matrix (Table 11-3). Note that the table is symmetrical on both sides of the diagonal row of X's, and one calculates values for one side only. Hand analysis is easier, however, if the values are filled out on both sides.

Cluster Analysis

For hand cluster analysis, taxa in the similarity matrix must be listed in one's best approximation of the final order of similarity. Rearranging Table 11-3, in which taxa are not so ordered, requires one to place taxa with the lowest similarity coefficients close together. Study of Table 11-4, which is Table 11-3 rearranged by similarity, shows C and F together (coefficient 2), D and H together (coefficient 4), and G and E together (coefficient 3) with B close to them (coefficients 4 and 5). This sorting is rough and will be refined as the analysis proceeds. Note that on a rearranged matrix, the lowest distance coefficients fall adjacent to the diagonal row of X's.

The cluster analysis method explained here is called the *unweighted pair-group method using arithmetic averages* (*UPGMA*) and is a commonly used distance method. This cross-averaging method is also known as average linkage.

The first step in hand cluster analysis is to find primary clusters, i.e., those which consist of only two taxa (Figure 11-3). Since distance mea-

TABLE 11-4
SIMILARITY MATRIX: MANHATTAN DISTANCE REARRANGED BY SIMILARITY

	Taxa							
	C	F	A	G	E	B	D	H
C	X	2	5	11	14	12	14	14
F	2	X	5	11	14	12	12	12
A	5	5	X	8	11	11	11	11
G	11	11	8	X	3	5	5	7
E	14	14	11	3	X	4	8	10
B	12	12	11	5	4	X	10	10
D	14	12	11	5	8	10	X	4
H	14	12	11	7	10	10	4	X

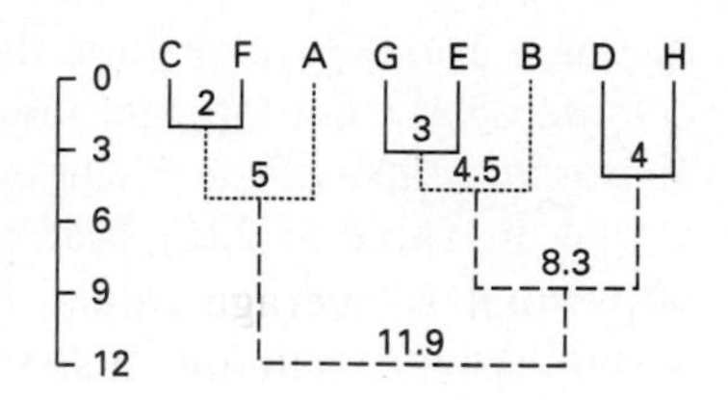

FIGURE 11-3
Phenogram. Manhattan distance of the demonstration insects. Primary clusters are indicated by solid lines, secondary clusters by dotted lines, and tertiary clusters by dashed lines. The scale on the left is a distance measure.

sures signify greater similarity with smaller numbers (i.e., shorter distances), the first primary cluster is found by locating the lowest number in the rearranged matrix (Table 11-4): C and F form a cluster at value 2, and G and E form a second primary cluster at value 3. Although taxa B and F show a difference of 4, F is already a member of a primary cluster. D and H, however, can and do form a third primary cluster at value 4.

Note that it is entirely possible for a given primary cluster to group at a distance far greater than that which separates members of other primary clusters or even that which separates some secondary or tertiary clusters. In any case, a primary cluster always consists of two taxa that are closer to one another than either is to any other taxon.

Secondary clusters contain more than two taxa but only one primary cluster. Although a given study may turn out to contain no secondary clusters, if they do exist, they must be delimited before the search for tertiary clusters begins.

Starting with a primary cluster such as C–F, one scans the columns or rows for C and F for an unclustered taxon with the lowest average distance value to C and F; here, that is 5 and 5 for taxon A. If A does not yet belong to another cluster, indicated by another value of 5 or less in the A column, it joins the cluster C–F to form a secondary cluster at an averaged value of 5 [$(5 + 5)/2 = 5$; Figure 11-3]. A comparison of pairs of values in the columns of the primary cluster G–E reveals that unclustered taxon B is closest with values 5 and 4. These two values are now averaged, and taxon B joins G–E to form a secondary cluster at the level 4.5. Although taxon G is most similar to the last remaining primary cluster, D–H, with values of 5 and 7, taxon G is already a member of a secondary cluster, and one must conclude that the primary cluster D–H is not part of a secondary cluster. While many taxa with sequentially higher distances may join a secondary cluster, no secondary cluster can contain more than one primary cluster.

Tertiary clusters are clusters of clusters. Locating all these tertiary clusters completes the analysis for this example. Each taxon in the example is now in one of three independent clusters, one primary and two secondary. Independent clusters are joined by cross-averaging values between each possible pair of clusters that are to be joined. Here there are three possibilities. The sure way to establish which of the three clusters in the sample will join at the lowest average value is to extract smaller matrices that cover each possibility from the original similarity matrix (Tables 11-3 and 11-4). In this case, pairs BEG and DH cluster at 8.33, the smallest average value of the three possible pairs (Table 11-5).

The two remaining clusters, ACF and GEBDH, are joined in the final phenogram by the same cross-averaging process. That is, the similarity coefficients of A with each member of group GEBDH are summed, the same sum is figured for C and F, and the average of the 15 similarity coefficients is calculated. The resulting average is 11.9, the level at which the last separate clusters are joined (Figure 11-3).

Note that the number of values to be averaged at any stage of clustering is the product of the numbers of taxa in the two groups that are to be clustered. By the time the final phenogram is complete, every number in the distance matrix has been used.

Occasionally a perplexing problem is presented by two taxa or clusters that are equally similar to a third but differ from each other by a greater value.

	P	Q	R
P	×	3	4
Q		×	3
R			×

$10/3 = 3.33$

To solve this problem, one departs from the usual dichotomous branching and makes a trifurcation, averaging all three values.

OVERVIEW OF DISTANCE METHODS

During the heyday of phenetics, numerous methods were developed and applied to actual problems of systematics. Other distance methods have been developed since then, especially for molecular data. The following

TABLE 11-5
MINIMATRICES: DETERMINING TERTIARY CLUSTERS

	A	C	F		A	C	F		B	E	G
B	11	12	12	D	11	14	12	D	10	8	5
E	11	14	14	H	11	14	12	H	10	10	7
G	8	11	11								
	104/9 = 11.56				74/6 = 12.33				50/6 = 8.33		

section is intended as a short introduction to some of these methods. Others have virtually disappeared from the literature, but ample descriptions or references can be found in Jardine and Sibson (1971), Sneath and Sokal (1973), Dunn and Everitt (1982), and Felsenstein (1988*b*).

Calculation of Similarity Coefficients from Character Data

Similarity coefficients are used in an attempt to measure the overall similarity between pairs of individual taxa or OTUs. They form several families, especially distance and correlation coefficients. However, only the distance coefficients have been widely employed; they measure dissimilarity but are complements of similarity coefficients. Association coefficients have been used primarily with binary (two-state) data; they have a long history (Sokal and Sneath 1963).

Euclidean Distance Also known as taxonomic distance, this coefficient measures the distance between two taxa in multidimensional space and is widely employed in phenetic studies. It is invariant when rotated as in ordination and is easily visualized, but it is inappropriate if one wants to measure evolutionary distance between taxa (although it has often been so interpreted). Figure 11-4 shows how the expression for Euclidean distance is derived. Connecting points j and k with a line creates a right triangle, to which the Pythagorean theorem, $c^2 = a^2 + b^2$, may be applied to find the distance c between j and k. The value for side a is the difference between the two character values on the X scale, and the value for side b can be figured in the same way on the Y scale. The Pythagorean formula for this purpose may be rewritten $c^2 = (Xj - Xk)^2 + (Yj - Yk)^2$. In this example, $c^2 = (9 - 4)^2 + (3 - 10)^2$, $c^2 = 25 + (-7)^2$, $c^2 = 74$, and $c = 8.6$.

Adding a third character, Z, requires a three-dimensional space and the addition of a Z axis to the taxonomic space. The formula then becomes

$$c^2 = (Xj - Xk)^2 + (Yj - Yk)^2 + (Zj - Zk)^2$$

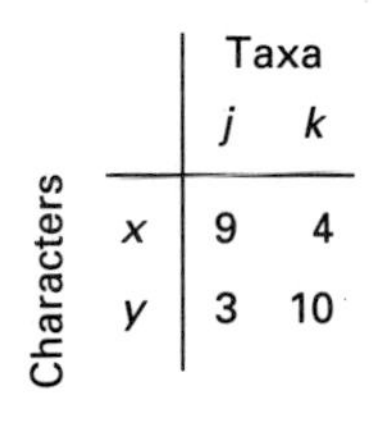

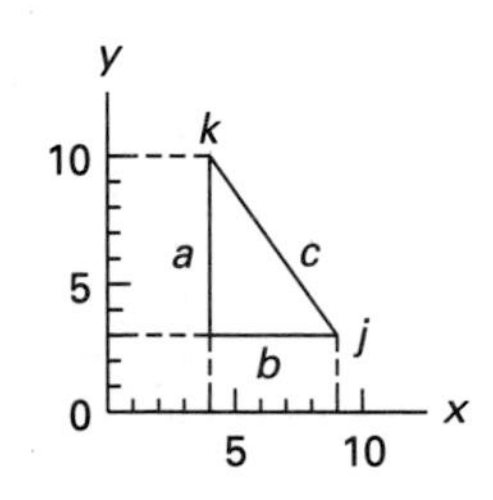

FIGURE 11-4
Derivation of a Euclidean distance coefficient.

For each character added, another dimension, represented by a term equivalent to $(Xj - Xk)$, must be included. The number of dimensions in a study is equal to the number of characters used.

The formula for taxonomic distance between two taxa j and k (or jk) for n characters is expressed as follows:

$$\Delta jk = \sqrt{\sum_{i=1}^{n} (x_{ij} - x_{ik})^2}$$

This formula is a direct derivation of the Pythagorean theorem modified to determine the distance between two points in hyperspace of an unlimited number of dimensions. The squared expression within parentheses is the value of each character dimension (i refers to each character). The sigma or summation symbol (Σ) instructs one to add up all the squared differences, the $i = 1$ under the summation symbol means to begin with the first character, and the n above the summation symbol means that one must continue through the last character. The entire expression tells one to take the square root of the sum of all the squared differences between each pair of character states for taxa j and k. No matter how complex or large a study may be, similarity coefficients are calculated by pairs of taxa; this expression is all that is needed for taxonomic distance similarity coefficients.

The formula for Euclidean distance may be simplified so that for all the characters of each pair of taxa j and k, the expression becomes

$$\sqrt{\sum_{i=1}^{n} (j - k)^2}$$

A related coefficient, Euclidean distance squared, expressed as

$$\sum_{i=1}^{n} (j - k)^2$$

has also been widely used. These coefficients are compared in Figure 11-5. The differences between all these distance measures (including Manhattan distance) are due to the use or nonuse of roots and squares.

Manhattan Distance The formula for this distance between each pair of taxa is expressed as follows:

(*a*) Euclidean distance

	A	B	C	D	E	F	G	H
A	X	3.6	2.6	4.4	3.9	2.2	2.8	4.1
B	3.6	X	4.2	3.5	2.0	4.0	2.2	3.2
C	2.6	4.2	X	4.9	4.9	1.4	3.9	4.7
D	4.4	3.5	4.9	X	3.5	4.5	2.6	2.0
E	3.9	2.0	4.9	3.5	X	4.7	1.7	3.5
F	2.2	4.0	1.4	4.5	4.7	X	3.6	4.2
G	2.8	2.2	3.9	2.6	1.7	3.6	X	2.6
H	4.1	3.2	4.7	2.0	3.5	4.2	2.6	X

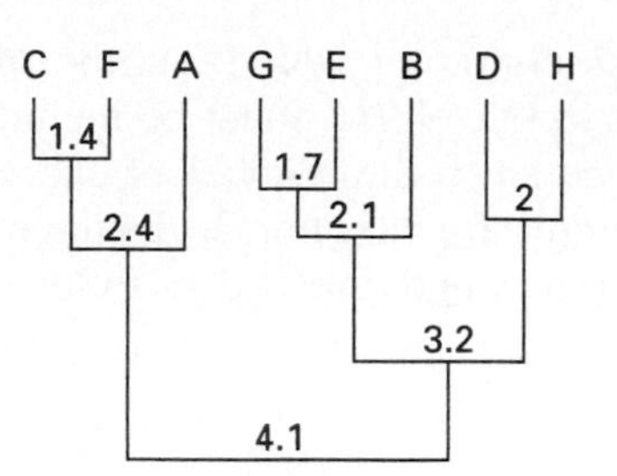

(*b*) Euclidean distance ^{2}

	A	B	C	D	E	F	G	H
A	X	13.0	7.0	19.0	15.0	5.0	8.0	17.0
B	13.0	X	18.0	12.0	4.0	16.0	5.0	10.0
C	7.0	18.0	X	24.0	24.0	2.0	15.0	22.0
D	19.0	12.0	24.0	X	12.0	20.0	7.0	4.0
E	15.0	4.0	24.0	12.0	X	22.0	3.0	12.0
F	5.0	16.0	2.0	20.0	22.0	X	13.0	18.0
G	8.0	5.0	15.0	7.0	3.0	13.0	X	7.0
H	17.0	10.0	22.0	4.0	12.0	18.0	7.0	X

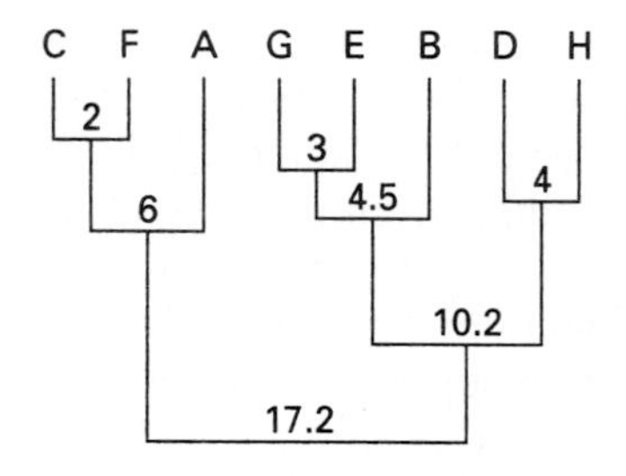

FIGURE 11-5
Euclidean distance (*a*) and Euclidean distance squared (*b*). Phenograms derived from the demonstration data comparing Euclidean distance and Euclidean distance squared. Compare the Manhattan distance phenogram shown in Figure 11-3. Differences in the three phenograms are due to the use versus the nonuse of squares and roots.

$$\sum_{i=1}^{n} |j - k|$$

It is simply the sum of the absolute values of the character state differences; i.e., the differences are all positive numbers. This measure differs from the Euclidean distance in that it neither squares the character state differences nor takes the square root. In fact, both distances are cases of a general class of distance functions, Minkowski metrics. The Manhattan distance is not invariant when rotated, and so it is inappropriate for ordination. However, it is interpretable as the evolutionary distance between taxa, because it measures character state differences as steps. It is therefore the distance used in numerical cladistics.

There have been variations on both measures. The most common has involved an averaging of the distance by the number of character states. This type of scaling has been used because the values of the distances increase with the number of elements used in the computation. Note that while using this procedure, pheneticists were counting characters, not character states, as they thought. Multistate characters can be transformed into two-state characters, increasing the number of characters but not the number of character states, without changing the results of distance measures. Dividing distance coefficients by the number of characters is meaningless. Its original purpose was to facilitate comparison of one phenogram with another, but this is seldom necessary.

Numerous other distance measures have been proposed (Sneath and Sokal 1973), including multivariate distances such as Mahalanobis's generalized distance, and they continue to be developed for specialized problems. However, none have achieved as wide use as the Euclidean and Manhattan metrics, nor are they likely to.

Correlation Coefficient Early in the history of numerical taxonomy, the *correlation coefficient* was widely used as a measure of similarity (e.g., Michener and Sokal 1957; Sokal and Michener 1958*b*). The expression for the correlation coefficient is

$$r = \frac{\sum (j - \bar{j})(k - \bar{k})}{\sqrt{\sum (j - \bar{j})^2 \sum (k - \bar{k})^2}}$$

Correlation values of 1 are supposed to indicate complete similarity, while values approaching 0 indicate greater difference. Negative correlations are also possible and indicate even less similarity. The process of clustering correlation coefficients by UPGMA is the same as that for distance values except that one works from higher to lower numbers.

Note that the expression above requires one to take an average of the codes for the characters of each taxon. Jardine and Sibson (1971) pointed out that such a value is biologically "absurd." While the phenogram using the demonstration data is reasonably similar to the results obtained with the use of distance coefficients (Figure 11-6), a simple reversal in coding of one character, say, PnEx, while it does not change distance analyses in any way, completely changes the correlation phenogram, especially in regard to the placement of taxon A. There is no reason for even the most dedicated pheneticist to continue to use the correlation coefficient.

Matching Coefficient When all characters are limited to two states, the simple matching coefficient can be used. If one compares two taxa by counting the pairs of matching character states and divides this sum by the number of characters, one obtains a matching coefficient (Sokal and Michener 1958*b*; Sneath and Sokal 1973). This coefficient has various weaknesses, and the same information can be obtained by using more efficient methods. The matching coefficient is now rarely used. Two-state (*a*) Correlation 1

	A	B	C	D	E	F	G	H
A	X	0.12	0.04	-0.12	0.30	0.06	0.31	-0.25
B	0.12	X	-0.27	0.23	0.77	-0.35	0.60	0.34
C	0.04	-0.27	X	-0.35	-0.35	0.77	-0.31	-0.44
D	-0.12	0.23	-0.35	X	0.38	-0.27	0.60	0.80
E	0.30	0.77	-0.35	0.38	X	-0.46	0.88	0.37
F	0.06	-0.35	0.77	-0.27	-0.46	X	-0.41	-0.37
G	0.31	0.60	-0.31	0.60	0.88	-0.41	X	0.53
H	-0.25	0.34	-0.44	0.80	0.37	-0.37	0.53	X

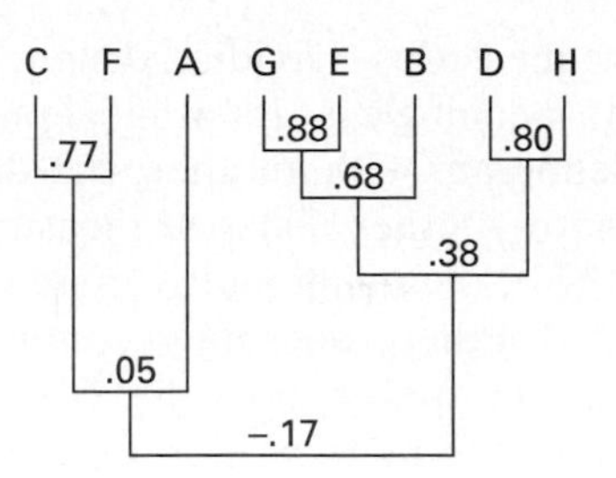

(*b*) Correlation 2

	A	B	C	D	E	F	G	H
A	X	0.36	0.10	0.14	0.42	0.41	0.53	0.10
B	0.36	X	0.00	0.26	0.77	0.00	0.65	0.39
C	0.10	0.00	X	-0.24	-0.24	0.65	-0.17	-0.42
D	0.14	0.26	-0.24	X	0.40	0.00	0.63	0.81
E	0.42	0.77	-0.24	0.40	X	-0.19	0.88	0.41
F	0.41	0.00	0.65	0.00	-0.19	X	-0.03	-0.08
G	0.53	0.65	-0.17	0.63	0.88	-0.03	X	0.58
H	0.10	0.39	-0.42	0.81	0.41	-0.08	0.58	X

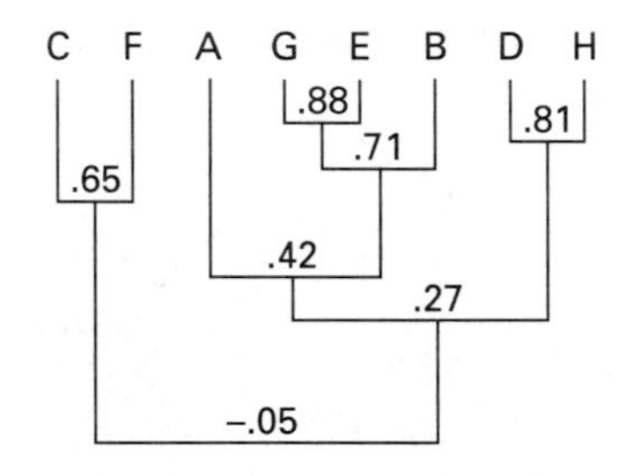

FIGURE 11-6
The correlation similarity coefficient. (*a*) Correlation coefficients and phenogram based on the data in Table 11-2. (*b*) Correlation coefficients and phenogram based on the same data but with the coding of character PnEx reversed (i.e., 0 = 2 and 2 = 0). Note the new placement of taxon A. This change in data has no effect on the similarity matrix or on a phenogram based on any distance coefficient.

data can best be analyzed using Manhattan distance, where unmatched pairs are counted.

Standardization

Because a character with many character states contributes more to a phenetic study than does a character with few and because numerical pheneticists treat all character states (characters) as having equal weight, a method known as character state standardization has been developed. The adjustment consists of taking the mean and standard deviation of the character state codes for each character in a data matrix and recoding them in standard deviation units with a mean of zero. The additional computation aside, standardization exaggerates the magnitude of character states that in fact show slight differences (possibly mere errors in measurement) and minimizes the contribution of highly distinct character states in a complex character. This procedure is no longer recommended.

Inference of Phylogenies from Distance Matrices

Cluster analysis has been the subject of an extensive and complex literature (Jardine and Sibson 1971; Sneath and Sokal 1973). The cross-averaging method UPGMA (average linkage clustering) used in our demonstration was introduced by Sokal and Michener (1958*a, b*) and remains the most commonly used method. Rohlf and Sokal (1981:462) found it to be typical of phenetic studies. In spite of its extensive use, there are serious problems with this method, as there are with any averaging method of cluster analysis.

The computer performs UPGMA analysis of the similarity matrix by recording as a cluster the two taxa with the lowest similarity coefficient in a distance matrix or the highest one in a correlation matrix, averaging the coefficient for the two clustered taxa, and treating the cluster so formed as a single unit in a new matrix that is shorter by a row and narrower by a column than the original. This process is repeated for the remaining most similar pairs of taxa and identified clusters until the matrix is used up and the phenogram is complete. (Rounding off of errors in the machine method may introduce minor discrepancies between it and the more accurate hand methods described earlier.)

Michener (1970) demonstrated that UPGMA cluster analysis can grossly distort relationships. Figure 11-7, drawn from data presented by Michener, is a shortest spanning tree on which a UPGMA cluster analysis is portrayed by contour lines around the ordinated taxa. The problem that concerned Michener is exemplified in the relationship of taxon B. The shortest spanning tree shows that B is connected by similarity to four other taxa but is most similar to taxon C. However, C is far more similar to taxon D, with which it is clustered. Sequentially, D, F, I, and E all join the CD cluster. Because each taxon is given an average value as it is formed, B is forced to cluster with its second most similar taxon, A. The fact that B is more similar to C is completely lost when UPGMA cluster analysis is used. A related weakness of the UPGMA method is illustrated by the question of where taxon G, the most distinct member of the group, might fit. The shortest spanning tree also clearly relates G to B.

The Wagner Distance Method

Farris (1972) developed a cladistic algorithm based on a taxon-by-taxon matrix of phenetic differences. A value is assigned to each comparison between pairs of taxa (OTUs). Such an approach is particularly suitable for data sets such as those derived from immunological comparisons and DNA hybridization. It can be used for a molecular datum that is expressed as a degree of difference between taxa. Its advantage over other

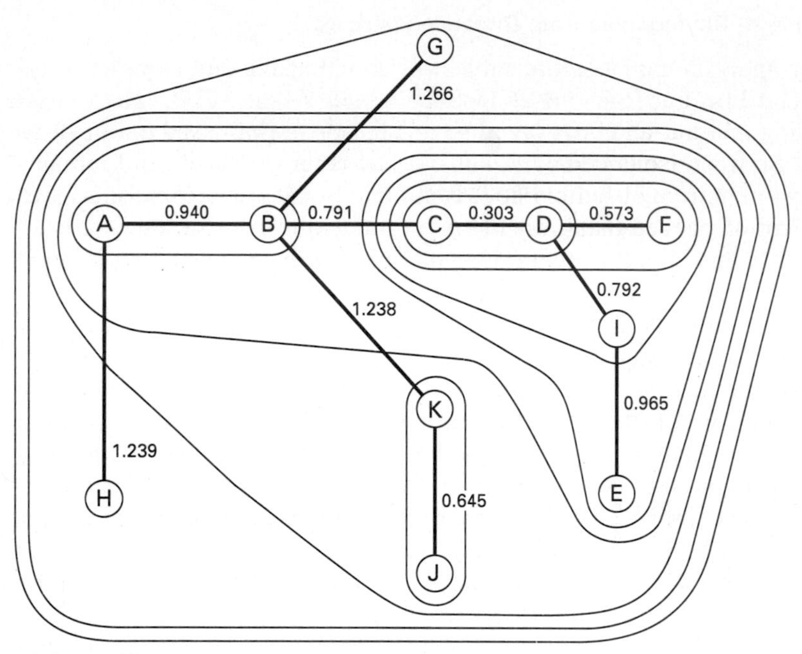

FIGURE 11-7
Shortest spanning tree with UPGMA cluster analysis, expressed as contour lines, of the *Hoplitis producta* group of bees, showing faulty phenetic relationships determined by UPGMA cluster analysis. (*Data from Michener 1970.*)

techniques for processing such data is that it does not make the assumption that the rate of evolutionary change is the same in all lineages. Further modifications of the Wagner distance method were suggested by Swofford (1981) and Farris (1981, 1985).

The trees calculated by the Wagner distance method are undirected trees. Three techniques for the rooting of such trees are available: (1) finding the nearest connection with an out-group, preferably a sister taxon, (2) assuming a uniform rate of change within the holophyletic group, and (3) using a minimization procedure (Farris 1972:658). Our example, using cladistically arranged data, roots trees by employing a hypothetical ancestor that is primitive in all character states. The use of different rooting methods may lead to rather different cladograms (Hillis 1985).

In the original Wagner distance method, synapomorphic as well as plesiomorphic character states were utilized in calculating the distance matrix. To make it a more strictly cladistic method, Farris (1977:835–836) calculates a coefficient of special similarity. He removes all character states that are not synapomorphies [i.e., all ancestral (plesiomorphic) character states] of any included pair of taxa, thus converting the phylogenetic method back into a strictly cladistic one.

Felsenstein (1982:395) discussed certain shortcomings of the Wagner distance method and compared it with other methods that employ pairwise comparisons. He concluded that there is no guarantee that pairwise methods reveal better topologies than other methods do. It must also be remembered that converting a data matrix into distances always entails a loss of information (Steel, Hendy, and Penny 1988).

Other Distance Methods

The construction of phylogenetic trees from molecular data is often based on a matrix of distances. Some types of biochemical comparison, such as microcomplement fixation, yield data directly in the form of distances. Nondistance data—allozyme frequencies and amino acid data—are usually converted to distances for purposes of analysis. The earliest and still most frequently used method was proposed by Fitch and Margoliash (1967), but about a dozen other methods for constructing phylogenetic trees from molecular distance data have since been proposed (Swofford 1981). All these methods attempt to fit a tree to a matrix of pairwise differences (or similarities) between taxa. The objective is to find a tree that comes as close as possible to predicting the distances of the matrix of differences (Fitch 1981).

In the Fitch-Margoliash method, a provisional tree is formed by phenetic clustering but is then modified by a series of trial-and-error rearrangements. In each trial a "length" (it may be negative) is assigned to each branch, and these lengths determine a matrix of tree-derived distances. The derived distance between a pair of taxa is the sum of the lengths of the branches that must be traversed in tracing a path on the tree from one member of the pair to the other. The "best" rearrangement is the one that gives the best fit of the tree-derived matrix to the data distance matrix (Farris 1981). The major shortcoming of the Fitch-Margoliash method is that a conscientious search for the best tree requires a great deal of computer time.

Methods proposed in recent years to avoid the inefficient search for the maximum-parsimony tree among all possible branching patterns include, in addition to the Wagner distance method (Farris 1972), the following: Tateno, Nei, and Tajima (1982) (modified from Farris), Li (1981), Sattath and Tversky (1977), Fitch (1981), and Saitou and Nei (1987). None of these authors claims that his or her method necessarily finds the correct tree, but all these methods are said to have special advantages. On the basis of computer simulations, Saitou and Nei claim that their "neighbor-joining method" and Sattath and Tversky's method are generally better than the other methods. This was confirmed by Sourdis and Nei (1988). Nevertheless, these uncertainties demonstrate that one should not automatically follow the findings of one particular numerical method but should also make use of all other available information for the construction of a classification. Distance methods are said to have two advantages over parsimony methods: They require much less computer time, and they usually give the branch length as well as the topology of the estimated tree.

Judging the Quality of Distance Methods

With the development of many innovative methods of assessing phenetic similarity, some of which produce different results from the same data, it became obvious that a method of quality assessment was needed. Cophenetic correlation analysis was introduced by Sokal and Rohlf (1962) to meet this need. Their original intent was to use the cophenetic correlation coefficient (rcs) as a measure of how well a given phenogram fits the original data. However, rcs measures only how well the cluster analysis fits the distance matrix. Farris demonstrated the relationship in a careful mathematical analysis (1969) and concluded that cophenetic correlation is not necessarily an accurate measure of the quality of a taxonomic result. Nonetheless, Farris (1977:831) and Swofford (1981:25) have used it as a measure of goodness of fit (see also Rohlf and Sokal 1981).

The greatest potential weakness of some of the distance methods is that they are based on the assumption of an equal rate of evolutionary change in all lineages. Rate constancy, for instance, was originally assumed by Sibley and Ahlquist (1983) in the interpretation of their DNA hybridization data. Unfortunately, in recent years so many irregularities in the timing of the molecular clock have been found for both enzyme and DNA changes that considerable caution must be exercised in constructing branching points on the assumption of rate constancy (Britten 1986; Gillespie 1986; Catzeflis et al. 1987). Fortunately, most modern distance methods do not require rate constancy.

Application of the UPGMA method is based on the assumption of an equal rate of change on all branches of the dendrogram. Other distance methods, such as the Wagner distance method, do not make this assumption.

CHARACTER METHODS

In contrast to distance methods, character-based methods of phylogeny inference use data about the states found in taxa as a starting point.

Hennig and his earlier followers constructed their cladograms by hand, but now many computer programs are available to aid the experienced taxonomist. However, Kitching (1985) found that even with complex data the manual method produced the trees that best reflected the total information (including life history data) on the included taxa. The beginner needs a special warning: Avoid an unnecessary reverence for the computer as a black box that unfailingly produces ideal solutions. The fact is that every computer program looks for a most parsimonious tree, one based on the largest possible groups of compatible character states, or the maximum-likelihood tree. Meeting these computational goals—parsimony, compatibility, or maximum likelihood—may produce results that only approximate biological reality. One should always use one's knowledge of the organisms to examine the reasonableness of the trees presented by a computer program.

The assumptions employed in a computer analysis may have been invalid. Too much weight may have been placed on relatively weak character states and not enough on the complex ones that are most likely to have been unique evolutionary innovations. Dollo's rule or *Darwin's principle* may have been violated. The best way to understand the limits of numerical analysis is to understand the process fully.

All students will greatly benefit from starting their exposure to cladistic analysis by first attacking a small problem by hand. This is why we have started with such an example. We have used the same data that demonstrated distance methods. Furthermore, we have added two computer methods, *clique* and *Wagner trees,* that illustrate some of the approaches used in phylogenetic analysis programs. None of these demonstrations is as sophisticated as current computer packages, but both can be used by a patient student without a computer, and both will serve to introduce a beginner to machine methods.

For further information, we recommend two volumes by Duncan and Stuessy. The first (1984) is a symposium in which many more advanced aspects of computer analysis are discussed, and the second (1985) reprints many important journal papers on the subject.

Another way to gain experience and to understand the differences among the various available methods is to study a particularly well done published analysis. We recommend the classification of the danaine butterflies by Kitching (1985). The author compared the results of two phenetic and three cladistic methods with one another and with a manual analysis. The use of a variety of phenetic and cladistic methods is particularly advisable when there is a deficiency of useful taxonomic character states. This is shown by Cutler and Gibbs (1985) for the phylum Sipuncula, where they "strongly recommend a synthetic approach whenever possible."

Character State Matrix

Character states that are to be used in a cladistic analysis are assembled into a data matrix in which characters are coded to indicate the sequence from ancestral (plesiomorphic) to derived (apomorphic). In hand analyses branched characters may complicate the analysis, but they actually are of some help in inferring the phylogeny, as the added assumptions can more clearly specify the most parsimonious phylogeny. The most primitive character state found in each character is coded 0, and increasingly derived character states are given sequentially higher numbers. In older computer programs it is more convenient to use linear or additive binary coding, as shown by Farris (1970); this is not necessary in the newest programs. Pimentel and Riggins (1987) provide a helpful discussion of coding methods.

The completed data matrix contains all characters in rough ranking from high—usually the more complex examples—to low quality and the character states of each character coded from primitive to derived.

It is helpful if the characters are given a mnemonic code (Table 11-1). For example, EyC would be suitable for eye color, and AbC for abdomen color. Since all character states will be entered onto the final phylogenetic tree, the use of such mnemonics helps readers understand the final phylogenetic hypothesis.

A Simple Example

Character-based phylogeny inference is best explained by an example. The method we will employ is nonnumerical cladistic analysis, which has been used by Hennig (1966*a*) and many of his followers. The group of eight imaginary insects introduced at the beginning of the chapter (Figure 11-1, A through H) are provided as subjects. Although they have character states that may be found on real insects, the demonstration data are to be considered strictly in isolation. Another imaginary insect related to the group of eight (Figure 11-2) is used to locate the various characters listed in Table 11-1. It also provides an out-group example on which primitive to derived sequences can be based. Each character state found in the study group in a derived form is provisionally considered to be primitive in the out-group insect, although the out-group insect has several derived features of its own. The fact that the study group is monophyletic is seen in the elongate body and the exserted head; other "hidden" features reinforce this conclusion.

Table 11-1, the same data set used for the demonstration of the distance method, lists the characters that differ in the members of the demonstration group in rough order of quality from high to low, with each character identified by a mnemonic code. The character states in each character are already listed in a primitive to derived sequence judged by comparison with the out-group.

Given that the study group is monophyletic, a drawing of what one knows about the cladogenesis as one begins a study most resembles a fan (Figure 11-8). The fan is a polytomy, and it is important to note that the members of a polytomy may be placed in any order. The first step in cladistic analysis is to find all characters that have a derived state in only one taxon. Table 11-2 shows that there are two such characters, both concerned with femoral spines: aFSp and bFSp. These are autapomorphies; when placed on the fan diagram, they do not group any taxa.

Several principles have been employed to establish from a given set of data the cladogram that is most likely to be the right one. One is *compatibility*. Here that tree is considered most probable which has the largest number of characters that are not in conflict with one another and with the tree (see below under clique). The second principle is *parsimony* (e.g., Wagner trees) (Chapter 9). The most parsimonious tree is the one that, while accounting for all character states in the data matrix, has the fewest character state steps. Because incompatible characters force an assumption of parallelism or reversal within one character or the other and thus require at least one extra character state step to arrive at the most derived state, the two principles—parsimony and compatibility—are related. Often but not always, the two methods dictate the same tree. Most frequently, especially in large studies, more than one equally parsimonious tree is found. Ranking characters by quality (weighting) helps avoid this problem. Neither the compatibility method nor the parsimony method is free of problems, and serious objections have been raised to both.

Because it is difficult to see repeated cladistic patterns in a data matrix, drawing individual trees for each character is often helpful. Figure 11-9 contains trees for all but the two autapomorphous characters, which are already on the fan tree in Figure 11-8. Here the first eight characters (Figure 11-9*a* through *h*) are of higher quality, for their derived states are unique to the study group. The remaining four characters (Figure 11-9*i* through *l*) are questionable because some have ambiguous states and some are too simple to be reliable.

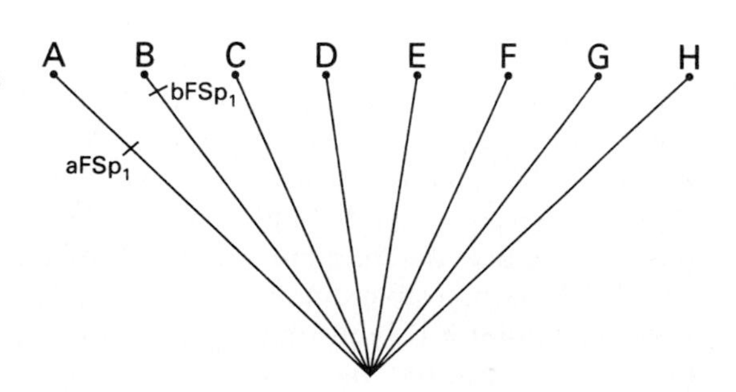

FIGURE 11-8
Fan cladogram showing what is known at the beginning of an analysis of the demonstration data in Table 11-2. Two autapomorphous characters have been added to the cladogram to show that characters unique to a single taxon add no cladistic information to the analysis.

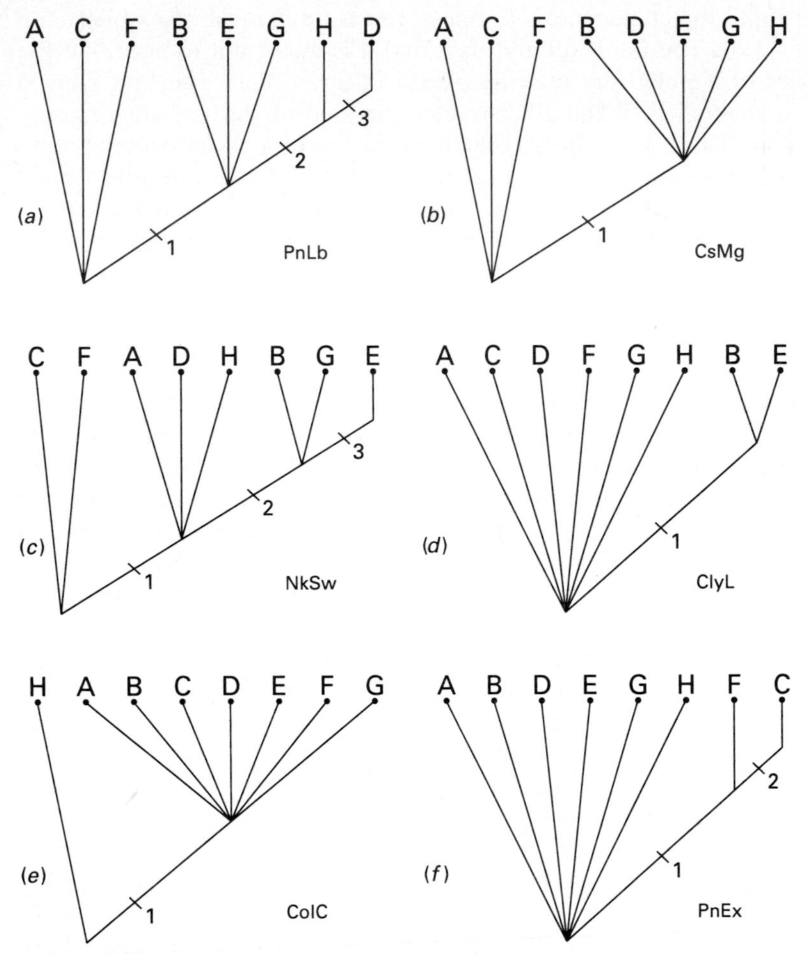

FIGURE 11-9
(*a* through *f*) Trees for 6 of the 12 characters in the demonstration data.

The most obvious pattern that emerges in scanning the character trees is found in PnLb, CsMg, and MmbL (Figure 11-9*a, b,* and *h*), which group taxa A, C, and F as primitive and the remaining five taxa in a holophyletic group. A good place to start is with these three characters. Figure 11-10 combines these three one-character trees. Two holophyletic groups emerge: GEBDH and DH. The three trees are completely compatible; because no character state had to be entered more than once, the tree has maximum parsimony, or least possible length.

At this point it is important to remember that apparent group ACF has been entered on the tree, and the remaining, weaker characters should be

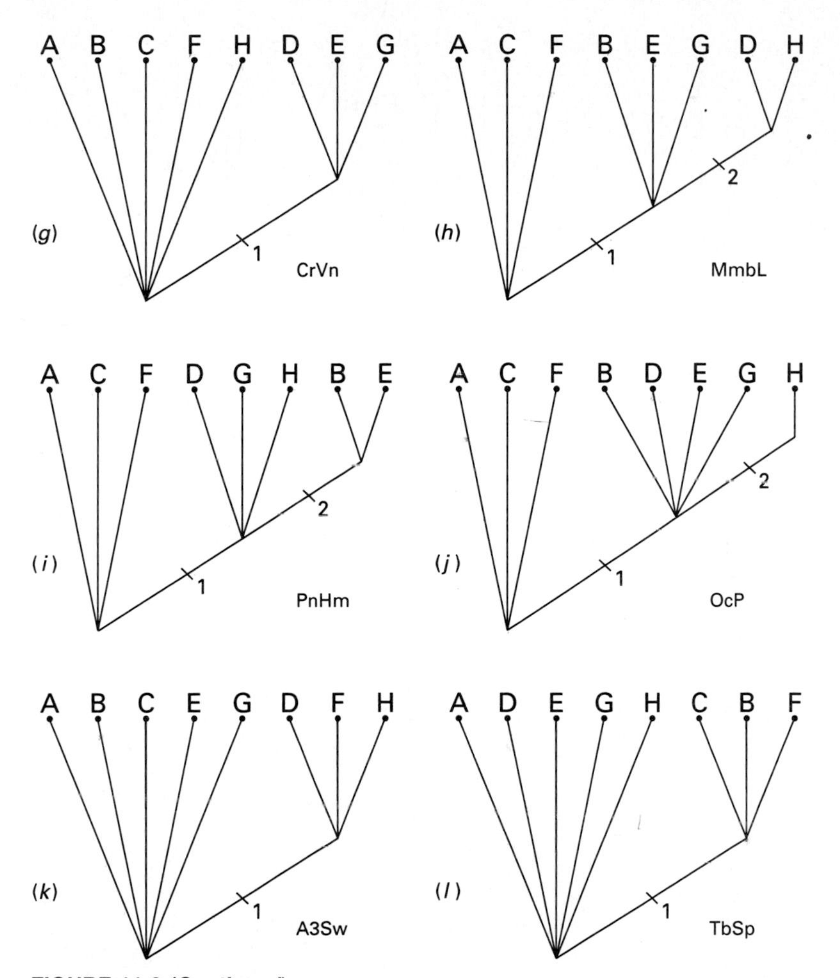

FIGURE 11-9 (*Continued*)
(*g* through *l*) Trees for the other 6 of the 12 characters in thc demonstration data.

reviewed to see whether any are also compatible. Of these, PnHM and OcP (Figure 11-9*i* and *j*) fit easily, as do the two autapomorphous characters aFSp and bFSp. The last two weak characters, A3Sw and TbSp (Figure 11-9*k* and *l*), are incompatible. All but four characters, then, are completely compatible and produce a parsimonious cladogram (Figure 11-10*c*).

Placing the incompatible characters on the tree is the next task. These incompatible characters contain character states that are parallel, convergent, or reversed, and placing them requires more steps on the cladogram than each single-character tree shows. If both primitive and

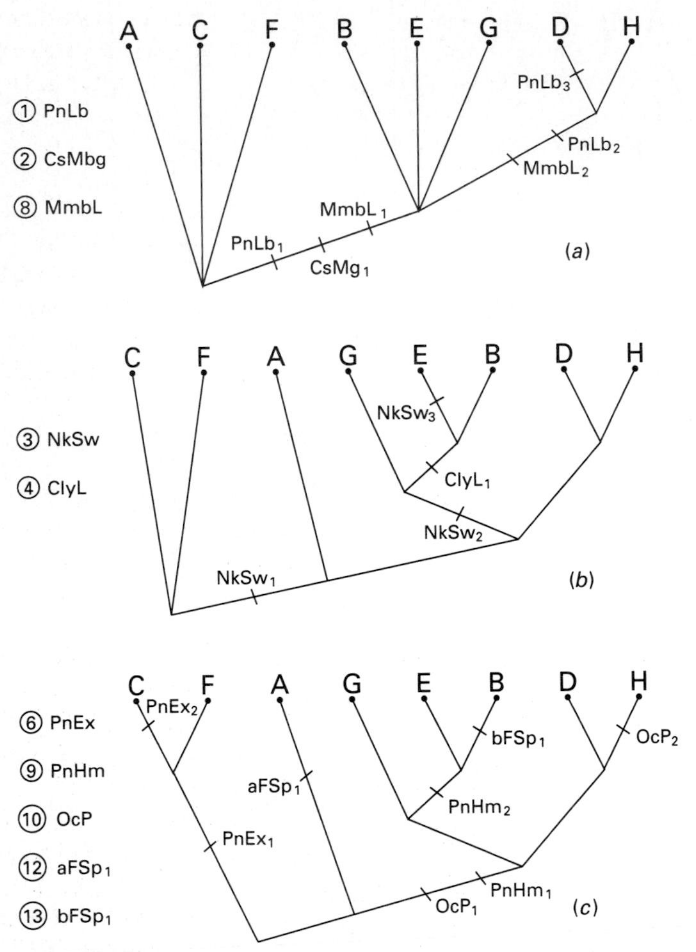

FIGURE 11-10
Three steps (*a* through *c*) in the construction of a cladogram for the demonstration insects: good characters.

derived states of a character appear both within and outside a well-established holophyletic group, it is probable that the derived state evolved more than once. Several numerical parsimony techniques are available for the analysis of homoplasies (see below).

Characters ColC and CrVn are unique to the study group and were originally judged to be reasonably complex. The judgment (one can still change one's mind), then, is that these two characters evolved only once but later were lost. The best procedure is to find the smallest holophyletic group that contains all the derived states of each character and postulate that the feature evolved at its base. With CrVn, the smallest such group is GEBDH; for ColC, it is the whole group. Tick marks are entered at the appropriate places on the cladogram. These character states are then reversed in the most parsimonious manner to make the taxa within the holophyletic groups more primitive. Instead of a tick mark, it is convenient to use an X for reversals so that they can be identified quickly in the completed cladogram. With character ColC, a single reversal subtending taxon H is all that is needed. With CrVn, two reversals are needed, one each subtending taxa B and H. These reversals have been entered on the completed cladogram shown in Figure 11-11*a*.

When reversals are postulated, it is wise to review what the result would be if each character state evolved more than once. With ColC, re-

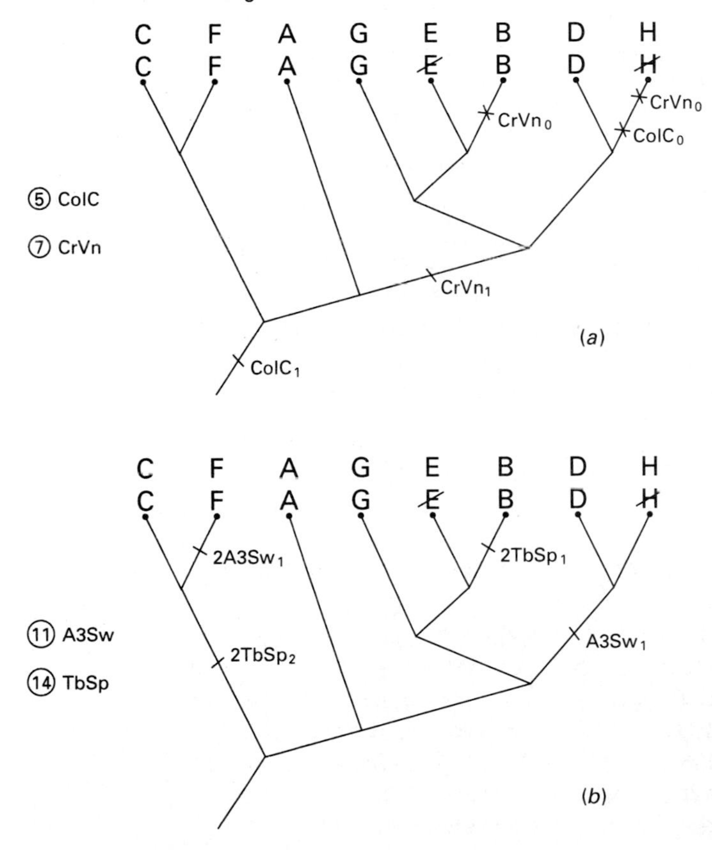

FIGURE 11-11
(*a*) Adding reversed characters to the cladogram. (*b*) Adding parallel characters to the cladogram.

versal is clearly the most parsimonious solution, for the character is entered with only two steps. If parallel evolution were postulated, the derived state ColC1 would have to have evolved four times, on internodes 1–2, 3–A, 4–5, and 7–G. (For an explanation of node numbers, see Figure 11-12.) By contrast, character CrVn took three steps as a reversal and would take three steps as a parallelism, at 5–G, 6–E, and 7–D. Both solutions are equally parsimonious. Which one is correct depends solely on the likelihood of single or multiple evolution of the derived character state. Such are the uncertainties of cladistic analysis.

Two characters remain—A3Sw and TbSp—both of which are weak. Though they have been placed into a primitive to derived sequence in the data matrix, evidence for the sequence is lacking, since all states have been found both within and outside the study group. A new tool, however, is available: the completed cladogram. The procedure is to postulate alternative possibilities and then accept the most parsimonious answer. It is useful to place a mark above the taxa that have a given character state and then count how many steps are required to place the character state on the tree compared with the number needed for the opposite assumption. Character state $A3Sw_1$, for example, needs only two steps, at 2–F and 4–5, to express the data matrix properly but requires

FIGURE 11-12
The completed cladogram of the demonstration data with all characters entered. For an explanation of the numbers preceding some character mnemonics and the character codes in parentheses on the stem of the tree, see the section on evolutionary systematic methods.

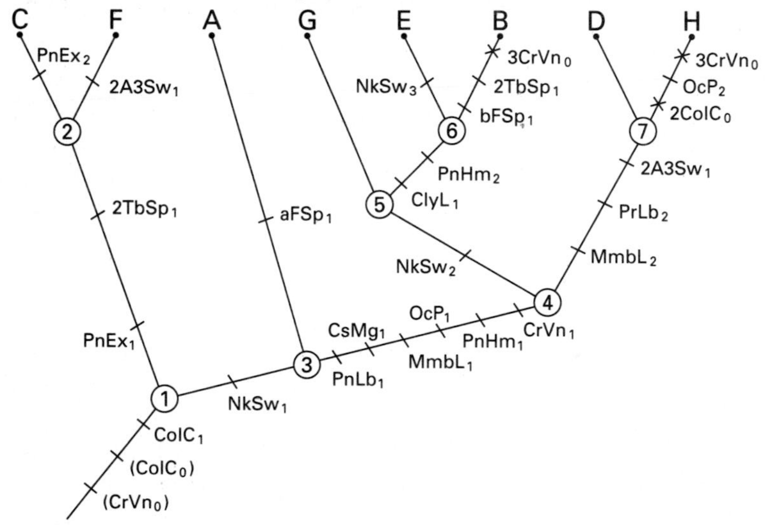

three steps, at 2–E, 3–A, and 4–5, if the coding is reversed. Similarly, $TbSp_1$ requires two steps, at 1–2 and 6–B, to be placed on the cladogram, while switching the primitive to derived sequence requires four steps, at 3–A, 5–G, 6–E, and 4–7. If the character code is preceded by the number of times each character state appears on the tree, it may easily be identified as a parallel character state on the completed tree. These two characters are entered on the cladogram shown in Figure 11-11*b*.

Sometimes both trials are equally parsimonious and no final decision about placement can be made. Such character states cannot contribute materially to the study and are best forgotten. Remember (see above) that numerical programs are available for the analysis of homoplasies. The completed tree with the character states, including clearly shown parallelisms, reversals, and autapomorphies, appears in Figure 11-12.

Problems Encountered in the Construction of a Cladogram

So-called phylogenetic trees have been proposed since the days of Haeckel. They were rarely based on a rigorous methodology; usually they were products of intuition based on a few character states. However, since a few key character states (e.g., notochord, feathers, three pairs of legs) often tell most of the story, many of these early phylogenies were fully confirmed when better methods were employed later. Other phylogenies were not confirmed because their authors had committed serious errors. It is necessary to call special attention to these errors because all of them can still be found in the recent literature. It is worth noting that many of these errors can be avoided or detected with machine analysis.

Blunders Since any numerical analysis consists of a long series of steps, errors such as the wrong copying of a number creep in easily. The procedure therefore has to be checked and double-checked. The best test is to translate the cladogram back into a data matrix and compare it with the original data matrix. If they do not match, a blunder has been committed somewhere.

Unsupported (Empty) Internodes Dichotomies are found in some trees that are not supported by any apomorphies. The result is an "empty" internode. The maintenance of such empty internodes is unjustified. When such an unsupported internode is eliminated, a trichotomy or sometimes even a polytomy will result. This is at least an honest representation of the available data.

The Unwitting Use of Plesiomorphic Character States A leftover from identification keys that is still found in some modern phylogenetic trees is the with-without alternative. One internode is labeled as being with a given character state, and its sister internode is labeled as being without it. Obviously, one of the alternatives is a plesiomorphic character state, which is inadmissible in a cladogram. Another variation on the same error is to label one branch as having five antennal segments, petals, or fin rays and its sister internode as having four of the same character states. In all such cases, since one of the character states is derived, the contrasting character state must be primitive. The elimination of primitive character states from such a tree may result in the presence of empty internodes on the tree. Diagrams that consist of contrasting character states of this sort on sister internodes are diagnostic keys rather than cladograms.

Even though the errors pointed out in the preceding paragraphs are well understood, they have appeared in recently published research, and continued vigilance must be exercised to avoid them.

The following sections describe situations in which paraphyletic groups may appear to be holophyletic.

Taxa without Apomorphies In a given study, it sometimes happens that no apomorphic character states can be found to give a character state to a terminal taxon. In the sample analysis, for instance, internode 5–G (Figure 11-12) has no such character states. In the literature, such taxa are sometimes referred to as character stateless taxa. When, as in the example, the terminal taxon is a single species, all that can be implied is that no difference has been found between taxon G and its ancestor at node 5.

An absence of apomorphic character states may have two causes. Either the distinguishing character states have so far been missed, as in a terminal sibling species, or the taxon is paraphyletic. For instance, the widely recognized Reptilia, a paraphyletic group, share all derived character states with one or both of their ex-groups (mammals and birds), while the character states not shared with the ex-groups (e.g., ectothermy) are primitive. Although useless for cladogram construction (genealogy), plesiomorphic character states—all contrary claims notwithstanding—are often highly useful for purposes of classification.

Unwise Choice of Character States The reliability of a cladogram depends on the information content of the character states used to construct it. The use of different sets of character states may lead to entirely different classifications even when they are based on the same algorithm. Recent examples include the proposal by some cladists that the lungfishes are the ancestors of the tetrapods, not the rhipidistians, as claimed by other cladists, and the claim by some cladists that the archosaurians rather than some synapsids are the sister group of the mammals.

Nearly every cladistic analysis contains a few hypotheses that inspire less than full confidence. Prominent among them are purported holophyletic groups for which the evidence consists of one character state or a few very weak ones: those, for example, which are demonstrably parallel, convergent, or simple or are reversals or losses. Often the use of such character states results in hidden paraphyly. The hypothesis may be correct, but further study is indicated. If nothing can be done to improve the hypothesis and it must be left in the study, it is wise to point out such places in the analysis and discuss alternative hypotheses. A major problem is to establish exactly which character states are unreliable. While some guidelines have already been given, the topic is further discussed below.

Incompletely Studied Taxa When a character state supports a holophyletic group on a cladogram, a reader naturally assumes that the character state is possessed by all members of the group or is present in the form of more derived modifications. However, it is often impossible to study every feature of every species in the study group. In small studies, character states may not be known because of a lack of certain immature forms or fresh specimens, or a particular morphotype or sex may not be available for some species. In large studies (classes within a phylum or orders within a class), these problems are compounded by the numbers of species involved. Attempting to check every species guarantees that the study will never be finished. In such classes, assumptions must be made about the distribution of some character states, and false assumptions may of course lead to false conclusions. Paraphyletic groups may appear to be holophyletic groups.

There is no perfect solution to such problems, but it is important that assumptions be clearly noted as part of the analysis. When one deals with very large taxa, it is helpful to indicate which species have been studied as samples of the larger group. Such information allows other systematists to verify the results and may suggest areas in need of investigation. Choice of species is important, too. Relatively primitive species are of course useful, for they are relatively similar to their sister groups. However, it is erroneous to consider all character states of a primitive taxon as primitive. For example, the platypus might be chosen as a primitive mammal, but this species, like many primitive organisms, has its own derived features, i.e., the poison spurs of the male and its duckbill. To represent all mammals by such closely related taxa as a cow, a goat, and a horse would not inspire much confidence. However, features common to the platypus, the opossum, and the shrew would be fairly certain to be present in the common ancestor of these animals.

Early accounts of cladistic analysis gave the impression, still held by some, that any cladistic analysis that is not completely dichotomous is a failure. Later accounts stressed that complete dichotomy is only a methodological goal, not a measure of success. Each bifurcation in a dichotomous cladogram represents not only an inferred ancestor but also an implied speciation event. However, not every speciation is accompanied by the production of character states that are useful in a cladistic analysis. The existence of sibling species is an extreme example. It follows, then, that a cladogram cannot always be resolved into dichotomous branches; polytomies are inevitable.

In any study in which the smallest taxonomic units are groups of species, some groups may lack convincing autapomorphies and others may even lack convincing synapomorphies; hence, polytomies may remain in any study. Hennig's advice (1966*a*:236) is apt here; he said that no one expects an author "to present the phylogenetic sister-group relations more accurately than they are known," and he warned workers about the need to point out weaknesses in their studies.

OVERVIEW OF CHARACTER-BASED METHODS

Analogous to computer methods for distance data, computer methods have been developed for character data. Various scientists have proposed methods that differ primarily in terms of whether they emphasize compatibility or parsimony. As with distance methods, different methods sometimes produce different results; often they produce ambiguous results (more than one possible solution is found). Unlike the results of distance methods, however, the results of cladistic studies and the distribution of character states suggested by a given solution can be reviewed, and informed judgments can be made about their value.

Some of the older programs required the use of large mainframe computers, but most of the algorithms referred to here can be solved using the newer programs on a modern desktop machine. A large variety of software packages now exist (see the Appendix at the end of this chapter), with new ones frequently becoming available. The relatively small fee for their use is repaid in time saved.

The first character-based methods were developed by Camin and Sokal (1965) and Edwards and Cavalli-Sforza (1964), with additional pathbreaking work done by Farris and his colleagues (Kluge and Farris 1969; Farris 1970). One of the major criticisms of pure cladistics was that it was interested only in cladogenesis, that is, in the location of the nodes in a cladogram, ignoring entirely all aspects of anagenesis. Wagner trees (Farris 1970), which consider the number of steps in a lineage, broke with the original cladistic tradition. Cladistic analysis thus became phylogenetic analysis. More important, the new approach permitted the development of algorithms that made it possible to incorporate quantitative characters. Since that time there has been a steady stream of improvements, modifications, and additions to these methods. Unfortunately, no textbook summary of these methods exists.

The earlier treatments are now obsolete and do not justify full explanation. Many currently employed methods are derived from Wagner's groundplan method but go considerably beyond the first numerical versions (Farris 1970) of it. They are parsimony methods in that the tree judged to be optimal is the one which allows for the simplest (most parsimonious) explanation of the observed distribution of character states. A different approach is used in some of the compatibility methods developed by Estabrook and his followers. A compatibility method is one in which that tree is selected which has the largest number of characters perfectly compatible with it. We will begin with compatibility methods.

Compatibility Methods

The best known method of compatibility analysis has been named *clique* by G. F. Estabrook, one of its early proponents (Estabrook 1972; Estabrook, Johnson, and McMorris 1975*a* and *b*; Estabrook and Landrum 1975; McMorris 1977). Meacham (1981) published a noncomputer account of this method, which has been adopted here with further simplifications. The system is based upon Le Quesne's observation (1969, 1972) that two two-state characters are cladistically incompatible only if their distribution among the taxa in the study group includes all four possible combinations of matches and mismatches. If any one of the four is not found, the two characters are mutually compatible. Le Quesne made no assumptions about which character states were ancestral and which were derived. For a further discussion, see Felsenstein (1982).

Since the method requires two-state characters, a multistate data matrix (Table 11-2) must be converted to a two-state data matrix. How this is done can be found by comparing the original data matrix (Table 11-2) with the two-state matrix shown in Table 11-6.

The number of characters in the data matrix—14—increases to 22 in the two-state matrix, in which each character in its derived expression is a true character state, not a character. Although increasing the number of required comparisons would seem to complicate the analysis, the reduction of characters to character states provides a balancing simplification.

TABLE 11-6
CLIQUE: TWO-STATE DATA

Clique character no.	Character	Character state	Taxon								
			Anc	C	F	A	G	E	B	D	H
1	NkSw	1	0	0	0	1	1	1	1	1	1
2		2	0	0	0	0	1	1	1	0	0
—		3[*]	0	0	0	0	0	1	0	0	0
3	PnEx	1	0	1	1	0	0	0	0	0	0
—		2[*]	0	1	0	0	0	0	0	0	0
4	ClyL	1	0	0	0	0	0	1	1	0	0
5	ColC	1	0	1	1	1	1	1	1	1	0
6	PnLb	1	0	0	0	0	1	1	1	1	1
7		2	0	0	0	0	0	0	0	1	1
—		3[*]	0	0	0	0	0	0	0	1	0
—	CsMg	1[†]	0	0	0	0	1	1	1	1	1
8	CrVn	1	0	0	0	0	1	1	0	1	0
—	MmbL	1[†]	0	0	0	0	1	1	1	1	1
—		2[‡]	0	0	0	0	0	0	0	1	1
—	PnHM	1[†]	0	0	0	0	1	1	1	1	1
—		2[§]	0	0	0	0	0	1	1	0	0
—	OcP	1[†]	0	0	0	0	1	1	1	1	1
—		2[*]	0	0	0	0	0	0	0	0	1
9	A3Sw	1	0	0	1	0	0	0	0	1	1
—	aFSp	1[*]	0	0	0	1	0	0	0	0	0
—	bFSp	1[*]	0	0	0	0	0	0	1	0	0
10	TbSp	1	0	1	1	0	0	0	1	0	0

[*]Autapomorphy.
[†]Duplicate of clique character number 6.
[‡]Duplicate of clique character number 7.
[§]Duplicate of clique character number 4.

First, since all autapomorphies (derived states found in single taxa) are compatible with all other character states, autapomorphic character states may be temporarily eliminated. (There are six autapomorphies in Table 11-6, four of which were part of multistate characters in Table 11-2.) Second, all character states with identical distributions are also compatible with one another. One example only of each such group of duplicate character states must be retained; the rest (six more two-state characters) may be eliminated. A new data matrix containing the 10 two-state characters remaining (Table 11-7) simplifies the generation of a compatibility matrix (Table 11-8). Even with data in a primitive to derived sequence, the clique method produces unrooted trees unless an ancestor, which is coded 0 for every character, is included in the data matrix.

The clique matrix is concerned only with compatibilities, not with rel-

TABLE 11-7
CLIQUE: ESSENTIAL DATA MATRIX

Clique character no.	Taxon								
	Anc	C	F	A	G	E	B	D	H
1	0	0	0	1	1	1	1	1	1
2	0	0	0	0	1	1	1	0	0
3	0	1	1	0	0	0	0	0	0
4	0	0	0	0	0	1	1	0	0
5	0	1	1	1	1	1	1	1	0
6	0	0	0	0	1	1	1	1	1
7	0	0	0	0	0	0	0	1	1
8	0	0	0	0	1	1	0	1	0
9	0	1	0	0	0	0	0	1	1
10	0	1	1	0	0	0	1	0	0

TABLE 11-8
CLIQUE: CHARACTER RANKING

Rank	Character	Compatibilities	Incompatibilities
1	3	8	1
2	2	7	2
3	4	7	2
4	7	7	2
5	1	6	3
6	6	6	3
7	5	5	4
8	8	5	4
9	10	4	5
10	9	3	6

ative degree of incompatibility. The matrix will always be symmetrical on either side of the diagonal row of X's, and only half the matrix must be calculated (Figure 11-13).

Compatibilities are found by comparing each character with each other character (row 1 with rows 2, 3, etc.) to see whether the combinations 00, 01, 10, and 11 exist for any pair. Since the hypothetical ancestor, coded 0 for all characters, is included, the combination 00 will always appear. If any one of the other three combinations is missing, the pair of characters is compatible. If all four combinations are found, the pair is incompatible.

C = Compatibility
I = Incompatibility

Clique character no.	1	2	3	4	5	6	7	8	9	10
1	X									
2	C	X								
3	C	C	X							
4	C	C	C	X						
5	I	C	C	C	X					
6	C	C	C	C	I	X				
7	C	C	C	C	I	C	X			
8	C	I	C	I	C	C	I	X		
9	I	C	I	C	I	I	C	I	X	
10	I	I	C	I	C	I	C	C	I	X
Compatibilities	6	7	8	7	5	6	7	5	3	4
Incompatibilities	3	2	1	2	4	3	2	4	6	5

FIGURE 11-13
Clique compatibility matrix.

It is convenient to picture the four corners of each cell in the matrix as representing the following chart of possibilities:

00

01

10

11

As a combination is found, a mark is placed in the appropriate corner of the cell. Any cell with four marks, then, indicates incompatible characters. The total number of compatibilities (C) and incompatibilities (I) for each character can be found by counting the row and column compatibilities for that character. The sum of compatibilities and incompatibilities will equal one less than the number of characters in the matrix (Figure 11-13).

Given the completed compatibility matrix, clique analysis locates groups of characters that are completely compatible with one another. These groups are called cliques. The clique with the largest number of compatible characters is the place where one begins constructing a dendrogram.

The ranked list of characters with the number of their compatibilities (Table 11-8, rank-compatibility test) provides a means of finding characters that cannot possibly be part of the largest clique in the study. When the rank of a given character exceeds the number of compatibilities of that character, the first clique must have been exceeded. In the study group, the first six ranked characters have six or more compatibilities. The next two have five, but their rankings of 7 and 8 place them below the critical level. The first six characters are the only possible candidates for membership in the largest clique in the character set. To test these six candidates, one constructs a compatibility diagram (Figure 11-14*a*). The six characters are arranged in a circle, and a line is drawn between each pair of compatible characters.

This compatibility graph connects each character with each other character. Thus, the group is the largest clique. Note that the clique contains more than these six characters. The autapomorphic characters belong to all cliques; moreover, the six duplicate characters also belong to this clique. The total number of characters in the largest clique is 18 of the original 22. Only four characters remain to bc grouped.

The largest clique does not always contain all the characters that pass the rank-compatibility test. When it does not, some lines will be missing in the compatibility diagram. If many lines are missing, it may be convenient to make an incompatibility graph that connects only the characters that are incompatible. Such a graph can be demonstrated with the four characters excluded from the first clique (Figure 11-14*b*). If the character with the most connections indicating incompatibility is eliminated, a second clique is found (characters 5, 8, and 10). Character 9 forms the third clique. None of these four characters had duplicates, but the six auta-

FIGURE 11-14
Clique (*a*) compatibility and (*b*) incompatibility diagrams.

(*a*)
Compatibility diagram

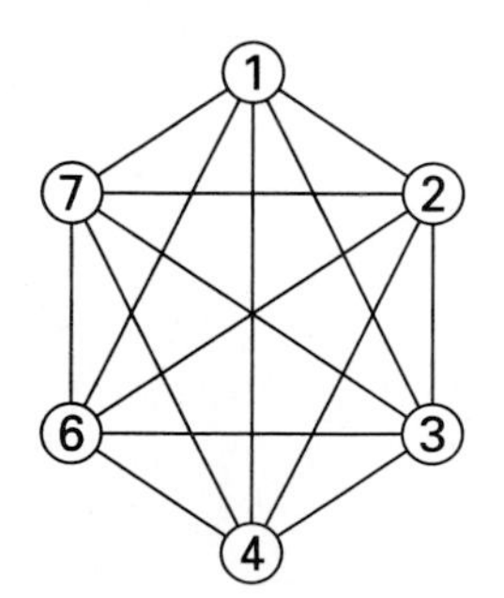

(*b*)
Incompatibility diagram

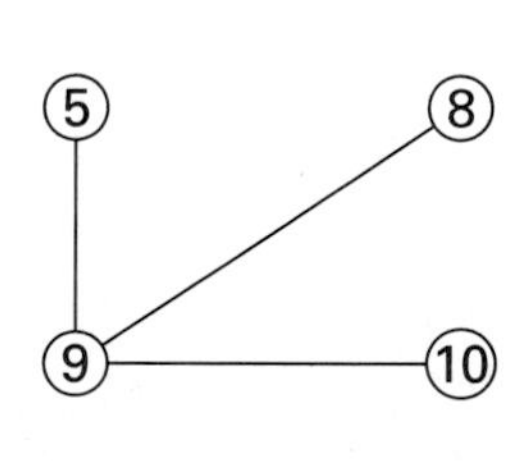

pomorphies belong to both cliques. Hence, the second clique consists of nine characters, and the third consists of seven.

When one forms the tree, cliques are considered in order, from largest to smallest. Here the tree will be completely formed by the first clique; the other two are of no value in cladistic analysis. In larger studies, more than one clique is usually needed to form a cladogram.

The first six characters that form clique 1 may be diagrammed as in Figure 11-15*a*. Because each character defines a different holophyletic group and all the characters are compatible, they can be combined in any order, and each addition will develop the cladogram. The process of forming the tree is shown in Figure 11-15*b*.

FIGURE 11-15
Clique cladogram construction. (*a*) The six distinct two-character trees which form the first clique. (*b*) Construction of the cladogram by "popping" taxa.

(*a*)

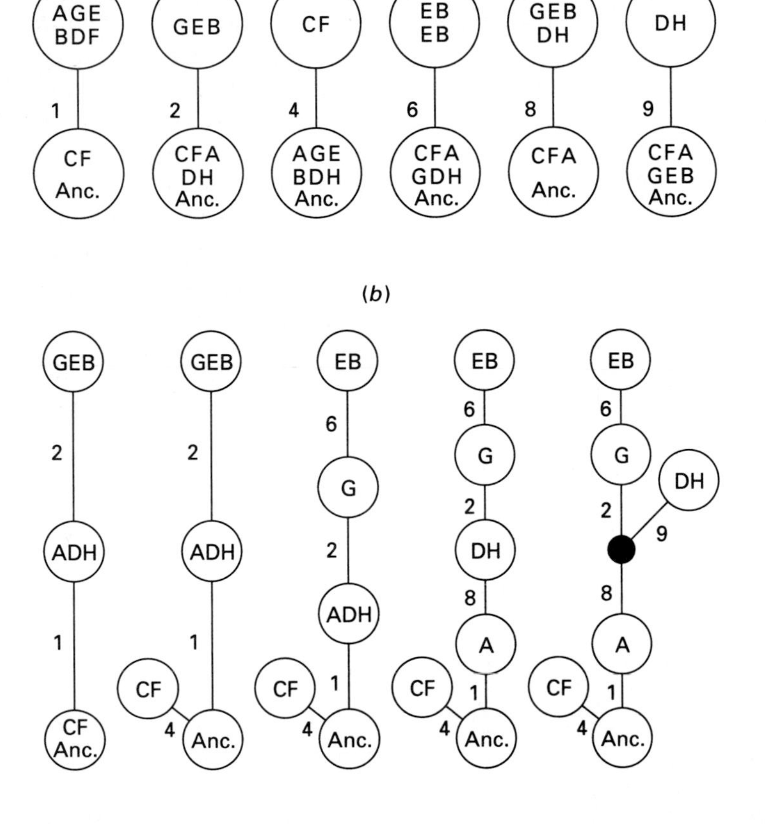

Combining character trees is called popping. The tree as formed is essentially complete, but some relationships are not clear. Pairs CF, DH, and EB remain, and two taxa, A and G, rest within the tree. All these except taxon G will be popped when the autapomorphous characters are returned to the tree. Taxon G has no character states of its own and is unresolvable. Characters of the second and third cliques are incompatible with the cladogram; they must be placed in the most parsimonious manner, as parallelisms or reversals.

Note that the clique method clearly identifies the weakest characters and does so more precisely than does the rule-of-thumb hand method we described first.

Parsimony Methods

Hennig and his early followers constructed their cladograms by means of trial and error. Only synapomorphies were employed; anagenesis was ignored.

Interestingly, when computer cladistics developed, it was based on a very different tradition. In the 1950s the botanist W. H. Wagner developed a *groundplan divergence method* (1961) in which not only the branching points of the phyletic lines but also the number of steps by which the terminal taxa advanced beyond the branching nodes were depicted (Figure 11-16). This manual method was used successfully in the 1950s by several of his students. Wagner consciously adhered to the principles of evolutionary taxonomy, stating that "it tries to estimate the amounts (grades), directions (clades), and sequences (steps) of phylogenetic divergence" (1980:173). "The tree is based upon taxonomy, not the reverse (taxonomy being defined here as phenetic classification, clustering by relative amounts of resemblance and difference)" (Wagner 1980:175).

A very early computer method was that of Camin and Sokal (1965). They constructed a compatibility matrix by placing the character states of each character on a pattern or tree of each other character and then

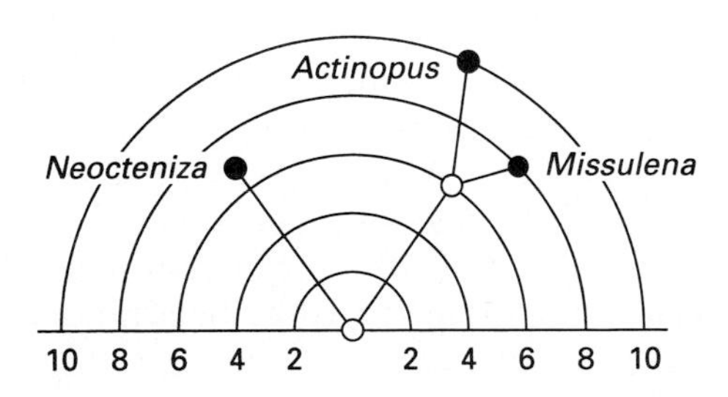

FIGURE 11-16
A phylogenetic analysis of *Neocteniza, Actinopus,* and *Missulena* using the Wagner groundplan divergence method. Concentric semicircles represent anagenetic divergence based on the number of synapomorphies and autapomorphies in each taxon. (*From Wiley 1981.*)

tabulated the number of extra steps (if any) required for each character–character tree combination. Character trees were then ranked by quality (most compatibilities and lowest number of extra steps) and combined in that order to produce a parsimonious tree. The original version had flaws, which have been corrected in modern computer programs.

Kluge and Farris (1969) later developed a computer method, formalized by Farris (1970), whose output is similar to that produced by the manual methods of Wagner. The algorithms are based on the construction of minimal or shortest spanning trees. Farris (1970) presented two algorithms, one to produce rootless trees and the other to produce trees with an ancestral root. The goal of these algorithms is the discovery of the most parsimonious trees, called Wagner trees. The two algorithms are not guaranteed to find the most parsimonious tree, and for most data sets they do not. The algorithms begin with a single taxon and sequentially add more taxa to the growing tree until all taxa have been added.

Many modifications and improvements of the methods of searching for parsimonious trees have been made in subsequent years [see Hendy and Penny (1982) and the documentation for the programs cited in the last section in this chapter]. They include rearranging the branches of a tree to see if a more parsimonious tree can be produced and the discovery of multiple parsimonious trees. Also, the initial algorithms dealt with characters whose states were arranged into a linear array (''ordered'' characters). Characters with unordered states, those obeying much more complex patterns of evolution, or those with the frequencies of states within the taxa specified can now be accommodated [see Fitch (1971), Hartigan (1973), Sankoff and Rousseau (1975), Swofford and Berlocher (1987)]. More up-to-date computer packages that use the newer methods are listed in the Appendix at the end of this chapter. The object of the short descriptions that follow is to convey an approximate idea of the steps in the search for a basic Wagner tree. The simpler version is from Whiffin and Bierner (1972); a more complex version is from Kluge (1971).

To produce a rooted tree with Farris's original Wagner algorithm, one can add to the data matrix an ancestral taxon (anc) whose character states are all primitive and are coded 0 (Table 11-9; ignore the section with ancestral nodes 1 to 7 for now). (See below for other methods of rooting.) From this matrix, a Manhattan distance similarity matrix is calculated (Table 11-10), which is identical to the Manhattan distance matrix used in the description of clustering methods except for the additional row and column for the ancestral taxon. The Wagner algorithm must start with the ancestral taxon, and taxa are added to the tree in the order of their distance from the ancestral taxon. Each new taxon is added at a position determined by its similarity to a taxon that is already in the tree,

TABLE 11-9
FARRIS-WAGNER TREES: DATA MATRIX

	Taxa									Ancestral nodes						
Character	**Anc**	**C**	**F**	**A**	**G**	**E**	**B**	**D**	**H**	**1**	**2**	**3**	**4**	**5**	**6**	**7**
NkSw	0	0	0	1	2	3	2	1	1	0	0	1	1	2	2	1
PnEx	0	2	1	0	0	0	0	0	0	0	1	0	0	0	0	0
ClyL	0	0	0	0	0	1	1	0	0	0	0	0	0	0	1	0
ColC	0	1	1	1	1	1	1	1	0	1	1	1	1	1	1	1
PnLb	0	0	0	0	1	1	1	3	2	0	0	0	1	1	1	2
CsMg	0	0	0	0	1	1	1	1	1	0	0	0	1	1	1	1
CrVn	0	0	0	0	1	1	0	1	0	0	0	0	1	1	1	1
MmbL	0	0	0	0	1	1	1	2	2	0	0	0	1	1	1	2
PnHM	0	0	0	0	1	2	2	1	1	0	0	0	1	1	2	1
OcP	0	0	0	0	1	1	1	1	2	0	0	0	1	1	1	1
A3Sw	0	0	1	0	0	0	0	1	1	0	0	0	0	0	0	1
aFSp	0	0	0	1	0	0	0	0	0	0	0	0	0	0	0	0
bFSp	0	0	0	0	0	0	1	0	0	0	0	0	0	0	0	0
TbSp	0	1	1	0	0	0	1	0	0	0	1	0	0	0	0	0

but it is placed on a line drawn from the internode subtending the nearest relative, thus generating an ancestral node.

When taxa in the example are ranked according to their distance from the ancestor, the list runs as follows: anc, A, C, F, G, H, B, D, E. The first segment of the tree consists of anc and the closest taxon, A (Figure 11-17). Taxa C and F are tied for second place, and either may be chosen. Here C is added to the tree by generating node 1 on the internode between anc and A (Figure 11-17*b*). Taxon F differs from C by 2 and from A by 5; it is added to the internode between C and node 1, generating new node 2 (Figure 11-17*c*). The process continues (Figure 11-17*d* through *g*) until the cladogram is complete. The order in which taxa E, B, and D, which are equally distant from the ancestor, should be added to the tree presents a problem. One should choose the taxon that is closest to a taxon already on the tree. Here E differs from G by 3, while B and D both differ from their closest relative by 4. For this reason, E is entered first, and either B or D may be next. Since B and D will join different internodes, no conflict is found.

The more complete tree that includes character states expands the data matrix (Table 11-9) to provide descriptions of hypothesized ancestors. Simultaneously, it expands the similarity matrix (Table 11-10) to provide distance data between hypothesized ancestors (new nodes) and taxa on other internodes. As shown on the data matrix (Table 11-9), A

TABLE 11-10
FARRIS-WAGNER TREES: SIMILARITY MATRIX

	Taxa									Ancestral nodes						
	Anc	**C**	**F**	**A**	**G**	**E**	**B**	**D**	**H**	**1**	**2**	**3**	**4**	**5**	**6**	**7**
Anc	X	4	4	3	9	12	12	12	10	1	3	2	8	9	11	11
C	4	X	2	5	11	14	12	14	14	3	1	4	10	11	13	13
F	4	2	X	5	11	14	12	12	12	3	1	4	10	11	13	11
A	3	5	5	X	8	11	11	11	11	2	4	1	7	8	10	10
G	9	11	11	8	X	3	5	5	7	8	10	7	1	0	2	4
E	12	14	14	11	3	X	4	8	10	11	13	10	4	3	1	7
B	12	12	12	11	5	4	X	10	10	11	11	10	6	5	3	9
D	11	13	11	10	6	9	11	X	4	11	13	10	4	5	7	1
H	10	14	12	11	7	10	10	4	X	11	13	10	6	7	9	3
1	1	3	3	2	8	11	11	11	11	X	2	1	7	8	10	10
2	3	1	1	4	10	13	11	13	13	2	X	3	9	10	12	12
3	2	4	4	1	7	10	10	10	10	1	3	X	6	7	9	9
4	8	10	10	7	1	4	6	4	6	7	9	6	X	1	3	3
5	9	11	11	8	0	3	5	5	7	8	10	7	1	X	2	4
6	11	13	13	10	2	1	3	7	9	10	12	9	3	2	X	6
7	11	13	11	10	4	7	9	1	3	10	12	9	3	4	6	X

differs from anc by $aFSp_1$, $ColC_1$, and $NkSw_1$ (Figure 11-18*a*). As a second taxon is added and node 1 is formed, these three character states must be sorted to see which belongs to A alone. Both A and C have $ColC_1$, but only A has $aFSp_1$ and $NkSw_1$. The hypothetical ancestor at node 1, then, differs from the original ancestor only by $ColC_1$. A new column is entered in the data matrix (Table 11-9) to reflect this difference. Treating node 1 as a new taxon, one adds a new row and column to the similarity matrix and calculates the distance of node 1 from each of the taxa. Finally, the other character states needed to describe taxon C are found in the data matrix and entered on the growing tree (Figure 11-18*b*). Figure 11-18*c* includes taxon F.

The process is repeated as before. Character states are entered on the tree at each step, character states of each node are entered in the data matrix, and the distances between the new node and other taxa and nodes are calculated. Note that character state $aFSp_1$ is a parallel character state but is correctly entered subtending taxon C. As the tree develops, the character state will automatically be located wherever it belongs. Similarly, reversed character states are also entered, at least in this example, in their correct positions. Such reversals become apparent when a taxon cannot be placed on the tree without a reversal. Only the first three steps, which demonstrate the system, are shown. The final tree is identical to the cladogram in Figure 11-12.

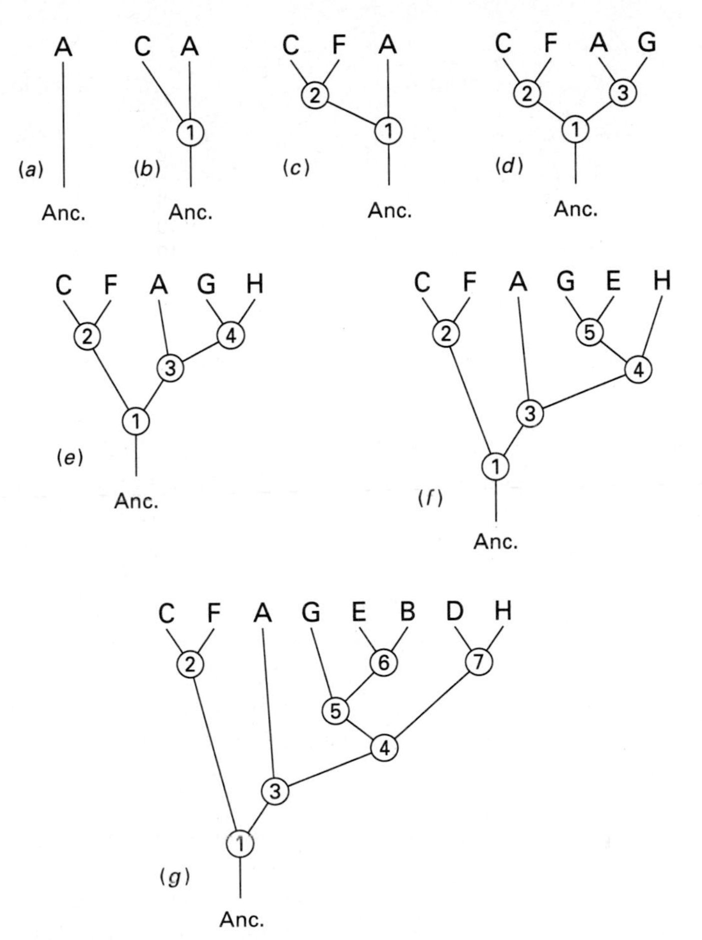

FIGURE 11-17
Wagner tree construction using the simplified method of Whiffin and Bierner (1972).

The final data matrix (Table 11-9) records the character states of each taxon and those inferred for each node, while the similarity matrix (Table 11-10) contains the distances between all combinations of taxa and nodes, accounting for parallel and reversed character states in the process.

As explained in Chapter 9, these computational difficulties are a very minor problem compared with other difficulties inherent in the methodology of cladistic analysis. To justify the most parsimonious (shortest) tree, earlier cladists tended to minimize the frequency of homoplasies, particularly reversals. Parsimony methods and their difficulties were dis-

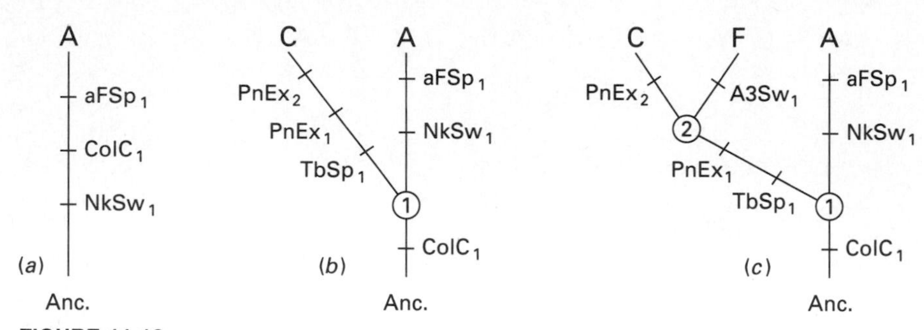

FIGURE 11-18
Wagner tree construction using the method of Farris; only the first three steps are shown.

cussed by Estabrook (1978), Anderson (1978), Panchen (1982), and Felsenstein (1982, 1988*b*). Numerical parsimony techniques that attempt to eliminate homoplasies were proposed by Farris (1970), Fitch (1971), Hartigan (1973), Sankoff and Rousseau (1975), Swofford and Maddison (1987), and others.

The clique compatibility method and the Wagner tree parsimony method just described have been explained simply so that they can be attempted by hand by a patient person with a data set that is not too large. It must be emphasized that they are very primitive forms of these algorithms, useful for helping a student appreciate how a computer program approaches the difficult problem of machine cladistic analysis.

Current programs are far more sophisticated; they employ many variations on compatibility and parsimony techniques and incorporate many enhancements. For example, after the first run, a parsimony method may eliminate the weakest character states and be run again. The process is repeated until no further improvements can be found. Various branches can be moved about to achieve a better fit of character states and a more parsimonious tree. In clique analysis, an incompletely resolved tree can be broken into smaller trees, and any smaller cliques not used in the first run can be tried on the partial trees. Parsimony may be restricted in such a way that Dollo's rule is not violated.

Maximum-Likelihood Estimation

Felsenstein (see his 1982 and 1988*b* reviews) has been the leading developer of statistical methods for inferring phylogenies. These methods have generally been applied only to molecular data. They presume that evolution follows a simple stochastic model; under this model, some patterns of distributions of character states among the taxa are more probable than others are. The tree for which the probability of observing the data is highest is chosen as the maximum-likelihood estimate of the phylogeny. Trees are produced that may have less than the best obtainable parsimony scores but greater biological reality.

These methods have been criticized primarily on the grounds that the models of evolution upon which they are built are incorrect oversimplifications of reality (Farris 1986). Even if this is true, the statistical methods should not simply be abandoned but should instead be more vigorously explored and improved; they will play an important role in the future of numerical systematics.

In his 1982 and 1988*b* papers, Felsenstein reviewed many current approaches to numerical phylogeny inference, and these articles are highly recommended. We especially applaud his discouraging of dogmatism and his desire for absolute certainty of results. Felsenstein ended his review with the statement "The adoption of a methodology that explicitly acknowledges uncertainty is a paradoxical necessity if phylogenetic inference is to be placed on a firm scientific footing" (1982:399).

Other Methods

Lake (1987) developed a method of phylogeny inference for nucleotide sequence data that is based on the fact that under a certain generalized model of nucleotide evolution, the relative frequencies of some patterns of character state distribution have a particular relationship to one another. Specifically, an algebraic combination of the frequencies of several patterns has an expected value of zero. This is referred to as the method of invariants. See Cavendar (1989) for a discussion of this promising method.

Other methods for the inference of phylogenies based on character data have been developed. Felsenstein (1988*b*) should be consulted for an overview and an entrance into the literature.

Weighting

Experienced taxonomists know that different character states can have highly unequal weight, that is, unequal amounts of information on relationship (Chapter 7). The pheneticists' failure to give differential weights to character states was a serious deficiency of their methods. Cladists as well as evolutionary taxonomists consider weighting essential for the improvement of classifications.

In addition to the nonnumerical methods discussed in Chapter 7, numerical methods of weighting are now available. Any current program will allow one to assign different numerical weights to characters a priori. Furthermore, Farris (1969, 1979*b*) described several methods by which weights can be calculated during the numerical analysis. Basically they require rerunning data through the program, which has given the best (most unambiguous) character states higher weight with each repetition, and giving homoplasious or seemingly irrelevant character states increasingly lower weights or removing them altogether. Usually, three to four runs will optimize the cladogram. In the final output, all character states are again given a weight of 1. Felsenstein (1981, 1982) described a maximum-likelihood approach to the weighting of character states and showed some similarities of Farris's (1969) character weighting by successive approximations to compatibility analysis.

WHICH METHOD IS BEST?

No taxonomist has the time to test one numerical method after the other in order to see which one produces the best, or most meaningful, phylogeny. However, in a remarkably large number of recent taxonomic revisions, the authors actually used a variety of distance and character-based methods to determine which would produce a phylogenetic hypothesis that best reflected the natural pattern of that group. In the course of this testing and comparing of methods, it has already become apparent that certain methods are inferior. It is hoped that eventually the number of proven methods will be narrowed to a few (Felsenstein 1988*b*).

APPENDIX: Computer Programs

A large array of computer programs are now available. They all have certain advantages and disadvantages. The NTSYS package does UPGMA phenetic clustering and can be obtained through Dr. F. James Rohlf, Department of Ecology and Evolution, State University of New York, Stony Brook, NY 11794. There are a number of packages for phylogenetic analysis. The most widely distributed ones are listed below.

- PHYLIP (Phylogeny Inference Package). Contains the broadest array of analyses, including distance methods, parsimony, compatibility, and maximum likelihood. The programs are slow but work on many different computers. They can be obtained through Dr. J. Felsenstein, Department of Genetics, University of Washington, Seattle, WA 98195.
- PAUP (Phylogenetic Analysis Using Parsimony). This is the most comprehensive package for parsimony inference of phylogenies; it is fast and very efficient at finding most parsimonious trees. It runs on MS-DOS, Apple Macintosh, and other machines. It can be obtained through

Dr. D. L. Swofford, Illinois Natural History Survey, 607 East Peabody Drive, Champaign, IL 61820.

• HENNIG-86. This is an efficient and very fast program for finding most parsimonious trees that run on MS-DOS–based machines. It can be obtained through Dr. J. S. Farris, Department of Ecology and Evolution, State University of New York, Stony Brook, NY 11794.

• MacClade. This program contains many tools for the manipulation and analysis of phylogenies and character evolution. It runs on Apple Macintosh computers. It can be obtained through David Maddison, Museum of Comparative Zoology, Harvard University, Cambridge, MA 02138.

Every few years new versions of these programs are issued, and new programs appear regularly. In some of the packages, particularly PHYLIP and PAUP, several versions are available that are especially suitable for particular data sets and particular computers (mainframe computers versus microcomputers). Some of these packages can be obtained free; for others a fee must be paid which is quickly repaid in saved computer time. The prospective user should inquire at the addresses listed above what the latest versions of these programs are. Christopher A. Meacham and Thomas Duncan, University Herbarium, University of California, Berkeley, CA 94720, have compiled a list of software that is of interest to systematists; the list will be updated to contain information about the latest versions and new programs.

PART **C**

METHODOLOGICAL ISSUES

The three chapters that constitute this part concern taxonomy as a whole rather than microtaxonomy and macrotaxonomy. They provide information about methods and principles relating to taxonomic collections (Chapter 12), taxonomic publications (Chapter 13), and the theory and practice of zoological nomenclature (Chapter 14).

CHAPTER 12

TAXONOMIC COLLECTIONS AND THE PROCESS OF IDENTIFICATION

All classification is based on the comparison of specimens that represent populations and species. We can determine the species-specific characteristics of an animal only by comparing it with members of other similar species, preferably its nearest relatives. An adequate comparative collection is therefore as indispensable to taxonomists as electron microscopes, Warburg apparatuses, and ultracentrifuges are to cellular and molecular biologists.

Collections must be borrowed from museums, which are repositories of systematic collections, or collected by the specialist. Commonly, both sources must be tapped. Borrowed material is usually insufficient in certain crucial areas and does not provide the biological information that is vital in modern taxonomic research. However, it would require many years of effort for a single collector to achieve the broad geographic scope seen in a museum collection accumulated during scores of years or even centuries.

SYSTEMATIC COLLECTIONS

Value of Collections

Museums play an important role as centers of documentation. They supply a permanent record of faunas and floras, particularly for localities where the biota has been destroyed by natural catastrophies or the activities of people. This is particularly true of the localized biota of streams, lakes, islands, and tropical forests. Museums contain a sampling of many areas that are inaccessible owing to their remoteness or for political reasons. Much of the material preserved in institutions can be replaced only at very high cost or cannot be replaced at all. Other material is of unique value because it forms the basis of published research. It may be needed again at a later period for verification of the original data or for renewed study in the light of more recent knowledge or new techniques. For a survey of the scientific significance of taxonomic collections, see *Biological Materials* (Mayr and Goodwin 1956) and Miller (1985).

A taxonomic revision of a given group can be done only if adequate material from the majority of included species can be assembled. It is therefore legitimate, indeed necessary, for curators of museums to build up their collections in groups that they or their associates plan to study in the future. The working taxonomist knows from sad experience how many revisions were started only to be set aside for lack of adequate material. Material in a collection may not be in use, but this does not mean that it is ready to be disposed. Collections are reference tools, just as necessary as the books in a library, which are not in continuous use yet must be available when needed.

Purpose of a Scientific Collection

For the old-style, typologically oriented taxonomist, a collection was an identification collection. When the taxonomist obtained additional specimens of species already represented in a collection, he or she considered them duplicates to be used for exchanges or given away. According to current thinking, biological classification consists in the ordering of populations (Chapter 6). Collecting, then, is the sampling of populations. Considering the great variability of most natural populations (Chapter 4), an adequate sample of every population should be collected and preserved.

But what is "adequate"? In making the right decision one must be guided by a number of facts. More material is needed in a species with strong individual and geographic variation than in a uniform species. More material is needed for studies of specific and subspecific characters than for studies of the characters of higher taxa. Birds, for instance, are on the whole more uniform in their anatomy than they are in their plumage characters. For this reason it would be illogical to preserve large series of spirit specimens of every species, as are needed for the study of the geographic variation of plumage characters.

The size of the sample thus depends on the objectives of the research. The presence or absence of an anatomic character that is diagnostic for a higher taxon is in most cases not subject to much variation, and one or two preserved specimens of a number of genera may be all the working material needed. However, when a population analysis is attempted, whether dealing with size, proportions, coloration, or polymorphism, large samples from numerous localities are needed. In fishes, for instance, where each population of a species may vary in some meristic characters (e.g., fin rays, vertebrae), a single specimen per locality would tell us very little. All phena of a species should, so far as possible, be represented in the collection.

Collecting and Research

Most taxonomists spend little time collecting new material. Similarly, most specialists have caught up reasonably well with material they themselves have collected. There are, of course, exceptions. Some big expeditions of the nineteenth century gathered material that has still not been fully worked out. Large oceanographic expeditions tend to accumulate far more material than the extremely small number of specialists can work up at once. However, even this material is generally processed to make it readily available to specialists.

Scope of Collections

Only a few large national museums attempt worldwide coverage in all groups of animals. Most museums restrict themselves to a geographic area and to certain groups of animals. It is important for the staff of a museum to have a clear-cut acquisition policy. Coverage that is too broad inevitably leads to shallowness and a failure to obtain the depth required for monographic studies.

There has been a steady and wholesome trend away from broad-purpose expeditions and faunal surveys and toward intensive collecting of specific families or genera. The late Admiral H. Lynes, for instance, who was especially interested in *Cisticola,* a genus of African warblers with 46 species, made a series of collecting trips to nearly every corner of Africa. He combined the collecting of specimens with a detailed study of the ecology, habits, songs, and nest construction of these birds. The result was that the genus *Cisticola,* formerly the despair of bird taxonomists, is now reasonably well understood (Lynes 1930). The work of J. Crane on fiddler crabs (*Uca*) is another example.

If a faunal survey is to be attempted, however, it should be directed to a natural geographic area, unhampered by blind dependence on political boundaries. Moreover, faunal surveys have had the best results when concentrating on a particular taxonomic group. The most ambitious single undertaking along these lines in this century was probably the Whitney South Sea Expedition, which operated under the auspices of the American Museum of Natural History in New York. This expedition visited practically every island in the South Pacific from the Tuamotus and Marquesas in the east to the Bismarck Archipelago, Palau, and the Marianas in the west, largely concentrating on bird collecting. It operated continuously from 1921 to 1934, and its work was continued by single collectors into 1940. The essentially complete collections made during this single expedition supplied the material not only for scores of detailed revisions but also for basic zoogeographic and evolutionary studies such as *Systematics and the Origin of Species* (Mayr 1942). The value of a single well-made collection is generally far greater than that of an equivalent number of casually collected specimens.

Where and How to Collect

A collecting trip must be carefully planned. All possible geographic information must be obtained beforehand, including the distribution of vegetation types, altitudes, and types of seasons as well as information on means of public and private transportation, the availability of health care, and so forth. In addition, previous collections from the area must be carefully analyzed, and existing type localities must be mapped. A plotting of collecting stations and, in particular, a mapping of species distributions will reveal the location of crucial gaps. If the study of geographic variation is a major objective, the periphery of the range of each species should be given particular attention because this is where geographic isolates and incipient new species occur most frequently. If a species shows seasonal variation, the collections should be spaced seasonally. The season during which sexual maturity occurs is relatively short in many animals, and so collecting should be done during that time. This is even more important if recordings of song and courtship, nests, egg proteins, embryos, or other materials are needed and can be obtained only during the breeding season. In the case of allopatric populations whose categorical status is uncertain (species or subspecies), a special effort should be made to collect in the intervening region to determine whether intergradation occurs.

Owing to the rapid recent increase of human populations, resulting in more intensive agriculture and drastic deforestation, many areas are in desperate need of immediate collecting before the localized faunas become extinct. *At the present time this task is far more urgent than is collecting in remote uninhabited areas.*

Innumerable techniques for the collecting of different groups of animals are described in standard manuals. New techniques are continually being developed, such as the use of mist nets for bird collection and "black light" (ultraviolet lamps) for insect collecting. Different kinds of traps, baits, poisons, and so on are well known to specialists, who generally share such information with beginners. Depending on the taxonomic group to be covered, one of the following books will be the most useful: Anderson 1955 (vertebrates); Beer and Cook 1957 (ectoparasites); Bianco 1899 (marine animals); British Museum 1936 (various groups); Kirby 1950*a* (protozoans); Kummel and Raup 1965 (fossils); MacFadyen 1955 (soil arthropods); Martin 1977 (mammals); Oldroyd 1958 (insects); Russell 1963 (marine animals); Steyskal, Murphy, and Hoover 1987 (insects and mites); Van Tyne 1952 (birds); Wagstaffe and Fidler 1955 (invertebrates); Williams, Laubach, and Laubach 1979 (mammals). Knudsen (1966) includes many references to special techniques.

Collections of Specimens

The classical image of a systematic collection is that of preserved whole specimens. These specimens may be dried skins of mammals or birds or insects mounted on pins. Most reptiles, amphibians, fishes, and invertebrates are preserved in alcohol or another liquid preservative.

In the field every effort should be made to collect unbiased population samples. One should avoid accumulating large numbers of "aberrations" or concentrating on the conspicuous sex in sexually dimorphic species. Not only adults but also an adequate sample of all stages of the life cycle (including larvae) and associated parasites should be collected. Sampling should be done in a way that provides study material not only for the describing taxonomist but also for the evolutionist.

Biological Information

Modern systematists need a great deal of additional information, much of which they have to secure themselves while studying a living organism either in its native environment or in the laboratory. Much of this information can be preserved in the form of permanent records: films of courtship displays and other aspects of behavior, recordings of the vocalizations of animals (tapes, sound spectrograms), and collections, photographs, or casts of the work of animals (nests, galls, spiderwebs, tracks).

Museums increasingly include aquariums, terraria, insectaries, aviaries, and the like among their facilities in order to permit study of and experimentation with living species. Indeed, several of the larger natural history museums have a laboratory floor or wing for their live collections and experimental work. Special laboratory facilities are needed for most modern molecular research. The importance of molecular characters is stressed in Chapter 7.

Preservation of Specimens

Techniques of preservation differ from one taxonomic group to the next, but the purpose is always to preserve specimens so as to make them least subject to deterioration through the action of insect pests, mold, oxidation or bleaching by sunlight, drying out, protein decay, and the like. The *manuals* and *handbooks* referred to under "Where and How to Collect" contain descriptions of specific methods. Properly preserved specimens of some groups are still in satisfactory condition 200 years after collection. With more and more species becoming extinct, the problem of "permanent" preservation is often raised. Some suggested methods, such as embedding in plastic, are too new to permit predictions about permanency. Preservation in alcohol raises problems with regard to sealing and appropriate containers, as discussed by Storey and Wilmowsky (1955) and Levi (1966).

Anatomical, Tissue, and Molecular Material

Collections of whole animals must be supplemented by material that permits anatomic, histological, cytological (chromosomal), and biochemical-molecular research. Such materials can be frozen or stored in liquid preservatives.

Numerous newer techniques that are useful in taxonomic analysis have been developed by molecular biologists in recent years; they include electrophoresis and DNA hybridization (Chapter 7). The methods by which the material for such studies must be preserved are described in the relevant literature (Hillis and Moritz 1990).

Labeling

A specimen that is not accurately labeled is worthless for most types of taxonomic research. The label is so important that some taxonomists joke that to other taxonomists, it is more important than the specimen. Many kinds of information are desirable, but by far the most important single piece of information is the exact locality of a collection. For example, certain land snails have racially distinct populations living as little as half a kilometer apart, and the collection locality must be stated with great precision. If the locality is a village, farm, hill, creek, or other feature not easily found on commercial or geodetic (e.g., U.S. topographic) maps, its position relative to a well-known place should be added on the label ("22 km NW of Ann Arbor, Mich."). The county or district name should be given for less well known localities. Labels on specimens collected in mountains should give the altitude, and those from oceans should give the depth. Ecological information is essential for such forms as plant-feeding insects and host-specific parasites.

Whenever possible, the label should be written in the field when the specimen is prepared. The replacement of temporary labels with later permanent ones is a potential source of error. However, this cannot be avoided with insects when labels are printed for entire lots of specimens. All essential data should be recorded on the original labels. Data recorded in a field book are frequently overlooked and may be unavailable if the collection is divided. The original label should never be replaced by a museum label. A certain number of mistakes are always made in the transfer. If a museum label is desired, it should be added to the original label. Labels attached to specimens preserved in alcohol or formalin must be corrosionproof. Writing must be resistant to fading or washing out.

The data in addition to locality that are needed depend on the given group. Most good bird collectors, for instance, record on the label not only the locality, date, and collector's name but also the sex (based on autopsy), the actual size of the gonads, the degree of ossification of the skull (important for age determination), the weight (in grams), and the colors of the soft parts (Van Tyne 1952). The little extra time required to make these records is more than compensated for by the increased value of the specimens.

CURATING OF COLLECTIONS

Every taxonomist sooner or later is given the responsibility of curating collections. This requires a great deal of expert knowledge but also, more important, a clear understanding of the function of collections, the different kinds of collections needed in various areas of taxonomy, and various policies concerning the use of collections. The journal *Curator* and particularly the *ASC Newsletter* (Association of Systematic Collections) publish a great deal of information on this subject. The Society for the Preservation of Natural History Collections publishes a biannual *Collection Forum*. Useful information can also be found in *A Guide to Museum Pest Control* (1987, Association of Systematics Collections, Washington, D.C.), Cato (1986), Dowler and Genoways (1976), Hungay and Dingley (1985), and Mayr (1971).

Preparation of Material for Study

Bird and mammal skins are ready for study as sent from the field by the collector; mammal skulls have to be cleaned. Some insects should never be placed in alcohol or other liquid preservatives; others are useless when dried. Invertebrates that are preserved in alcohol or formalin are usually ready for study as preserved. Microscopic slide mounts or slides of parts of organs may have to be prepared in the case of smaller forms. Instructions can be found in textbooks of microscopic technique, including Bradbury (1984), Geligher and Key (1964), Gray (1973), Humanson (1979), Pearse (1980), and Postke et al. (1980). Most insects are pinned, and the wings are spread if they are taxonomically important (or beautiful), as in butterflies and moths and some grasshoppers. Species can be identified in many groups of insects only through study of their genitalia. Microscopic slides or dry or liquid mounts of the genitalia may have to be prepared. For information, consult Tuxen (1958), Scudder (1971), and Matsuda (1976). Special techniques are needed for the preservation of protozoans (Corliss 1963).

Housing

Research collections should be housed, like research libraries, in fireproof buildings that are reasonably dustproof. More and more museums keep their collections in air-conditioned buildings. Buildings originally designed for public exhibitions are usually unsuitable for the housing of research collections. Rapid changes in temperature and humidity are detrimental to museum cases and specimens. Photographs and films should be stored in air-conditioned rooms. Storage cases for specimens should be built sufficiently well to be insectproof and, ideally, dustproof. Various firms now construct steel cases that satisfy all these requirements. It must be remembered that insectproof museum cases reduce the labor of curating.

Cataloging

The method of cataloging depends on the group of animals. With the higher vertebrates, where collections consist of a limited number of specimens, it has been traditional to give each specimen a separate number and catalog it separately. This very time consuming procedure has been challenged, and it has been suggested that the methods of entomology and malacology be adopted for vertebrates (see below). All the specimens collected at a given locality or district or on one expedition are entered in the catalog together. This greatly facilitates the subsequent retrieval of distributional data and the preparation of faunistic analyses. Cataloging is usually done after the specimens have been identified at least as far as the genus. This permits a permanent reference to the contents of the collection long after it has been broken up and distributed in the systematic collection or even dispersed to other institutions.

Catalog entries of vertebrates usually contain the following items:

1 Consecutive museum number
2 Original field number
3 Scientific name (or at least generic name)
4 Sex
5 Exact locality
6 Date of collecting
7 Name of collector
8 Remarks

In groups for which the collections consist of large numbers of specimens, for example, insect collections, where additions of 100,000 specimens per year are not uncommon, it is customary to catalog accessions by lots, with each lot consisting of a set of specimens from a given locality or region. Lot numbers in turn may refer to the collector's diaries or to other sources of information on each collection. It is also customary to note whether a lot was received as a gift or by purchase or exchange; the names of the collector and donor are always given.

When museums and their collections were small, curators often had rather elaborate card-filing systems that facilitated the retrieval of all sorts of information, such as the collecting station and the name of the collector. Maintenance of such files is so expensive in terms of clerical and curatorial time that they have been abandoned in most museums where the emphasis is on research rather than providing information to the public. A properly organized and well-curated collection is a reference catalog in itself and permits rapid information retrieval.

Card methods have been replaced by *electronic data processing* (*EDP*). Numerous methods of employing computers in cataloging have been described in the volumes of *Curator,* particularly in a special computer issue [Vol. 30(2), 1987]. A reasonably up-to-date review of the field is provided by Sarasan and Neuner (1983).

The maintenance of catalogs and card files is time-consuming and should never be carried to the point where it interferes with work on collections. A list of the accessions is indispensable, however, since it often allows the recording of additional information on localities which cannot be entered in full on the specimen labels.

Arrangement of the Collection

As far as possible, the collection should be arranged in the same sequence used in a generally adopted classification. The sequence of orders and families is reasonably standardized in many classes of animals. Whenever it is not to be worked out as a separate collection, unidentified material is placed with the family or genus to which it belongs (see "Identification"). The contents of trays and cases should be clearly indicated on the outside. When specimens are of highly unequal size (e.g., fish), storage in a strict taxonomic sequence is very wasteful of space. Very large specimens may have to be stored separately.

Curating of Types

The names of species are based on type specimens (Chapter 14), which are thus official standards. Being virtually irreplaceable, these specimens must be curated with special care. Whenever doubt arises about the zoological identity of a nominal species, only reference to the type can resolve the doubt.

Many descriptions by classical authors are equally applicable to several related species. Early entomologists rarely if ever referred to the structural detail of the genitalia, which is now indispensable for diagnostic purposes in most groups of insects. Only reference to the type can establish the basis for the classical name. The curator must make every effort to ensure the safety of these irreplaceable specimens. Types are customarily deposited in large collections in public or private institutions which have come to be recognized as standard repositories of types.

When conducting an authoritative revision of a given genus, a specialist should be able to see all the existing types. If many of them are in a single institution, the specialist should travel there for examination. He or she should obtain scattered types by mail loan. Modern curators are quite liberal in lending type specimens to qualified specialists. The number of recorded losses of shipments has been small, and if there is a real need for the replacement of a lost type, a neotype can be designated, as permitted by the Code in specified circumstances (Chapter 14). Ideally, types should be housed in a separate collection to facilitate rapid removal in case of emergency and to avoid constant handling in a general-study collection. They should be clearly labeled in distinctive colors. If not previously cataloged, they should be numbered individually to facilitate reference to them in the literature and retrieval from the collection. Since many types are those of synonyms, an index by genera and another by species will save much time in locating the desired type. It is not economical to include in type collections all sorts of pseudotypes, that is, specimens that are not holotypes, lectotypes, neotypes, or syntypes (Chapter 14).

There is much to recommend the arrangement of type collections alphabetically according to the given specific name. A type collection is a reference collection rather than a classification, and ease of reference should determine which system is adopted. When number systems are adopted, too many errors are usually committed for them to be practical.

Type specimens assume such an important role in the taxonomy of lesser-known groups that many workers believe that no individual has the right to retain a type in a private collection after his or her study has been completed. Specialists sometimes donate their collections to a public institution but retain them on permanent loan until the end of their active research period. Retaining such types includes the obligation to send out specimens on loan if requests are made by other specialists.

Exchange of Material

In the days when taxonomists considered most of their specimens duplicates, exchanges were very common. However, selecting appropriate material for exchanges and keeping the required records are time-consuming activities, and exchanges are no longer popular except among private collectors. A specialist doing a monograph on a certain genus or family can always borrow material from other institutions and return it when the studies have been completed. Exchanges are most desirable in groups where series of unlimited size can be obtained and where material is offered from areas that are not readily accessible to an institution (e.g., intercontinental exchanges). Exchanges are least defensible when they lead to the dispersal of biologically important population samples. Exchanges are sometimes necessary to build up complete identification collections. When an exchange is carried out, generosity is always the best policy. It is generally not advisable to insist on exchanging specimen for specimen except when institutional policies or other unusual factors demand it. Many specialists give away excess specimens of a large series as open exchanges, not necessarily expecting a return.

Expendable Material

Improperly preserved or inadequately labeled specimens are usually valueless. Collections would be better off without them, yet they are much less of a burden on a collection than they would be if the curator were asked to find and eliminate them. The curator must make sure that each specimen is not an unlabeled type or a specimen which is unique or of historical value. The most efficient way to eliminate useless material is to ask specialists to pull out such specimens when scrutinizing the material during a revision.

Loans

Modern curators are very generous in lending specimens to qualified experts. The axiom that systematic collections are the general property of science, not of a specific institution or curator, is being acknowledged more widely. Every loan, however, involves loss of time and effort, and the borrower should reimburse the lender for this outlay. Research grants often include an item for such purposes, covering not only the costs of postage but also those of selecting the specimens, recording the loan, and getting the material packed for shipment. The modern curator, being essentially a research worker, must delegate these tasks to hired clerical help.

The borrower also has well-acknowledged obligations. A request for the loan of specimens should be as specific as possible, including a statement of the reason for the request and some indication of the length of time for which the material is needed. The beginner may be unable to borrow certain material except through a loan to his or her institution or to a well-known colleague or the beginner's major professor. In such cases any laxity in carrying out the conditions of the loan reflects not only on the beginner but on the sponsoring individual or institution. If the borrower is unable to complete the studies in the time designated, the person or institution that made the loan should be informed. The lender should never be placed in the embarrassing position of having to write and ask about the status of the study. The borrower should not ask for material that is not actually needed or ask for material that can be studied easily by traveling to the museum.

If a specialist has agreed to identify a collection provided that he or she receives certain specimens, the specialist should make sure that the terms of the agreement are well understood and should return to the lender a list of the specimens which he or she has retained. All types and unique specimens must be returned to the lender in such cases. With anatomic material, it is understood that dissection, the purpose for which the specimen was originally collected, will partially or entirely destroy it. Those borrowing such material are obliged to preserve a pictorial record of the dissection.

To cut down on loans, institutions sometimes make temporary or permanent transfers of collections. For instance, an inland museum that owns a small collection of marine invertebrates that are not used for exhibition or instruction may transfer it to a large museum that is active in marine research. However, excessive consolidation of taxonomic collections not only creates monopolies but has another inherent danger: The concentration of so much irreplaceable material in a single institution makes it exceedingly vulnerable to destruction in the event of a catastrophe. Long-term loans to leading specialists are a better solution in most cases.

IDENTIFICATION

A number of very different activities are usually combined under the word *identification*. They all involve the identification of previously unidentified material, but for very different purposes. (See Chapters 1 and 6 for the distinction between identification and classification.)

Sorting of Collections

All material gathered on collecting trips and expeditions must be sorted and at least tentatively identified before it can be incorporated into a collection. The first rough sorting of freshly collected material is often done in the field. An entomologist may keep specimens of different species from different hosts in separate containers. Collections made during oceanographic expeditions are often roughly sorted immediately, in part because different kinds of animals may require different methods of preservation. After the specimens have been properly preserved and labeled, the usual practice is to segregate unstudied material down to orders and, whenever possible, to families or even genera. Such material is then available to the specialist, who can undertake precise identification. Even such a tentative identification requires skill and experience. When large collections are involved, the establishment of a special sorting organization is advisable. The Smithsonian Institution in Washington maintains the Smithsonian Oceanographic Sorting Center (SOSC), which is charged with the preliminary sorting of the material gathered by the Indian Ocean Expedition and other oceanographic expeditions. When identified to family or genus, the material is shipped to the specialists who have agreed to study it.

The ease of sorting newly received collections depends on the nature of the material (larvae or adults, microscopic in size or not) and the maturity of the taxonomy of a given group and in a given region. In birds, for instance, identification at the species level is virtually never a problem, although subspecific identification may be difficult. In less well known groups for which no recent revisions, keys, or manuals are available, definite identification may be possible only down to the level of the family or at best to that of the genus.

Linnaeus thought that every zoologist should know every genus. (Only 312 genera of animals were known to him.) At a time when the collections and the number of described species were still small, taxonomists attempted to identify every specimen, even when it belonged to groups about which they had no special knowledge. It is now realized that this is a very wasteful approach. When accessions are sorted in a modern zoological museum, they are identified only to the level (order or family) where they become available to the specialist. It is far more economical and important for a taxonomist to devote time to the preparation of new monographs and keys than to attempt identification down to the species level in groups with which he or she is not familiar.

Moreover, identifications by nonspecialists are often erroneous. Most large collections contain numerous examples of misdeterminations. The original specimen in a series might have been quite authentic in such cases, but other specimens were subsequently added without critical analysis and without determination labels by experts. Such misdeterminations cause more trouble than does leaving the material unidentified until it can be studied by an expert.

Determination Labels

Sooner or later all material in collections is seen by a qualified taxonomist or specialist who is able to identify it to species. Each specimen or each series should be labeled when such identification is made. The determination label should give the scientific (generic and specific) name and author, along with the name of the determiner and the year in which the identification was made. With this information entered for every specimen, the authenticity of the determination is established, and its dependability can be readily evaluated at any subsequent date on the basis of progress which may have been made in the study of the group during the intervening years. In bird and mammal collections these names are usually written in pencil so that they can be changed easily if there is a change of nomenclature.

Identification of Individual Specimens

The taxonomist is frequently called upon to identify a particular specimen or species. If such identification is highly important for a special research project in applied biology or experimental zoology, a taxonomist will gladly make every effort to identify the species on which the research is based. However, it is not the job of the taxonomist to undertake the routine identification of ecological collections or archaeological material. Such identification work is the responsibility of the ecologist or archaeologist who wants the material identified.

There are federal and state agencies charged with the responsibility of identifying economically important animals. These agencies employ specialists, to each of whom a particular group is assigned. The research taxonomist, by contrast, is not responsible for routine identifications. Nothing reduces the productivity of a research museum more than attempting to fill miscellaneous identification demands by the public.

Process of Identification

Even a rank beginner trying to identify a specimen can usually tell that it is a bird, a spider, a grasshopper, or a butterfly and can then immediately go to keys and manuals for the appropriate zoological groups. The real beginner, however, will have trouble with all but the most common kinds of animals. When in doubt about the order to which an animal belongs, the beginner should try the simple keys given in general textbooks and handbooks. Even the advanced student may encounter unusual species or immature or exotic forms that cannot be placed on sight in the proper family or order. However, modern works are generally available which provide family and subfamily keys to assist in this stage of identification.

New handbooks, keys, field guides, and bibliographies are published every year, and any listing, such as that provided by Mayr (1969:114–115), is soon out of date. Every natural history library can provide titles. This is also true for faunal works covering specific countries in Asia, Australasia, Africa, and South America. There is hardly a region in the world not covered by a reasonably modern fauna of mammals, birds, butterflies, or other better known groups.

Identification is vastly more difficult when no convenient keys and manuals are available. The beginner is advised not to attempt it. If a monograph or technical revision of recent date is available (see below for how to find such literature), the specimen is run through the keys, the description of the appropriate species is checked character by character, the specimen is compared with any illustration that may be given, and the recorded geographic distribution is checked. If all these points agree, the identification is considered as tentatively made, subject to comparison with authentic specimens and provided that no additional related species have subsequently been described. For further details on these steps, see the following sections. When no recent monograph or revision is available, no one but a specialist should attempt a determination. Even a specialist will not waste time trying to identify single specimens except under the exceptional conditions specified above.

MATERIAL FOR REVISIONARY OR MONOGRAPHIC WORK

The major occupation of the taxonomist, except perhaps in the case of a specialist in a very well known group of animals, is the preparation of revisions. The early part of the process—the determination of the species to be classified—resembles the process of identification only in some ways. The details of this procedure are described in Chapters 5 and 6.

However, before the actual work of revising can be started, the taxonomist must gather the needed specimens and the literature. Let us assume the taxonomist wants to revise a certain tribe of beetles from South America. After the taxonomist has completed a first examination of the material available in his or her own collection, the next task is to write to the major museums whose scope includes the Coleoptera of South America and ask for a loan of their material. During such correspondence the taxonomist may discover that another taxonomist has also started to revise this group; it will then be necessary to negotiate with the other specialist about how to divide the task. In the case of large collections rich in types, the taxonomist will have to make arrangements to visit these museums rather than ask for a loan of the material.

The tracing of literature is usually far more difficult than the tracing of the available material. One may start with a general biological bibliography such as Smith and Reid's *Guide to the Literature of the Life Sciences* (8th ed. 1972) and Besterman's *Biological Sciences* (1971). An excellent source is R. W. Sims's *Animal Identification: A Reference Guide* (1980). For vertebrates, one can consult Blackwelder's *Guide to the Taxonomic Literature of Vertebrates* (1972); for insects, Gilbert and Hamilton's *Entomology: A Guide to Information Sources* (1983). There are also many very specialized bibliographies such as Skarbilovich's *Bibliography of Russian Literature on Nematodes* (1985) (605 pages!). Every better library will help the beginner become familiar with the available bibliographic resources.

When there is no recent monograph or revision, the most recent catalog for the group should be consulted. The catalog will give literature citations pointing to the descriptions of all species known up to the time of its completion. Some catalogs furnish even more information, e.g., complete bibliographies under each genus and species, lists of synonyms, and geographic distribution. Taxonomic research is greatly facilitated by a good catalog, because the catalog brings together the most significant published references to the group being studied.

Reference to Current Bibliographies

Catalogs are inevitably out of date soon after they are published. This difficulty may be partially compensated for by the issuance of supplements. Nevertheless, it is not at all unusual to find that even the most recent catalog is 20 years old. In some of the major insect orders no general catalog has been prepared since 1900, and some groups have never been cataloged from a world standpoint.

Fortunately, there is an unusual bibliography of the literature in systematic zoology, a great reference work entitled *The Zoological Record*. It is indispensable for taxonomic work and has appeared every year from 1864 to the present. Each new scientific name is given, together with a reference to the place of publication and the type locality. The names are arranged alphabetically under families, but a systematic arrangement is followed for families and higher groups. Current numbers are available separately by purchase or subscription, and complete sets are held by major libraries.

The Zoological Record, formerly published by the Zoological Society of London in cooperation with the British Museum (Natural History), is now published by Biosis (Philadelphia). The following 20 sections of *The Zoological Record* are published separately and can be obtained as a group or singly each year: (1) Comprehensive Zoology, (2) Protozoa, (3) Porifera, (4) Coelenterata, (5) Echinodermata, (6) Vermes, (7) Brachiopoda, (8) Bryozoa, (9) Mollusca, (10) Crustacea, (11) Trilobita, (12) Arachnida[1] and Myriapoda, (13) Insecta, (14) Protochordata,[2] (15) Pisces, (16) Amphibia, (17) Reptilia, (18) Aves, (19) Mammalia, (20) List of New Generic and Subgeneric Names. These sections can be ordered from Biosis, 2100 Arch Street, Philadelphia, PA 19103–1399.

The most common method of using *The Zoological Record* is to start with the most recent volume and work back to the date of completion of the most recent catalog or revision. The genus or other group in question can be located in the table of contents of the section devoted to the particular class of animal. New names, synonymies, distributions, and in some instances even biological references are given. If the citation is not clear because of its abbreviated form or if the exact title of the publication is important, reference can be made to the bibliography of papers arranged by author at the beginning of the section. For special needs, there is an elaborate subject index that covers various phases of morphology, physiology, ecology, and biology.

Some groups of animals have never been cataloged or monographed. This is especially true among insects. For these groups, it is necessary to work through the entire *Zoological Record* back to Volume 1 (1864).

The best annual review of the taxonomic literature prior to 1864 can be found in the *Berichte über die wissenschaftlichen Leistungen* in different branches of zoology, including entomology and helminthology, published

[1]To Arachnida are added Merostomata, Pantopoda, Pentastomida, Tardigrada, Myriapoda, and Onychophora.

[2]Photochordata are treated together with Pogonophora, Enteropneusta, Graptolithina, Pterobranchia, and Phoronidea.

in Wiegmann's *Archiv für Naturgeschichte* (Berlin, 1835 et seq.). Additional important bibliographic aids covering this early period of zoology include Engelmann (1846), Agassiz and Strickland (1848), and the catalog of scientific papers published by the Royal Society (1800–1863). Sherborn's *Index Animalium* (1758–1800, 1801–1850) gives a complete list of generic and specific names proposed up to 1850, and Neave's *Nomenclator Zoologicus* (1939–1940, 1950, 1966, 1975) lists all generic names for the period 1758–1965.

Nominal Species and Zoological Species

Not all names (*nominal species*) found in the literature represent different zoological species. Many individual variants have been erroneously described as separate species. Chapters 4 and 5 give instructions on how to determine which names refer to valid species and which refer to intraspecific phena.

Equally troublesome are cases where similar (or not very similar) species masquerade under the same name. The correct determination of the actual zoological species in the group to be revised is the most important basic step in taxonomic research. It requires consulting original descriptions (or improved redescriptions), surveying authentic specimens, and, in any authoritative revision or monograph, studying the actual type specimens.

Original Description

Although the secondary literature is often a great help, reference should always be made to original and more recent authoritative descriptions. Unless this is done, one cannot be sure that misidentifications in the secondary literature have been found and corrected. Original descriptions are located by means of catalogs, monographs, the *Zoological Record,* or other bibliographic sources, as described above.

Copies of the original descriptions may be difficult to find. Even the largest libraries are not complete, and the average university library is often found to be wanting. This is not so much a reflection on the caliber of libraries as it is evidence of the extent and diversity of scientific publications throughout the world. Although largely confined to a half dozen languages, taxonomic papers are published in practically every country. The sheer size of this literature poses a problem for libraries with limited budgets. The situation is further complicated because the priority rule places a premium on earlier works. No taxonomic work published since 1758 becomes out of date if it contains new names, and as a result of limited editions, losses through the years, and other factors, too few copies exist to supply all biological libraries. Microforms and facsimiles of some very old works are available at major libraries, and specialists may arrange for photocopies of the pages of periodicals they have to consult frequently.

The search for original descriptions is often complicated. Helpful, if not essential, to the searcher are full familiarity with and use of all locally available scientific libraries, reference to the *Union List of Serials* to locate publications in other libraries for interlibrary loan, extensive use of microfilm services and other copying devices, accumulation of reprints by purchase or exchange with other workers, and accumulation of photocopies.

Descriptions are the foundation of taxonomy, since only the printed word is relatively permanent. Types may be lost, and the original authors are available for only a brief span of years to pass on "their" species.

The original description should be read several times, first to obtain a general impression or mental picture of the actual specimen the original author studied. Then the characters that the original author or subsequent authors considered particularly important should be extracted and checked against the specimens in question. Finally, any comparative notes given by the original author should be checked. Such comparative characters are the most useful clues to identification.

Original descriptions are normally the court of last appeal for purposes of general identification. However, many of them are inadequate, particularly those published prior to 1800. The value of a description is in direct proportion to the judgment of the author and the author's ability to select significant characters and describe them in words. Also important are the extent and nature of the material available to the author at the time of description. For these reasons, descriptions given in a thorough and authoritative monograph of recent date are usually more usable than original descriptions are.

Illustrations are often as valuable as or more valuable than original descriptions, particularly when there are language difficulties. In such popular groups as birds or butterflies, many monographic works contain colored plates, which are often a great help in the rapid identification of specimens. Colored plates are not always well reproduced, however, and there are many opportunities for error if too much dependence is placed on them.

If the original description is accompanied by an illustration, the characters of illustration and those of description may conflict. It can sometimes be proved that the artist did not have access to the type specimen and used another in the belief that it agreed with the type. Such discrepancies are not infrequent in the works of early authors.

Comparison with the Type

Type specimens are the most authentic source but are much too valuable to be used for routine identifications. Ideally, in the course of a monographic study of a group, all type specimens should be reexamined. At this time the significant characters are usually known and can be checked using the same technique and the same interpretation of characters that are applied to the rest of the material.

In work with subspecies it is not always necessary to have type specimens for comparison if there is no question as to the identity of the species. However, a series of specimens from the type locality (*topotypical specimens*) is desirable to provide information on the characters and variability of the subspecies.

CHAPTER **13**

TAXONOMIC PUBLICATION

No piece of research is complete until its results have been published, because advances in every area of science are made known through publication. Even though a number of books give advice on various aspects of scientific writing, none address taxonomy. The objective of this chapter is to fill this void. Much has to be considered when one is preparing a manuscript for publication. Authors who neglect our recommendations or the more broadly established conventions of taxonomy are likely to frustrate their peers and damage their own reputation. A careful study of this chapter should benefit not only the beginner but the experienced taxonomist.

THE STRATEGY OF PUBLISHING

A research worker in any branch of science must ask at regular intervals whether he or she is pursuing the most productive line of research. Should the worker shift emphasis or switch to a different group of organisms? What sort of research is most needed in the particular group of organisms in which the worker specializes? A judicious balance of several approaches is often the most rewarding strategy.

A major taxonomic revision may require years of work prior to publication. The question inevitably arises whether there should be preliminary papers and whether the specialist should deal with other taxa to maintain a publication schedule. Obviously there is no excuse for padding a bibliography by publishing essentially the same results more than once. On the other hand, the fieldwork and research required for the preparation of a comprehensive revision may result in biological, evolutionary, and biogeographic findings that are better published separately in journals that specialize in those areas.

Some taxonomists find it difficult to complete their manuscripts. They should realize that no research is ever "complete." There will always be areas that have not yet been collected and collections in remote museums that have not yet been studied. The same is true in experimental research, where results are published even though more and perhaps slightly different experiments could well be conducted. The number of specialists is limited, and all specialists have a real obligation to publish their results as soon as a reasonably reliable summary of their research is feasible.

DOCUMENTATION AND INFORMATION RETRIEVAL

Every year many new authors start publishing. Tracking their output has become a formidable task. Various systems for the retrieval of this new information have been devised, most based on the resources of major libraries. Most major libraries have shifted their records to computer data banks that can be accessed by a variety of programs, some of which facilitate a literature search. Also, the animal taxonomist is fortunate in having a relatively complete printed record of all new names and taxonomic revisions in *The Zoological Record* (Chapter 12). It is essential for a modern systematist to be familiar with the resources and systems of the nearest major library; a periodic check of relevant journals and new acquisitions at that library is also useful. The day when everything a systematist needed could be contained in a personal library is long gone.

Every effort should be made by the taxonomist to facilitate the task of information retrieval and to avoid any act that might increase its difficulty. Name changing in violation of the principles laid down in the Preamble of the Code of Zoological Nomenclature (Chapter 14) and unnecessary splitting or lumping on the generic level are indefensible; so are changes in a widely adopted taxonomic sequence without a compelling reason and the proposal of new species not connected with complete revisions.

KINDS OF PUBLICATIONS

Taxonomic publications range from a short description of a new taxon that covers only part of a page to lengthy monographs and/or handbooks that may run to several volumes. These publications include identification manuals as well as revisionary works and new classifications. Some works stress the nomenclatural aspects; others stress life history, distribution, or illustration. There is a time in the history of the study of each group of organisms in which one kind of publication or the other is most useful. Specialists must fully understand the function of each kind of publication when selecting the form of their next publications in the field. The following comments on these different types of taxonomic publication may be helpful to the beginner.

Many different titles are used for more comprehensive taxonomic publications. Unfortunately, the meaning of such titles is not the same in different groups of organisms. Words such as *checklist, handbook,* and *review* likewise have different meanings in different fields. Mayr (1969:260–264) has discussed in considerable detail the nature of taxonomic papers published under such headings as *synopsis, review, revision, catalog, monograph, atlas, fauna, manual, handbook,* and *field guide*. In view of the lack of standardization of these terms, this material will not be discussed here. Instead, some general recommendations will be presented.

Description of New Taxa

The isolated description of new subspecies, species, and genera, divorced from revisional or monographic work, is the least desirable form of taxonomic publication, except in well-known groups. The preparation of such a description does not permit as careful a comparison of all related species as does a revision; it often results merely in the addition of still another name to the already long list of nominal species. The final reviser must then search out the type—often quite inaccessible—and will lose valuable time getting it properly identified. The description of isolated new species in poorly known groups of animals is usually a handicap to subsequent workers. Such isolated descriptions are justified only when names are needed for biological or economic work in progress or when a group has been revised recently and the new species can be readily fitted into the classification.

Theoretically, providing an adequate description of even a single species requires much of the work basic to the preparation of a revision, and authors could carry their work to that point with a little extra effort. All too frequently, however, the isolated description is based on a superficial acquaintance with previous work. Many more synonyms are created through isolated descriptions than through more substantial revisions.

Catalogs and Checklists

A *catalog* is essentially an index to published taxa arranged so as to provide a complete series of references for both zoological and nomenclatural purposes. Species and genera are often listed in an alphabetical sequence, because most catalogs are nothing but uncritical (prerevisionary) listings of nominal species. Valuable catalogs are thorough and comprehensive, and their preparation is therefore a highly technical task that requires great patience and an intimate knowledge of bibliographic resources and methods.

A *checklist* provides a convenient source of reference for the correct names of specimens and the arrangement of collections. Checklists vary greatly in elaborateness. A list of names deserves to be called a checklist only if a careful distinction is made within it between valid names and synonyms. A checklist in ornithology is a very careful, critical revision, usually with extensive synonymy and a detailed presentation of the geographic distribution of each species and subspecies. A checklist in entomology is generally only a list of names, often in alphabetical sequence. Compare, for example, J. L. Peters et al. (eds.), 1931–1987, *Checklist of Birds of the World,* Vols. 1–16 (Cambridge, Mass.: Museum of Comparative Zoology), and A. Stone et al., 1965, A catalog of the Diptera of America north of Mexico (Agricultural Research Service, U.S. Dept. of Agriculture, Washington, D.C.), 1696 pp.

Checklists are most useful in the better-known groups of animals. A checklist in which species are critically evaluated and carefully distinguished from synonyms qualifies as "primary zoological literature," whereas an uncritical listing of names or nominal species does not.

Revisions and Monographs

Revisions are taxonomic papers dealing with all the species of a species group, genus, or higher taxon. They provide a critical analysis of all nominal species and descriptions of previously undescribed species. Revisions vary greatly in completeness of treatment, ranging from synopses and reviews that merely summarize the literature to genuine monographs.

Monographs are complete systematic publications. They involve a full systematic treatment of all species, subspecies, and other taxonomic units and require a thorough knowledge on the part of the author of comparative anatomy, the biology of the species and subspecies included, the immature stages in groups that exhibit metamorphosis, and detailed distributional data. For the student of evolution such monographic treatises are the most rewarding type of taxonomic publication. They permit a detailed treatment of geographic variation, relationships, and distributional history. Generalizations on the structure of species, modes of speciation, the nature of taxonomic categories, and the like are based on such monographs. They have the disadvantage of requiring more complete material than do other kinds of taxonomic papers. Unfortunately, in the present state of our knowledge of many groups, especially among the invertebrates, few taxonomic papers can justify the title monograph. Monographs can be done more frequently for vertebrates.

Atlases

These publications provide complete illustrations of the species of a taxonomic group, reflecting the inadequacy of the printed word in conveying a mental picture of the general facies of an animal. The idea of an atlas also grew out of the need for taxonomic data that are strictly comparable from one species to another. Since the purpose of an atlas is purely taxonomic, semidiagrammatic drawings are commonly used.

Faunal Works

In a fauna the animals of a circumscribed region are listed and described. Thus, the scope is dictated by the limits of a geographic region rather than by the limits of a higher taxon. The area covered by a fauna may be as small as a township or as large as a continent.

In the case of well-known organisms, *local faunas* give the local naturalist an opportunity to provide ecological and behavioral information, often based on many years of fieldwork; some faunas are produced by teams of local naturalists.

Faunas covering large areas are highly valuable as a first survey of the animal life of a region. Examples include the following:

Fauna of British India. Taylor and Francis, London. Many volumes covering most groups of animals, published from 1888 on.

Biologia Centrali-Americana, 1879–1915, parts 1–215. Dulau and Co., London.

Fauna SSSR. Zoologicheskii Institut Akademii Nauk SSSR (more than 130 volumes published).

Faunistic reports on the whole are among the least efficient kinds of taxonomic papers and often amount to collections of isolated descriptions. The time required to run down all the locality records in a local fauna can be spent far more productively in preparing a revision or a manual.

Revisions of Higher Taxa

As a consequence of the increasing interest in macrotaxonomy, numerous works have appeared in recent years which deal exclusively with genera and higher taxa. Frequently they contain an evaluation of the characters which have been used by the author to infer relationship and phylogeny and to determine ranking. Such taxonomic works include the following:

Ax, P. 1984. *Das Phylogenetische System*. Stuttgart–New York: G. Fischer.

Corliss, T. O. 1984. The kingdom Protista and its 45 phyla. *Biosystems* **17:**87–126.

Ehlers, U. 1985. *Das phylogenetische System der Platyhelminthes*. Stuttgart: G. Fischer.

Ehrlich, P. R. 1958. The comparative morphology, phylogeny, and higher classification of the butterflies (Lepidoptera: Papilionoidea). *Univ. Kansas Sci. Bull.* **39:**305–378.

Greenwood, P. H., D. E. Rosen, S. H. Weitzman, and G. S. Myers. 1966. Phyletic studies of Teleostean fishes with a provisional classification of living forms. *Bull. Amer. Mus. Nat. Hist.* **131:**341–455.

House, M. R. (ed.) 1979. *The Origin of Major Invertebrate Groups*. London and New York: Academic Press.

Michener, C. D. 1952. The Saturniidae (Lepidoptera) of the Western Hemisphere. *Bull. Amer. Mus. Nat. Hist.* **98:**339–501.

Newell, N. D. 1965. Classification of the Bivalvia. *Amer. Mus. Novitates,* **2206:**1–25.

Nielsen, C. 1985. Animal phylogeny in the light of the trochaea theory. *Biol. J. Linn. Soc.* **25:**243–299.

Simpson, G. G. 1945. The principles of classification and the classification of mammals. *Bull. Amer. Mus. Nat. Hist.* **85:**i–xvi, 1–350.

The best survey of the classification of all living organisms is Parker, S. P. (ed.). 1982. *Synopsis and Classification of Living Organisms,* Vols. 1 and 2. New York: McGraw-Hill.

Evolutionary and Biological Publications

It is inefficient to include too much heterogeneous information in a single publication, particularly in a taxonomic monograph. Such works are usually read only by a few specialists; any information of broader biological interest published there will not come to the attention of the general biologist and is therefore bound to be ignored. Almost every author of a taxonomic revision makes ecological, evolutionary, or zoogeographic discoveries which may be of considerable importance to other biologists. Instead of burying these findings in the introduction to a monograph, the author should publish them in general journals such as *Ecology, Evolution,* or *Systematic Zoology,* where these articles will draw attention to the monograph. The fact that taxonomists have increasingly adopted this policy partly explains the noticeable improvement in the general image of taxonomy in recent years. No other group of biologists can make such important contributions to all problems relating to the diversity of the organic world. Taxonomists have an obligation to make their knowledge broadly available.

In addition to these by-products of monographs and revisions, more and more taxonomists publish short contributions to evolutionary or behavioral biology or to ecology as a result of their ad hoc investigations. Every volume of *Systematic Zoology* contains papers dealing with cytological or chemical attributes of species (and higher taxa), distribution patterns, behavior, taxometrics, variability, endemism, extinction, serology, geographic variation, speciation, rates of evolution, climatic rules, allometry, and so on. However, this is only one of a large number of journals that publish the results of evolutionary research conducted by taxonomists. The results of the research of animal systematists are also used in major books on evolutionary biology. Three are listed here as examples:

SIMPSON, G. G. 1953. *The Major Features of Evolution*. New York: Columbia University Press.

MAYR, E. 1963. *Animal Species and Evolution*. Cambridge, Mass.: Harvard University Press.

FUTUYMA, D. J. 1986. *Evolutionary Biology*, 2d ed. Sunderland, Mass.: Sinauer Assoc.

Theory and Methods of Systematics

Publications devoted to this area constitute a growing part of the taxonomic literature. They are cited in all the chapters in this book, particularly Chapters 6 through 11, and are listed in the bibliography.

Journals devoted to systematics and classification include the following:

Cladistics (Vol. 1, 1985)
Cladistics
Meckler Publishing Corporation
11 Ferry Lane West
Westport, CT 06880–5808

Journal of Classification
Classification Society of North America
Springer
175 Fifth Ave.
New York, NY 10010

Systematic Zoology (1952)
Society of Systematic Zoology
P.O. Box 368
Lawrence, KS 66044

Systematics Association
Special Volumes, 1(1978)–34(1988)
London: Academic Press
Taxon (1951)
International Bureau for Plant Taxonomy and Nomenclature
Tweede Transitorium
Uithof, Utrecht
The Netherlands
Zeitschrift fur zoologische Systematik und Evolutionsforschung (1963)
Verlag Paul Parey
Spitalerstrasse 12
D2000-Hamburg 1
Germany
Zoological Journal of the Linnaean Society of London (Vol. 48, February 1969) (formerly called *Journal of the Linnaean Society of London*)
Academic Press Inc. Ltd.
P.O.B. 6208
Duluth, MN 55806–0208

MAJOR FEATURES OF TAXONOMIC PUBLICATIONS

Most taxonomic publications contain a set of major components which deserve thorough discussion. Many excellent (and a few outworn) traditions have developed in the past 200 years of taxonomic research. Every taxonomist must know what they are and why certain standardized procedures should be followed to facilitate quick retrieval of information.

Descriptions

The chief objective of a description is to aid in the subsequent recognition of the taxon involved. It was realized at an early date that different kinds of descriptions approach this goal in a different manner. Linnaeus distinguished clearly between the general *descriptio* (character naturalis) on the one hand and the polynominal *differentia specifica* (character essentialis) on the other (Svenson 1945). The latter contained the essential characters by which the species is distinguished from its congeners. It corresponds to what is nowadays called a *diagnosis*.

The functions of the two kinds of descriptions—the general description and the diagnosis—are by no means identical. The diagnosis serves to distinguish a taxon from other known similar or closely related ones. The general description has a broader function. It should present a general picture of the described taxon, giving information not only on characters that are diagnostic in relation to previously described species but also on those which may serve to distinguish the species from unknown species. It should also provide information of interest to others besides taxonomists.

Linnaeus and many subsequent taxonomists have stressed the extreme practical importance of a short, unambiguous diagnosis. The diagnosis can only rarely be combined successfully with the general description. The latter, in turn, no matter how exhaustive, cannot always provide an adequate substitute for a type specimen or, in many cases, for illustrations. In describing animals the taxonomist should achieve two objectives: diagnosis and delimitation. *Diagnosis*[1] is the art and practice of distinguishing between things; *delimitation* is the art and practice of setting limits to things. Both enter into taxonomy; they are essentially different and their complementary roles should be clearly understood (Simpson 1945). Although the formal diagnosis in taxonomic work sometimes assists in the delimitation of a taxon, this function is performed mainly by the general description. The terms *diagnosis* and *description* may then be used as follows:

Diagnosis: A brief listing of the most important characters or character combinations that are peculiar to the given taxon and by which it can be differentiated from other similar or closely related ones.

Description: A more or less complete statement of the characters of a taxon without special emphasis on those characters which distinguish it from coordinate units.

The direct comparison of a species (or other taxon) with other specifically mentioned species (or other taxa) is usually called a *differential diagnosis*. Such comparisons with other species are of great practical help to students who lack specimens of the newly described form. Moreover, it forces the author to review all the evidence for and against the establishment of the new taxon (Rensch 1934) and thus ensures that the diagnostic characters of the new form are mentioned. If the nearest relatives are rare or poorly known, it is also helpful to make a comparison with a well-known, not so closely related species.

Original Description The description published at the time a new species, genus, or other taxon is proposed is called the *original description*. It has two primary functions. The first, as stated above, is to facilitate subsequent recognition and identification; the second is to make the new name available by fulfilling the requirements of Articles 10 through 20 of the Code.

[1]Ultimately from the Greek διαγιγνωσκειν, meaning "to distinguish between two (things)."

The importance of preparing a proper description cannot be overemphasized. The describer is forced to rely on words to convey meaning, yet words, no matter how carefully chosen, can rarely give an accurate and complete mental picture of the appearance of an organism. Nevertheless, the description should enable a subsequent worker to identify specimens without reference to the type. This goal can be achieved by the careful worker, particularly when the description is properly coordinated with illustrative material.

A good description requires on the part of its author (1) a thorough knowledge of the group of organisms concerned, (2) a knowledge of structure and terminology, (3) an ability to evaluate differences and similarities, (4) an ability to select and emphasize the important, (5) a full understanding of the precise meaning and use of the words and grammar of the language employed, and (6) a concern for the future worker. Ferris (1928:102) stated, "If [the describer's] work of recording the data has been properly done those data are available for re-examination and re-evaluation. His conclusions can be checked, they can be extended or modified or rejected as appears desirable, all without the necessity of recourse to his types."

In the less well-known groups, much of the taxonomist's time is spent in comparing and contrasting one description with another. This task is difficult in any circumstances, but it is easier when the descriptions approximate one another in style, arrangement, and form. This does not mean that a completely standardized description is always possible, because the factors which influence the order of presentation, form, and style vary from group to group. Within a particular group, however, much can be done to standardize descriptions and thus increase their effectiveness and utility.

Style The style generally used in descriptions as well as diagnoses is telegraphic and concise. It is usually characterized by the elimination of articles and the use of verbals (gerunds and participles) instead of verbs and by the selection of adjectives and nouns with explicit meaning. It also involves careful punctuation and adherence to a logical sequence of presentation. The most common form is a series of noun phrases. Thus the telephonic-style statement "The head is one-third longer than it is wide, the antennae are shorter than the body, and the outer antennal segments are serrate" becomes "*Head one-third longer than wide; antennae shorter than body; outer segments serrate.*"

Describing features that change during ontogeny, such as the ornamentation of ammonite shells, presents more complicated problems. A series of sentences describing such ontogenetic change might read: "The shell is smooth early in an individual's life, but as it grows, it develops broad, flat ribs separated by interspaces that are narrower than the ribs themselves. These ribs cross the venter, which becomes slightly flatter as the animal ages, without interruption." Made telegraphic, the passage runs: "*Surface smooth at first, later with broad flat ribs separated by narrower interspaces, ribs crossing venter without interruption, in adult slightly flattened.*" The descriptive style in the italicized statements has lost none of the precision or clarity of the full sentences, but the statements are only about half as long and can be read and understood more quickly.

Sequence of Characters The recommended sequence of characters is different in a diagnosis as opposed to a description. In a diagnosis it is customary to present characters in the order of their diagnostic importance (or what the author regards as the order of importance). This will facilitate rapid recognition. In a full description the material should be arranged in a standardized natural order, for instance, describing the body parts from anterior to posterior, first on the dorsal surface and then on the ventral surface. The details may be varied to fit the groups yet still maintain a natural and readily comparable order. For instance, the sequence of presentation for a dorsoventrally flattened animal group would be different from that for either a laterally compressed or a radially symmetrical group because of the different methods of orientation during study. The standardized sequence of characters helps ensure that nothing important is overlooked and that the description is comparative. It is very frustrating to use a taxonomic paper in which half a dozen species are described independently of one another, with details given, for example, of the antennae of one species, the pronotum of a second, and the elytra of a third. Such a procedure makes comparison impossible. Authoritative monographs usually adopt a standardized sequence of characters, and subsequent describers should follow it as far as possible. When one is starting out, it is useful to adopt the format used by a competent colleague as a standard and alter it as needed to fit the case at hand.

The utility of a description may be increased by the use of devices that enable the reader to locate quickly the particular characters being looked for. One such device is the use of paragraphs to break up the description according to main body divisions (e.g., in insects: head, thorax, abdomen, wings, genitalia). Where paragraphing is undesirable, the same effect may be gained by italicizing the key words. If the author has followed a natural sequence of presentation, either method will permit the reader to become oriented quickly at a particular point without having to read the whole description.

Dictated Descriptions When preparing revisions, many modern taxonomists dictate the descriptions into an automatic recording device to speed the tedious task of describing. When both hands are needed for viewing specimens through the microscope, dictating equipment with a foot pedal and a desk microphone is useful. This way the describer has a checklist of characters in standardized sequence which can be followed in dictating the descriptions of each taxon. The time saved by this method is considerable.

What to Include in a Description An exhaustive description of an organism would fill many volumes, as evidenced by works on the morphology (physical anthropology) and anatomy of the human species. It is therefore obvious that even a so-called detailed description of a taxonomic species is highly selective and in the nature of an expanded diagnosis. Furthermore, in these days of steadily rising printing costs, it becomes increasingly important not to include trivia. The amount of detail that should be included in a description depends on the group concerned and the state of knowledge about that group. Excessively long descriptions obscure the essential points; excessively short ones omit pertinent data. While the diagnosis serves to distinguish a species from other known species, the description should be detailed enough to anticipate possible differences from undescribed species. The description should therefore be more detailed in poorly known groups, because it is impossible to predict which characters will distinguish a new species from those which are still undiscovered. It should pay particular attention to character complexes that are variable in the given group. By contrast, subspecies in a well-known species of birds may differ from one another in so few characters that an extensive description would amount to a repetition of the species description. In such a case the description may not differ at all from a diagnosis—for instance, "Like subspecies *alba* but larger, upper parts blackish gray, not ash gray" (followed by a tabulation of the measurements).

As far as is practicable, descriptions should include all characters, both positive and negative, which are known to be useful or potentially useful in distinguishing other taxa at the same categorical level. However, characters of higher categories should be omitted except where they are anomalous or where the rank of the taxon is in doubt. For example, the description of a subspecies of Song Sparrow should not include reference to characters that are typical of all Song Sparrows (or worse, of all sparrows). Violation of this rule not only is uneconomical but distracts attention from the essential features of the taxon being studied.

Beyond these generalizations, there is little to guide the describer other than good judgment. More than almost any other aspect of taxonomy the description provides a permanent record of the author's ability to observe accurately, record precisely, select and interpret intelligently, and express the facts clearly and concisely (Cain 1959*b*).

The description should include a statement of the differences between the sexes or, if only one sex is available, a frank statement of that fact (e.g., "female unknown"). The characters of immaturity also should be discussed, along with the larval stages. Available behavioral, ecological, and other biological data should be presented. In the case of sibling species, such information is often more important than are morphological characters.

Whether the description should be based exclusively on the type is a disputed point. Those who favor this method argue that all too often it has eventually turned out that the original material—and consequently the description—was a composite of several species, which makes it very difficult to disentangle the characters of the various species. These researchers argue that it is much safer to restrict the description to the type and have it followed by a discussion of the variability of the rest of the material.

Others believe that this kind of treatment favors the erroneous typological view that the type has a special significance as far as the characters of the species are concerned. They prefer the description to be a composite drawn from a consideration of the entire material (Simpson's *hypodigm*) and propose to mention at the end (or in parentheses within the description) by what characters (if any) the type specimen differs from the rest of the material.

Actually the proponents of both methods agree that (1) the entire variability of the species material should be described and (2) it is advisable to mention the special features of the holotype. Different authors may use different methods to achieve these objectives.

Description of Coloration Differences in coloration are among the most important diagnostic characters in many groups of animals. A detailed description of the general pattern of coloration and the precise tone of the various colors is therefore essential in many taxonomic groups. Subspecific differences in birds, mammals, and butterflies are often largely a matter of coloration. Many attempts have therefore been made to standardize color descriptions, since descriptions such as rufous and tawny do not necessarily suggest the same shade of color to every taxonomist. It is for this reason that color keys are widely used in taxonomy. Those of Ridgway (1912), Maerz and Paul (1950), and Villalobos-Dominguez and Villalobos (1947) are highly recommended. When fine shades of color are involved, a direct comparison with topotypical material is advisable. Measuring devices (spectrophotometers) provide objectivity and standardization in such comparisons and permit quantification as needed for statistical evaluation (Selander et al. 1965).

Numerical Data The recording of a set of precise measurements is an integral part of a well-rounded description. If the new form differs from its relatives in its proportions, such proportions should also be recorded (Chapters 5 and 7). Exact data should be given of numerically variable features of structure or pattern, such as numbers of spots, spines, scales, and tail feathers. The reasons for including such data are stated in Chapter 5.

Descriptive Treatment A full descriptive treatment of a species may take the following form:

Scientific Name and Its Author
Bibliographic reference to place, date, and author of original description
Type (including type locality and repository)
Synonymy (if any) (see below)
Diagnosis and differential diagnosis (brief statement of essential differences from nearest relatives, see above)
Description
Measurements and other numerical data
Range (geographic)
Habitat (ecological notes) and horizon (in fossils)
Discussion
List of material examined

Because the type is the name bearer, all information relating to the type should be given immediately after the name. (Types of each synonym should be handled similarly.) Some systematists still include the type data immediately after the description, as if the type were the basis of the description. It is important that the type locality and especially the museum or other repository in which the type is deposited, with the specimen number for all new species, be clearly and fully recorded.

If the species is new, the form of the description is as follows:

X-us albus, new species
Type (including statement of type locality and repository)
Diagnosis and so forth as in a redescription

Illustrations Illustrations are in most instances vastly superior to verbal descriptions. Anything that can be made clearly and sufficiently visible in a picture should be illustrated. The value of illustrations is recognized in the International Rules, since a scientific name given to a published illustration was held to be valid (prior to January 1, 1931) even when not accompanied by a single word of description [Article 12*b* (7)]. Such a naming of illustrations was customary in the days of Linnaeus. In our day, however, sound taxonomists always present a diagnosis and full description (often with illustrations); this has been mandatory since 1931. See below for a discussion of illustrations.

Redescriptions The redescription of poorly described forms is an extremely important element of revisional and other taxonomic work. In the present state of knowledge about many animal groups, it is of greater importance than is the description of new forms.

The specimen or specimens on which a redescription or illustration is based should be clearly indicated because if the species has been wrongly identified, a new species may be proposed for *X-us albus* Jones, not Smith. In such a case the type specimen of the new species is the specimen (or is selected from among the specimens) on which the redescription or illustration was based.

If a good description, correctly and adequately stated, is readily available in the literature, it is wasteful to republish copies of it again and again. A bibliographic reference is sufficient.

Description of a Higher Taxon Much of what has been said about the description of species taxa is equally true for higher taxa. However, the description of a new higher taxon traditionally stresses diagnostic features. Citation of the type species (in case of a genus) or the type genus (in case of a family) reduces the amount of descriptive material that must be included.

In the case of higher taxa of vertebrates it is good practice to utilize characters of the skeleton that would be diagnostic in fossils. In mollusks and other groups of invertebrates that have a fossil record, the modifications of the hard parts, used for diagnosis, are usually the same in recent forms and in fossils.

The form of presentation is as follows:

X-us, new genus
Type species
Diagnosis, description, list of included lower taxa, discussion

Summary Recommendations for the preparation of descriptions can be summarized as follows:

1 The taxonomic characters should be treated in a standardized sequence.

2 The most easily visible characters should be featured.

3 A direct diagnostic comparison with the nearest relative or relatives should supplement the description.

4 Since words alone can seldom give an adequate picture of the diagnostic characters of a taxon, appropriate illustrations should be provided whenever possible.

5 The description should provide quantitative data supplemented with information on geographic range, ecology, habits, and the like.

6 Species that are poorly known or are in poorly known genera should be fully described.

7 The formal description should be followed by an informal discussion of the variable characters.

8 The description should be accompanied by full information on the type specimen and other material before the author.

9 Characters that are common to all members of the next higher category should be omitted from the description.

Synonymy

Different names given to the same taxon are *synonyms*. The correct establishment of *synonymies* is perhaps the most important task in the early taxonomic analysis of any group of organisms. All other tasks, such as the elaboration of a classification and the preparation of keys, depend on the correctness and completeness of the synonymies. A complete synonymy of every species and genus is therefore a necessity when a higher taxon is monographed or revised for the first time or when the previous treatment has become obsolete.

Unfortunately, the exhaustive preparation not only of a complete synonymy but also of a listing of all references to previous publications with all possible binominal combinations (in case of generic transfers) has become the misplaced ideal of scholarship for some taxonomists. The expense of printing endlessly repeated names and references (often a separate line is needed for each reference) is altogether out of proportion to the benefits conveyed. Friedmann (1950) illustrated in Part II of Ridgway's *Birds of North America* what this system leads to in a taxonomic group with a rich literature. Other examples are provided by monographs on mammals and fishes.

In the better known groups of animals it has now become customary to list in synonymies only names that were not at all or not correctly listed in the previous standard treatments. For birds, for example, in Peters's *Checklist of the Birds of the World* (1931–1987) synonyms are not listed that can be found in previous standard works, such as *Cata-*

logue of Birds of the British Museum (1873–1892), the *Handlist of Birds* (1896–1910), for American species Hellmayr's *Catalogue* (1918–1944), and for Palearctic species Vaurie (1959–1965). More recent checklists of birds do not repeat synonyms that are correctly cited in Peters. Only genuine synonyms are listed, not new combinations. There is perhaps no other section of a taxonomic paper in which economies can be achieved more easily.

For many groups of animals, particularly insects, there now exist only alphabetical lists of nominal species. Obviously, the first revision in such a group must give a complete synonymy. However, is is not necessary even in poorly known groups to repeat synonyms correctly placed in the last previous standard treatment, even if it is otherwise out of date. Since manuscript names (*nomina nuda*) have no existence in nomenclature, their listing in synonymy is confusing and should be avoided.

New synonymy is most usefully cited with the following sequence of data: (1) scientific name (in its original form), (2) author, (3) date of publication, (4) reference, (5) type locality, and (6) present location of type (optional). An example follows:

Oncideres rhodostictus Bates

Oncideres rhodosticta Bates, 1885, *Biol. Cent.-Amer., Coleopt.* **5**:367. [Lerdo, Mex.; British Mus. (Nat. Hist.)]

Oncideres trinodatus Casey, 1913, *Mem. Coleopt.* **4**:352 [El Paso, Tex.; U.S. Natl. Mus.] NEW SYNONYMY.[2]

The above form is recommended for a revision of a well-cataloged group. In poorly known groups, as described above, a full synonymy (i.e., a list of scientific names, incorrect and correct) may be required. This should include all references which have nomenclatural or zoological significance, arranged chronologically under the actual name (correct or incorrect) by which the author actually referred to them.

A full bibliographic synonymy, as used in a first revision of a poorly known genus (in a publication with no literature cited), would appear as follows:

Oncideres rhodostictus Bates

Oncideres rhodosticta Bates, 1885, *Biol. Cent.-Amer., Coleopt.* **5**:367 [type: Lerdo, Mex.; British Mus. (Nat. Hist.)]; Linsley, 1940, *J. Econ. Entomol.* **33**:562 (synon., distr.); Linsley, 1942, *Proc. Calif. Acad. Sci.*

[2]This synonymy was published as new in *J. Econ. Entomol.* **33**:562, 1940. Its use as an example here and elsewhere in the present discussion is not to be interpreted as a nomenclatural change. The words NEW SYNONYMY are usually printed in caps or small caps to draw the reader's attention.

(4) **24:**76 (distr.); Dillon and Dillon, 1945, *Sci. Pub. Reading Mus.,* no. 5:xv (key); Dillon and Dillon, 1946, *loc. cit.* **6:**313, 382 (revis.).

Oncideres putator; Horn (not Thomson, 1868), 1885, *Trans. Amer. Ent. Soc.* **12:**195 (key, distr.); Schaeffer, 1906, *Can. Ent.* **38:**19 (key).

Oncideres cingulatus; Hamilton (in part) (not Say, 1826), 1896, *Trans. Amer. Ent. Soc.* **23:**141 (distr.).

Oncideres trinodatus Casey, 1913, *Mem. Coleopt.* **4:**352 [type: El Paso, Tex.; U.S. Natl. Mus.].

Oncideres sp.; Craighead, 1923, *Can. Dept. Agr. Bull.* (n.s.) **17:**132 (larva, hosts).

Oncideres pustulatus, Essig (not LeConte, 1854), 1926, *Insects of Western North America,* p. 460, fig. 368 (habits, distr.).

Note that type locality and location of type are recorded for all genuine synonyms.

Many authors use the convenient device of a semicolon inserted between the specific name and the author [*X-us albus* Smith; (not Brown)] to distinguish between a misidentification, which has no nomenclatural status, and a homonym [*X-us albus* Jones (not Brown)], which has. Article 51 of the Code specifically forbids the use of a comma between a name and an author.

The above synonymy might appear in an abbreviated checklist as follows:

Oncideres Serville, 1835

1. *rhodostictus* Bates, 1885	So. Calif. to Texas
trinodatus Casey, 1913	No. Mex.
	L. Calif.

The usefulness of a checklist containing a terminal bibliography may be increased by giving page references to original descriptions which facilitate the library search of later authors—thus, *rhodostictus* Bates, 1885:367.

If it is desirable to indicate the various combinations under which each name has appeared, this can be accomplished by taking the oldest specific name and following it through its various combinations, then the next oldest, etc., as follows:

Megacyllene antennata (White)

Clytus antennatus White, 1855, *Cat. Coleopt. Brit. Mus.* **8:**252 [type: "W coast of America"; British Mus. (Nat. Hist.)].

Cyllene antennatus; Horn, 1880, *Trans. Amer. Ent. Soc.* **8:**135 (descr., syn. distr.); Craighead, 1923, *Can. Dept. Agr., Bull.* **27,** p. 33 (larva, biol.); Hopping, 1937, *Ann. Ent. Soc. Amer.* **30:**411, pl. 1 (revis.).

Megacyllene antennata; Casey, 1912, *Mem. Coleopt.* **3:**348, 351 (descr.).

Arhopalus eurystethus LeConte, 1858, *Proc. Acad. Nat. Sci. Phila.* 1858:82 [type: Sonora, Mex.; Mus. Comp. Zool., Harvard]; LeConte, 1859, in Thomson, *Arcana Naturae,* p. 127, pl. 13, fig. 9.

In this example, the semicolon between the specific combination and the author's name has again been used, this time to distinguish between a new combination (*Cyllene antennatus*; Horn, 1880) and the original combination (*Clytus antennatus* White, 1855).

Generic synonymy is handled much like specific synonymy, except that for new synonymy or full bibliographic treatment, the generic type (and its subsequent designator, if any) is also cited. (Paleontologists commonly also include the type locality and type location.) The synonymy of the genus *Dicrurus,* as cited in Vaurie's (1949) revision of the Dicruridae, is an example.

Dicrurus Vieillot, April 14, 1816, Analyse d'une nouvelle ornithologie élémentaire, p. 41. Type, by subsequent designation, *Corvus balicassius* Linnaeus (G. R. Gray, 1841, A list of the genera of birds, ed. 2, p. 47).

Edolius Cuvier, Dec. 7, 1816, Le règne animal, vol. 1, p. 350. Type, by subsequent designation, *Lanius forficatus* Linnaeus (G. R. Gray, 1855, Catalogue of the genera and subgenera of birds, p. 58).

Drongo Tickell, 1833, *Jour. Asiatic Soc. Bengal,* vol. 2, p. 573. Type, by monotypy, *Drongo caerulescens* Tickell = *Lanius caerulescens* Linnaeus.

Chibia Hodgson, 1836, *India Rev.,* vol. 1, p. 324. Type, by subsequent designation, *Edolius barbatus* J. E. Gray = *Corvus hottentottus* Linnaeus (G. R. Gray, 1841, A list of the genera of birds, ed. 2, p. 47).

Bhringa Hodgson, 1836, *India Rev.,* vol. 1, p. 325. Type, by original designation and monotypy, *Bhringa tectirostris* Hodgson.

Bhuchanga Hodgson, 1836, *India Rev.,* vol. 1, p. 326. Type, by subsequent designation, *Bhuchanga albirictus* Hodgson (Sharpe, 1877, Catalogue of birds in the British Museum, vol. 3, p. 245).

Chaptia Hodgson, 1836, *India Rev.* vol. 1, p. 326. Type, by monotypy, *Chaptia muscipetoides* Hodgson = *Dicrurus aeneus* Vieillot.

Dissemurus Gloger, 1841, Gemeinnützige s Hand- und Hilfsbuch der Naturgeschichte, p. 347. Type, by monotypy, *Cuculus paradiseus* Linnaeus.

Musicus Reichenbach, 1850, Avium systema naturale, pl. 88, fig. 9. Figure of generic details, no species included, cf. Bonaparte, 1854,

Compt. Rend. Acad. Sci. Paris, vol. 38, p. 450. Type, by tautonymy, *Dicrurus musicus* Vieillot = *Corvus adsimilis* Bechstein.

Dicranostreptus Reichenbach, 1850, Avium systema naturale, pl. 88, fig. 12. Figure of generic details, no species included. Type, by subsequent designation, *Edolius megarhynchus* Quoy and Gaimard (G. R. Gray, 1855, Catalogue of the genera and subgenera of birds, p. 58).

Keys

Keys facilitate the identification of specimens by presenting diagnostic characters in a series of alternative choices. The worker finds the correct name of his or her specimen by making the appropriate choice in a series of consecutive steps.

The procedure involved is somewhat analogous to that of the physician, who, using a series of questions and examinations, arrives by means of a process of elimination and confirmation at the diagnosis of the ills of a patient, or to the elimination method used in culture identification of bacteria.

The use of keys is old indeed. Voss (1952) gives an interesting history of the development of keys in systematic biology. Pankhurst (1978) provides a good treatment of the principles and practices of identification methods. In the Aristotelian procedure, keys were an instrument of classifying logic. Typologically an object was either *A* or not-*A*. The use of a key typologically for classifying purposes is misleading in view of the polythetic nature of many taxa. However, if a key is used as a purely pragmatic device, the taxonomist can cope with the existence of a polythetic taxon easily by keying it out repeatedly. This would be fatal in a classification but does not constitute a weakness in an identification procedure.

Keys are also a tool for taxonomic analysis since in their preparation one must select, evaluate, and arrange taxonomic characters. In this sense keys are an integral part of taxonomic procedure as well as a means of presenting findings.

The construction of keys is laborious and time-consuming, involving the selection and sifting of the most useful and most clearly diagnostic characters. Ideal key characters apply equally to all individuals of the population (regardless of age and sex); are absolute (two scutellar bristles versus one scutellar bristle); are external, so that they can be observed directly and without special equipment; and are relatively constant (without excessive individual variation). Unsuitable key characters include those which require a knowledge of all ages and stages of a species (e.g., "sexual dimorphism present" versus "sexual dimorphism absent," "male larger than female" versus "male smaller than female," "fall molt complete" versus "fall molt partial"), relative characters without an absolute standard (e.g., "darker" versus "lighter," "larger" versus "smaller"), and overlapping characters ("larger, wing 152 to 162" versus "smaller, wing 148 to 158").

Most sets of data permit the choice of several characters for the various primary and secondary divisions of the key. Here the writer must exercise judgment in selecting the most satisfactory characters at the various levels. If torn between the phylogenetic and utilitarian approaches to the problem, the author should remember that the primary purpose of a key is utilitarian; diagrams, lists, numbers, or order of subsequent treatment will take care of phylogeny. However, when making a key in a poorly known group (with many undescribed species), it is useful to arrange the key so that closely related species key out near one another. This facilitates the subsequent insertion of new species as well as the decision whether a species is new. Some study material permits the construction of a key that can present relationships without interfering with its main function of ensuring identification. Key construction is a procedure for which computer programs are now available (Dunn and Everitt 1982:110).

A good key is strictly dichotomous, not offering more than two alternatives at any point.[3] Alternatives should be precise. Ideally, the statements should be sufficiently definite to permit identification of a single specimen without reference to other species. In any event, identification should be possible without reference to the opposite sex or to immature stages, which should be treated in different keys when dimorphism is exhibited. If a key character neatly separates the species of a genus into two groups, except for one or two species that are intermediate or variable, it is legitimate to include these variable species in both subdivisions. A given species name may thus appear repeatedly in a key. The procedure which gives the quickest and most unambiguous identification should be adopted. Ordinarily new species should not be designated as such in a key. Also, it is customary to omit authorities from specific names in keys unless they are not mentioned elsewhere in the article.

The style of keys is telegraphic, like that of descriptions, and the phrases are usually separated by semicolons. Even though the primary contrasting characters of each couplet may be diagnostic and definitive, supplemental characters are desirable in the event that the primary characters are not clearly discerned or the specimen has been injured or

[3]If it is impossible to work out a key that permits the identification of all species, it is advisable to indicate this clearly and to key out as groups those species which cannot be diagnosed by key characters.

mounted in an unsatisfactory manner. Generalized distributions are often helpful. One of the most satisfactory methods for assembling data for the construction of a key is shown in the example of the method and the subsequent analysis given in Table 13-1. This example is oversimplified to demonstrate the method more clearly.

Several types of dichotomous key are used in taxonomic papers. By far the most common is the bracket key; the other is the indented key. The indented key has the advantage that the relationship of the various divisions is apparent to the eye. It has the disadvantage, especially in a long key, that the alternatives may be widely separated and that it is wasteful of space. For these reasons, it is generally used only for short keys, keys to higher taxa, and comparative keys (keys which not only serve the purposes of identification but also treat the same comparative characters at each level for each group). An *indented key* based on the hypothetical data given in Table 13-1 might be as follows:

A. Wings opaque
  B. Antennae serrate
    C. Eyes entire .... *completa*
    CC. Eyes emarginate .... *emarginata*
  BB. Antennae filiform
    C. Legs red .... *rufipes*
    CC. Legs black .... *nigripes*
AA. Wings clear
  B. Tarsal segments linear
    C. Antennae black .... *smithi*
    CC. Antennae red .... *ruficornis*
  BB. Tarsal segments bilobed
    C. Antennae black .... *californica*
    CC. Antennae yellow .... *flavicornis*

TABLE 13-1
ARRANGEMENT OF KEY CHARACTERS*

Name of species	Wings	Antennae	Antennal color	Eyes	Tarsal segments	Leg color
smithi	*clear*	filiform	*black*	entire	*linear*	black
completa	*opaque*	*serrate*	black	*entire*	linear	black
emarginata	*opaque*	*serrate*	black	*emarginate*	linear	black
rufipes	*opaque*	*filiform*	black	entire	linear	*red*
nigripes	*opaque*	*filiform*	black	entire	linear	*black*
flavicornis	*clear*	filiform	*yellow*	entire	*bilobed*	black
ruficornis	*clear*	filiform	*red*	entire	*linear*	black
californica	*clear*	filiform	*black*	entire	*bilobed*	black

*Characters used in examples are italicized.

The *bracket key,* which is used almost exclusively by most taxonomists, has the advantage that the couplets are composed of alternatives placed side by side for ready comparison and that it uses less space because it is not indented. When properly constructed, it can be run forward or backward with equal facility by following the numbers, which indicate the path that the various choices follow. This type best fulfills the diagnostic purpose of a key. Its main disadvantage is that the relationship of the divisions is not apparent to the eye. An example based on the same data previously used follows:

1.		Wings opaque	2
		Wings clear	5
2.	(1)	Antennae serrate	3
		Antennae filiform	4
3.	(2)	Eyes entire	*completa*
		Eyes emarginate	*emarginata*
4.	(2)	Legs red	*rufipes*
		Legs black	*nigripes*
5.	(1)	Tarsal segments linear	6
		Tarsal segments bilobed	7
6.	(5)	Antennae black	*smithi*
		Antennae red	*ruficornis*
7.	(5)	Antennae black	*californica*
		Antennae yellow	*flavicornis*

Pictorial keys deserve mention among various other kinds of keys designed for special purposes. The pictorial key is of value for field identification by nonscientists. During World War II, for example, malaria crews based their control operations on the results of field identifications of anopheline mosquito larvae (Figure 13-1). The fact that critical characters were illustrated as well as described made the keys usable by medical corps personnel and engineers as well as entomologists. Other examples of pictorial keys are one by Corliss (1959) of the higher taxa of Ciliates and one of the rotifer genus *Ptygura* by Edmondson (1949). Pictorial keys have also been employed in field guides to vertebrates and flowering plants.

Those who employ pictorial keys must remember at all times the utilitarian purpose of keys and the characters used in them. The character on which a given taxon "keys out" may be of no particular biological or phylogenetic significance for the particular taxon. It is simply the character which gives the best assurance of correct identification. Keys are not phylogenies.

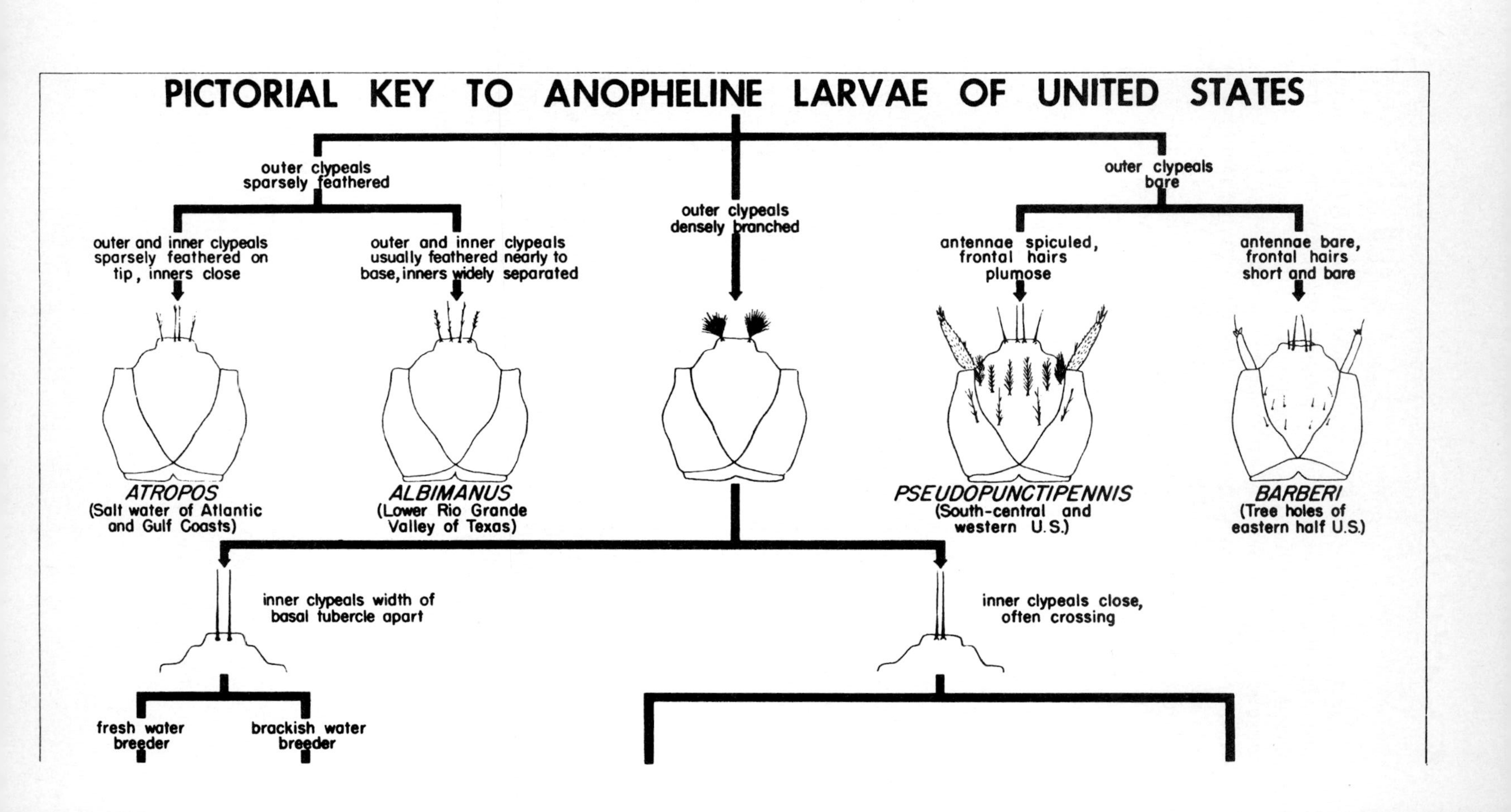
PICTORIAL KEY TO ANOPHELINE LARVAE OF UNITED STATES
outer clypeals sparsely feathered
outer clypeals densely branched
outer clypeals bare
outer and inner clypeals sparsely feathered on tip, inners close
outer and inner clypeals usually feathered nearly to base, inners widely separated
antennae spiculed, frontal hairs plumose
antennae bare, frontal hairs short and bare
ATROPOS
(Salt water of Atlantic and Gulf Coasts)
ALBIMANUS
(Lower Rio Grande Valley of Texas)
PSEUDOPUNCTIPENNIS
(South-central and western U.S.)
BARBERI
(Tree holes of eastern half U.S.)
inner clypeals width of basal tubercle apart
inner clypeals close, often crossing
fresh water breeder
brackish water breeder

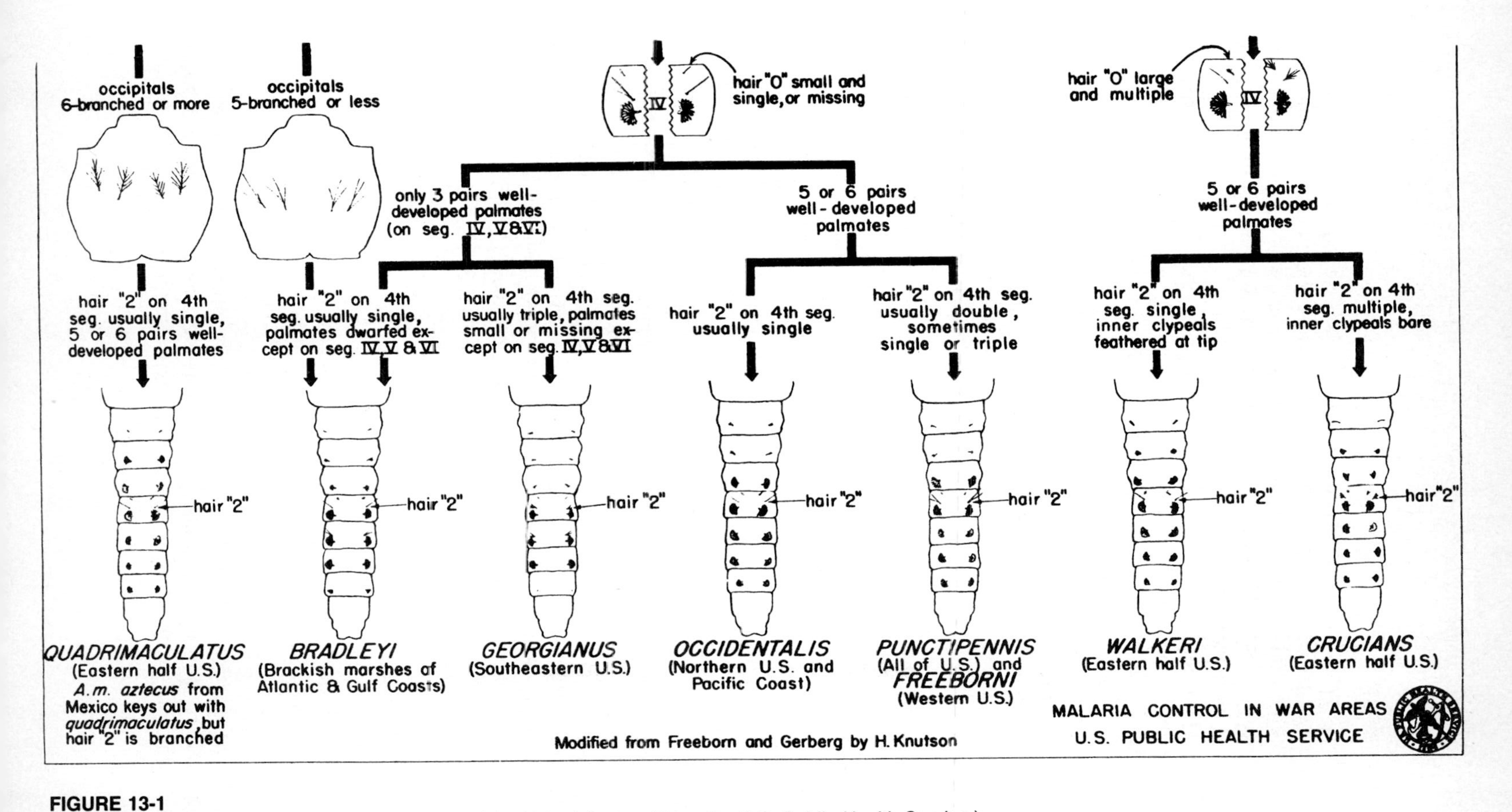

FIGURE 13-1
Pictorial key to larvae of anopheline mosquitoes of the United States. (*From the U.S. Public Health Service.*)

Illustrations

The old saw is true: One picture is worth a thousand words, especially in taxonomic papers. It is often impossible to describe a complex structure, say, the genital armatures of an insect or the palps of a male spider, adequately in words. Descriptions without illustrations are of limited value in many organisms.

The taxonomist must decide in each case what kind of illustration would be most useful. Except for special purposes, an accurate but somewhat diagrammatical line drawing is usually the most informative illustration for a taxonomic publication. A scientist who lacks artistic talent should not be discouraged, because clear diagrammatic drawings are satisfactory for scientific purposes and in some cases are superior to artistic drawings. Ferris (1928) called this type of drawing drafting and expressed the opinion that any conscientious scientist can learn to do it. Several books have been published on the subject, including Kuhl (1949), Papp (1976), and particularly Zweifel (1988) and the Council of Biology Editors handbook *Illustrating Science* (1988).

Sketches of bilaterally symmetrical animals made with a pencil on tracing paper can be "corrected" for symmetry by tracing one half on the other. The original outline can be sketched freehand, but at least with microscopically small organisms, this can be done more quickly and accurately by mechanical means. The camera lucida, or drawing tube, projects the microscope image on a piece of paper by means of prisms and a mirror. While looking in the microscope, one sees the specimen superimposed on the reflection of the paper. By carefully adjusting the light, one can draw an outline with both the specimen and the pencil point clearly in view. Another method of obtaining the outline is the direct projection of an image on a screen or paper by a microprojector attached to the microscope. A photographic negative can be enlarged onto matte finish printing paper; the outline is inked directly on the photograph, and the photographic emulsion then is washed off. Some workers prefer to sketch freehand on crosshatched paper, guided by a grid in the eyepiece of the microscope.

Maps, often an important kind of illustration in taxonomic papers, are best done using available outline maps (base maps). After mounting the map on bristol board, one attaches a transparent acetate overlay. Specific collection localities can be indicated on the overlay by symbols, and the general range can be indicated by an area of light cross-hatching.

Reproduction Planning is necessary in the preparation of figures. Especially for journal publication, an author must design illustrations with the place of publication in mind. Otherwise important information may be lost because the final illustrations are printed too small or had to be rearranged by the editor to save space. Line drawings (black and white only, not gray) should be made no more than twice as large as they will finally appear. If the original is very large, it can be reduced by photocopying. (The line on reduced photocopies of such large originals may need strengthening.) The sharpness and clarity of all line work can be improved if the original is reduced about one-half. However, with much more reduction, finer lines begin to disappear and stippling becomes muddy. Figures that require halftone reproduction (all photographs and pictures involving different shades of gray rather than pure black and white) should be produced at about the same size at which they are to be printed, because the halftone process necessary for the reproduction of photographs automatically results in some loss of detail. More yet is lost when reduction is necessary.

In the layout of an illustration, consideration must be given to the format and the size of the page of the publication in which it is to appear. Correct proportions for the original figure can be obtained by expanding on a diagonal line through a rectangle drawn to page or column size. Usually, room must be left at the foot of the finished illustration for the legend.

Once the proportion is determined, all symbols and letters to be used can be applied in a size large enough to be easily read and small enough that the published figure does not resemble a billboard. The lettering on a line drawing, for example, should be about twice as large as the type in which the text of the journal is set. After reduction and printing, the resulting labels will be easy to read yet unobtrusive. Obviously, symbols and letters on photographs which will not be reduced must be about the same size as the journal type. On dark backgrounds, it is sometimes wise to use white (reversed) symbols and letters. Legibility is paramount. The shapes of symbols and letters on gray backgrounds, which will become less definite when halftones are made, must be particularly clear and simple. Hand lettering is rarely satisfactory, but various mechanical lettering guides will produce legible labels that can be reduced or enlarged to proper size by means of xerography or photocopying. Transfer and wax-backed printed letters and symbols suitable for labels are now widely available in a range of typefaces and type sizes, greatly simplifying the production of illustrations.

Editors are limited by the size of the printed page, the need for captions, and the amount of space available. At many scientific journals, editors are volunteers with research objectives of their own. Such editors appreciate a thoroughly professional author whose illustrations are ready for publication without alteration. All magnifications included in figure legends should be tentative, because it is difficult to tell in advance how much reduction a particular editor will find appropriate. A labeled scale line on the illustration avoids these problems.

When many illustrations are used, grouping is often required for economical use of space. Line drawings may be assembled onto bristol board and rubber-cemented down, with the labels applied after the most economical arrangement has been found. Paper edges of individual drawings will not show; if they do, they can be whited out by the printer. Trimmed edges of photographs, however, do show. When several photographs are to make up one illustration, all the edges should be cut clean and square with a paper cutter and the pictures should be mounted flush with one another. The printer can mask out narrow straight white lines between pictures without difficulty. It is not wise to combine photographs and line drawings in one illustration, because sharp reproduction of both black-and-white lines and gray areas will require the shooting of two negatives and the subsequent registering of separate layers of film. Such hand work is expensive and introduces the possibility of error. Moreover, line work can be printed on any kind of paper whereas halftones require a glossy, smooth-finished paper. Many books and journals are now printed on glossy paper for that reason. Others use glossy paper only when required, and some do not use them for any reason. It is clear that an author needs to be familiar with the practices of the journal which will publish his or her research results.

Graphs and charts involving curved lines or straight lines that are neither strictly horizontal nor strictly vertical are reproduced as line drawings; in preparing them for publication, the same rules as to size, proportions, and lettering apply. However, such figures should be drawn on plain white paper or blue-lined coordinate paper only. The camera will pick up printing in any other color, and it will become part of the figure.

Photography for scientific publication requires special approaches and setups. Hints on equipment and lighting are given by Blaker (1977).

One should retain a complete and correct copy of any manuscript submitted for publication. Unless the journal instructs otherwise, only one or two sets of good photocopies of illustrations should be sent with the manuscript. The originals are kept in case they will need simple alteration for publication, and they are sent when the paper is approved and the editor requests them. A good photograph of an illustration, made to the correct size, facilitates review.

References and Bibliography

References in taxonomic papers are generally printed in footnotes, in parentheses in the text, or in a terminal bibliography. Footnotes are useful when only a few references are involved and when repeated reference to the same bibliographic item is unnecessary. However, since they are costly to handle in typesetting and printing and may add materially to publishing costs, parenthetical references are preferred. When references are numerous, they are most frequently handled in a terminal bibliography. In most cases this bibliography should be as short as is consistent with its purpose, and the items included should be chosen selectively. Frequently the value of the terminal bibliography can be increased greatly by including parenthetical comments on the nature of the subject matter covered. Unverified references can be included when they are necessary for completeness, but they should be clearly marked.

Bibliographic items should receive full citation, including author, date, title, publication or publisher, volume, and pages, in the manuscript. They may be edited to suit journal requirements when the final draft is prepared. The references to the terminal bibliography can be made by enclosing an author's name and date (sometimes also page) in parentheses. Two or more references to publications by a single author in the same year can be designated by appended letters (Smith 1940*a*; Smith 1940*b*). The author-date system of bibliographic reference now in general use is far better than the straight numbering system which is sometimes used. The number system tells nothing about the reference; moreover, the author-date system permits the addition of references during the preparation of the manuscript without the necessity of renumbering all references beyond the point of insertion.

The designation *Bibliography* implies completeness of coverage of the subject. *Literature Cited* indicates restriction of references and is self-explanatory. For illustrations, see the bibliography at the end of this book.

An example of footnote style[4] is given here. In the typescript, footnotes are often entered beneath a marginal line in the text wherever they occur rather than at the bottom of the page, because in the final publication pagination will be entirely different from the pagination of the typed pages.

There are numerous exceptions, especially in government documents, but a majority of literature citations will fit into one of the simple styles illustrated here. It is standard practice to list the year of publication immediately after the author's name, since this sequence agrees with that of the author-date system. The full title should be given in all but the briefest footnotes, because readers obtain valuable leads this way. Abbreviations of journals should follow standard works such as the *World List of Scientific Periodicals,* although one should always refer to the style of the journal for which the manuscript is intended. It is increas-

[4]W. M. Wheeler, 1889. *Amer. Nat.* **23**:644–645.

ingly common to eschew abbreviation in references in the interest of clarity; little space is saved by abbreviation, and typographical errors can render the abbreviated title of a journal incomprehensible.

FORM AND STYLE OF THE TAXONOMIC ARTICLE

Much scientific literature is of a rather ephemeral nature, except for a few classical papers. This is not true in taxonomy, where publications (particularly if they contain new names) are consulted for many generations, back to 1758 and earlier. Taxonomists must try to give their publications a quality which lives up to this timelessness. Innumerable details require attention, and many of them escape the beginner's attention. There are various guides to the preparation of scientific papers (Trelease 1951; Hurt 1949) as well as style manuals, such as *A Manual of Style* by the University of Chicago Press, the *U.S. Government Printing Office Style Manual,* the *Council of Biology Editors Style Manual,* and the *American Museum Style Manual* (the latter pays particular attention to the practices of the taxonomist). There are also some useful hints in Appendix E of the International Code (Chapter 14). Before submitting a manuscript to a journal, one should study the instructions to authors which nearly every journal has and the particular style it prefers for citations, bibliography, and so forth. This will save a great deal of retyping.

Title

The title is the first part of a paper encountered by the reader, although it is often the last item to be added in the preparation of a paper. Its bibliographic prominence and significance warrant much care in its selection. The title should be long enough to be specific about the contents of the paper but brief enough for easy indexing. Short words are preferable to long ones. The most important nouns should be near the beginning of each series of words. The title should contain key words which in indexing will classify the article. This is particularly important now that some abstracting services base their entries entirely on titles. Punctuation should be avoided unless it is essential to meaning. Among the essential elements of a title are (1) a clear indication of the field involved (taxonomy, morphology, ecology, etc.), (2) the scientific name of the taxon treated, (3) indications of the order and family by means of scientific names (which may be in parentheses), or, rarely, a well-known common name, and (4) the geographic area, fauna, or locality. The following are examples of good titles:

A Taxonomic Revision of American Leafhoppers (Homoptera, Cicadellidae)

A Checklist of the Birds of Alabama
Geographic Variation of *Hippodamia convergens* in Southern California (Coleoptera, Coccinellidae)
Two New Species of Wood Rats (*Neotoma*) from the Rocky Mountain Region

The following are a few examples of poor titles for taxonomic papers. On the basis of the principles enumerated above, the objections to these titles are obvious.

New Hymenoptera
Notes on Mammals
The Western Biota
A Collecting Trip to Texas
Additions to the Fauna of Nebraska
Studies in the Mollusca
A New Acanthiza

Titles need not be as bad as these, however, to cause difficulties for catalogers, abstractors, reviewers, and other bibliographers. Authors have no cause for complaint that their work is being overlooked if it masquerades under an incomplete, ambiguous, or misleading title.

Author's Name

The author's name follows the title. Bibliographic problems are simplified if an author always uses the same form of his or her name. The entomologist Laporte sometimes published under the name Laporte, sometimes as the Comte de Castelnau, and the bibliographic confusion which resulted persists in the modern literature. Women taxonomists who begin publication before marriage frequently avoid confusion by continuing to publish under their maiden names or by a system of hyphenation, e.g., Dorothy McKey-Fender. It is customary in America to omit degrees and titles from the author's name, although they are used in many European journals. The author's institutional address and the correct postal address, if different, should follow the author's name in order to give credit to the supporting institution and facilitate correspondence. The mailing address can also be given at the end of the article or in a footnote.

When more than one author is involved, the order of names is determined by the nature of the contribution each author has made. When the work has been more or less equally shared, the names are usually arranged in alphabetical order. When the work has been disproportionately divided or there is a marked discrepancy in age or experience, the name of the "senior" author appears first. If the same coauthors publish several papers in a single year, it is advisable to change the sequence of names to facilitate ease of reference.

Abstract

Most journals require that the author provide an abstract, usually a single paragraph including brief statements of all new information covered in the work. The abstract is the source of references for many indexing and bibliographic services and is checked by many busy readers so that they can decide whether to read the full paper. For these reasons it needs to be a thorough, accurate, and completely clear—though very brief—summary of the content.

Introduction

Every taxonomic paper should include an introductory paragraph stating the scope of the paper and, where pertinent, the reasons for the study as well as the nature of the studied material. Frequently a brief historical review is appropriate. These features serve to orient the casual reader and the new student of the group and to refresh the minds of other workers in the field.

Acknowledgments

Acknowledgments may be included in the introduction when they can be treated as part of the natural sequence of exposition. Some authors place them in a footnote appended to the author's name; others, at the end preceding the summary. Many journals have specific methods of handling acknowledgments, as a glance at a recent number will demonstrate.

Giving proper credit is one of the most important responsibilities of the scientist. Acknowledgment should be made of all unpublished observations, determinations, and data derived from others. If an unpublished statement is quoted, the author of the statement should be allowed to prepare it especially for that purpose. Previously published data should never be utilized in such a way as to appear original.

Acknowledgment should be made of borrowed and donated specimens that have been studied. The methods of doing so vary with the amount of material received from any one source and the general plan of presentation of the paper. A way can be found for such acknowledgment even in complex cases.

Photographs, drawings, and other illustrative material lent or donated by others should be credited. Credit should also be given to artists and photographers for their work whether or not they received payment for their efforts. Good drawings or photographs are scientific contributions on a par with descriptive work and are frequently far more accurate and useful.

Full credit should be given to the collector, who, after all, is the real discoverer of the material. Assistance in outlining a research program (including the help of a major professor or senior colleague) should be acknowledged, as should help in the preparation of a manuscript by means of critical reading.

Finally, acknowledgment should be made of financial grants or institutional aid, such as the use of laboratory facilities and libraries. Frequently such help is a primary factor in making a particular taxonomic research project possible. Grant-giving agencies usually stipulate that the code number of the grant be recorded in all publications resulting from the grant.

Methods Used and Materials Studied

In a revision or monographic work it is desirable to include a statement about methods utilized and collections, specimens, or other materials studied. This enables the reader to evaluate conclusions and judge the thoroughness of the work. Standard methods for measuring, mounting, staining, and special preparations may be referred to by name and reference. Only new methods need to be described in detail.

Body of the Text

The material that constitutes the body of the text will, of course, depend on the scope and objectives of the particular paper. However, a complete systematic paper includes (1) a delimitation of the highest included taxon (family, tribe, etc.), (2) a key or keys to all intermediate taxa treated (genera), (3) synonymies and descriptions of taxa of intermediate rank (genera), (4) a statement of the type species, (5) comparison with other genera, (6) keys to the species of each genus, (7) synonymies and descriptions of each species, and (8) statements about type localities and location of types, general distribution, hosts and other significant biological data, comparisons with other species, and so on. For details on the preparation of descriptions and keys, see ''Keys'' in this chapter.

Summary

A summary is usually unnecessary in a strictly taxonomic paper. If required, it should be brief but not in telegraphic style. It should be written as a series of short paragraphs and should be specific, not written in broad general terms.

PREPARATION OF THE MANUSCRIPT FOR PUBLICATION

Aside from matters involved in the actual organization and construction of a taxonomic paper, some points should be kept in mind to facilitate editorial handling after the paper has been submitted for publication. Editors are much more likely to accept readily (and publish quickly) papers which are in good form and require minimal editing. Most publications have special form requirements, and much editorial time can be saved by careful study of the instructions given in the journal in which the paper is to be published. The sections of the *Council of Biology Editors Style Manual* (1988) dealing with biological journals are very helpful.

Typing

Manuscripts submitted for publication must be typed on standard (8½ by 11 in or 8 by 10½ in) white paper, double-spaced (some publications require triple spacing) and with at least a 1-in margin on all sides for adding proof marks and for editing. If approximately the same number of lines are typed on each page, the editor can conveniently estimate the space needed to print the manuscript. However, some editors require that pages end with completed paragraphs. Pages should be numbered consecutively in the upper right-hand corner.

Inserted pages are numbered alphabetically (e.g., 65*a*). Whole sheets should be used for insertions, regardless of the length of the inserted matter. When it becomes necessary to cut and rearrange, sheets should be assembled by pasting, not by pinning. Some journals request two copies of each manuscript, the original and a photocopy; some need three copies to facilitate review.

All tabular material should be typed on separate sheets, since it is usually set in a typeface different from that used for the text.

Underlining

Underlining is the standard convention for showing that material should be printed in italics. In a taxonomic manuscript submitted for publication, underlining should be limited to the scientific names of genera and species which appear in the text. New names should not be underlined, because the editor will usually mark them with a wavy line to indicate boldface. Indications of style or sizes of type for titles, headings, subheadings, side headings, and the like should be left to the editor. In general, marks which the author makes merely interfere with the editor's work, though marginal notes about the relative rank of headings may be helpful, and longer manuscripts should be accompanied by a table of contents that uses indentation to indicate the rank of headings.

Legends and Text Citations to Illustrations

Titles and legends should be self-explanatory. The list of these titles should be typewritten, double-spaced on separate sheets (several titles can be placed on a single sheet), and assembled in numerical order at the end of the manuscript, after the bibliography. A short identifying title and number must be placed on each illustration for purposes of identification. In production, titles and legends go to the typesetter with the rest of the manuscript, whereas illustrations are sent to a photographer. The printer may never see the original drawings.

The place of insertion of the illustrations should be marked in the manuscript. Illustrations are usually numbered starting with each article, but some journals number plates consecutively throughout a volume. In any event, a new series of figure numbers or letters should be used on each plate. Many journals designate figures with Arabic numbers and plates with Roman numerals. All figures should be referred to in the text by number.

Revision of the Manuscript

Few authors have sufficient mastery of the English language to write directly in final form for publication. Many competent scientists find it necessary to revise page after page not once but many times. T. D. A. Cockerell was an example of the former type of writer, while by his own testimony Charles Darwin was an inveterate reviser.

Trelease (1951) recommended careful reading of the manuscript 10 times, each time for one of the following: (1) consistency, (2) sentences, (3) clarity, (4) repetition, (5) connectives, (6) euphony, (7) punctuation, (8) style, (9) accuracy, and (10) length. Authors of taxonomic papers seldom follow the details of this recommendation, but most papers would benefit from more revisions than they receive. It often helps to put a manuscript aside for a while before the final revision is made. It is always advisable to have other persons read a manuscript before it is submitted for publication. A fully corrected copy of the manuscript should be kept by the author for use if the original is lost.

Word Processing

The use of a word processor greatly facilitates manuscript preparation, since it allows cut-and-paste revisions and thorough rewriting to be done before the manuscript is committed to paper, i.e., printed. Moreover, any number of original copies can be printed. The savings in time and effort are enormous.

Some word processors, however, have features that are best avoided in preparing a manuscript for publication. Right justification (lines whose right as well as left margins are even, as seen in this and most books) is possible as well as the "ragged right" margin characteristic of most typewritten material. Right justification is to be avoided in manuscripts, for it impedes the task of estimating the length of the manuscript in type. Moreover, it is harder for the editor and typesetter to read and thus increases the possibility of error. Some editors and publishers prefer that "soft" hyphens be avoided in a manuscript, i.e., that only essential hyphens be included and that no words should be divided at the end of a line. Right justification and word dividing are defended by those who value a neat manuscript, but clarity and ease of reading are much more important to editors, reviewers, and typesetters than mere neatness is.

Although many computer printers can produce italic type for scientific names, this can easily be missed by the typesetter. It is far safer to stick to standard underlining when italics are required.

If a letter-quality printer is available, a word-processed manuscript can be translated into camera-ready copy before it is submitted to an editor. Moreover, a manuscript can be submitted on floppy disks or go directly to the printer over the telephone by means of a modem. The use of this technology for the printing of research results has a number of built-in problems. If one's work is to be, both in matter and in manner, of sufficiently high quality to survive the scrutiny of generations of workers, it is likely to benefit from the improvements suggested by editors, reviewers, and proofreaders. Each will catch errors that even the most competent and meticulous scientist is bound to make.

For those unfamiliar with word processing, several excellent handbooks are available. A good one is Flegelman and Hewes (1983), which promotes no particular computer or word-processing program but does discuss features to be found in such programs and strategies for their use.

Proofreading

Most scientific journals permit the author to read proof before publication. A few journals place the entire burden of proofreading on the author and hold the author responsible for typographical or other errors which may pass undetected. In any event, proofreading, whether of manuscript or of type, is a very important part of an author's scientific responsibility. The scientific value of a paper can be greatly lessened by unfortunate typographical errors. Such errors are sometimes obvious to the reader; others are insidious and may be misleading.

In general, proof is submitted to the author to permit the elimination of printer's errors. Author's errors are the author's own responsibility, and most publications charge authors for corrections other than printer's errors. Changes in proof are costly and therefore should not be made unless they are necessary and the author is willing to assume the cost of the change.

Proofreading cannot always be done satisfactorily by one person. It is advisable to supplement personal reading by having someone else read slowly from the original manuscript while the proofreader (preferably the author) carefully reads the proof. Some authors read the manuscript into a tape recorder and play it back if no friend is available. Special attention should be given to punctuation, spelling of scientific names, numbers, and dates of all kinds. When corrections are necessary, they should be made according to the standard system of proofreader's marks that is given in most dictionaries and style manuals.

Prior to the electronic revolution in the printing industry, authors were able to correct galley proofs of their papers, each of which was equivalent to about three pages in the final publication. In galleys, author's changes, though discouraged, were possible if one wished to pay for them. Now, however, type is set on specialized computers—actually elaborate word processors—that have as an end product a master (i.e., camera-ready copy) for offset printing. The result is that authors now commonly see only the page proofs of their articles. While these new methods can reduce printing costs, changes in page proof, except for the correction of printer's errors, are prohibitive in cost because they require not only resetting but sometimes extensive remakeup of pages. It is essential that the final form of the manuscript, after revision and change resulting from editors' and reviewers' comments, be as close to perfect as possible. Some publications submit corrected page proofs to authors. In such cases proofreading cannot be restricted to individual words but must include much more. Modern typesetting machines cannot change a single letter in a word but must reset at least the whole line, while phototypesetters often rerun several lines, changing hyphenation as they go. The author should carefully check everything which has been rerun. It is also advisable to read the top and bottom lines of every page, because in the process of correction, mishaps most often occur in these lines. Corrected proof should be returned at once to the editor or printer to avoid delay in publication. The printing of an entire issue of a periodical can be held up by a single tardy author.

Reprints

Reprints are usually ordered when proof is returned to the editor. Papers that deal with general principles require more reprints than do those dealing with an author's limited specialty.

A good reprint collection is particularly valuable in taxonomic research, where so many publications are only one or two pages long. With the help of modern copying methods, an author can build up rapidly and inexpensively a specialized library of the entire literature on a taxon (except for a few very popular groups).

Letters requesting reprints should be specific. Most authors resent a request for "a set of reprints of your papers," except in unusual circumstances. Few authors have an inexhaustible supply of their papers, and the majority prefer to distribute their limited stock to those who will make the greatest use of them. Authors who wish to send all their reprints are free to do so. The receipt of requested reprints should be acknowledged; in some cases, especially when the expense of shipment is significant, the recipient should offer to reimburse the author for postage.

CHAPTER **14**

THE RULES OF ZOOLOGICAL NOMENCLATURE

The zoologist deals with an enormous number of items; each species, genus, and higher taxon is a different thing. It would be impossible to refer to them if each one did not have a separate name. The term *nomenclature* comes from the Latin words *nomen* (''name'') and *calare* (''to call'') and means literally ''to call by name.'' Through nomenclature, labels are provided for taxa at all levels to facilitate communication among biologists. The scientific names for species of organisms and the higher taxa in which they are placed form a system of communication, a language; these names must fulfill the same basic requirements as any other language.

Beginning with Linnaeus, an elaborate system of rules and provisions has evolved that decrees how the naming of animals should be done and how one can determine the correct name of a species or higher taxon. Underlying these very technical rules are certain basic principles or concepts that unfortunately are not understood by all taxonomists. Neglect of the conventional rules and ignorance of the underlying principles are the major reasons for the unsettled nomenclature of many groups of animals. This chapter calls to the attention of the working taxonomist the major rules and the most frequent difficulties in their application. This is necessarily a rather streamlined presentation; a far more detailed account is given in Mayr (1969). Every taxonomist who introduces new names into the literature or who wishes to determine synonymies should own a copy of the *International Code of Zoological Nomenclature* (see below). The present treatment is meant to be a commentary on the Code and should be used with the Code in front of the reader.

THE MOST IMPORTANT PROPERTIES OF TAXA NAMES

What are the important requisites of any system of communication, scientific nomenclature included? There are many, but three seem to be especially significant.

Uniqueness

A classification is a filing system, an information retrieval system. The name of an animal, like the index number of a file, gives immediate access to all the known information about that particular taxon. Every name has to be unique because it is the key to the entire literature relating to that species or higher taxon. If several names have been given to the same taxon, there must be a clear-cut method of determining which of them has validity. In zoological nomenclature, priority usually decides in cases of conflict.

Universality

Scientific communication would be very difficult if there were only vernacular names for animals; specialists would have to learn the names of taxa in innumerable languages in order to communicate with each other. To avoid this, zoologists have adopted by international agreement a single language, a single set of names for animals, to be used worldwide.

Stability

Any change of a well-established name is likely to produce confusion and impede information retrieval, since its name is the key to the information about a particular organism. Therefore, even the principle of priority can be set aside by the International Commission on Zoological Nomenclature in cases where stability is threatened. For a discussion of the concept of stability, see Mayr (1969:335–340, 344).

These three major objectives of the communication system of taxonomists are singled out for attention in the Preamble of the Code: "*The object of the Code is to promote stability and universality in the scientific names of animals, and to ensure that each name is unique and distinct. All its provisions are subservient to these ends.*"

The Preamble stresses another vitally important principle: None of the provisions of the Code "restricts the freedom of taxonomic thought or action." This means that no taxonomist shall be forced to accept a particular classification or the delimitation of a taxon or be guided in his or her original choice of the type of a taxon or be compelled to recognize a subjective junior homonym (Article 10*g*) for other than zoological reasons.

THE INTERNATIONAL CODE

The valid rules of zoological nomenclature are contained in an authoritative document entitled the *International Code of Zoological Nomenclature*. The original code and its revision were the work of the International Congress of Zoology. Now that these congresses are defunct, their role has been taken over by the General Assemblies of the International Union of Biological Sciences (IUBS). See Mayr (1969:299–301) for a history of the rules of nomenclature.

The International Congresses of Zoology have appointed a judicial body, the International Commission on Zoological Nomenclature, which interprets the rules and can publish opinions on controversial issues. In particular, the Commission can by plenary powers set aside priority or other provisions of the Code when they violate the principle of stability enunciated in the Preamble. Articles 76 through 82 specify the duties of the Commission. The latest edition of the Code was published in February 1985 and can be ordered from the International Trust for Zoological Nomenclature, 56 Queen's Gate, London SW7 5JR, or from the University of California Press, Berkeley and Los Angeles.

EVOLUTION OF THE THEORY OF NOMENCLATURE

Many provisions of the Code are the results of compromise between conflicting principles. Such compromises go back to Linnaeus, whose nomenclature embodied on the one hand the principles of Aristotelian logic and on the other hand methods for solving practical problems. The conflicts between stability and priority, taxonomic freedom and nomenclatural rigidity, and typification through type fixation and typification through type restriction are examples of other conflicts that have required compromises. Equally important are certain changes in the last 150 years in the basic concepts of taxonomy (Mayr 1989). These changes include the concept of taxa as populations rather than types, the nature of the type as a name bearer, the categorical status of infraspecific names, and finally the application of certain basic legal principles to the laws of nomenclature, such as the impropriety of a retroactive application of laws and the stabilizing effect of statutes of limitation.

All good law is living law. This is as true for the Code as it is for all other codes of law. The International Rules have prevented the development of chaos in the naming of animals and have helped greatly to standardize taxonomic procedures. No change of these rules should be undertaken lightly, yet the rules should not become rigid, losing contact with the conceptual evolution of taxonomy. It is the rules that will have to be adjusted to the conceptual development of taxonomy, not the reverse.

An unresolved difficulty is posed by the drastically different needs in regard to different taxonomic groups. The needs that exist in popular groups such as birds and mammals, in which a limited number of species are referred to by their scientific names hundreds or thousands of times annually, are very different from the needs in an obscure group of arthropods attended to by a single specialist, with many species not mentioned in the literature more frequently than once every 30 to 50 years. Different problems exist for the parasitologist, who must make tentative assignments of stages in the life cycle of parasites (e.g., cercariae) even though the connection with the adult stages can be established only by experiment, and for the paleontologist, who may be forced to adopt form genera until the true taxonomic status of the objects placed in those form genera is fully established. Dissension and controversy inevitably result when specialists in one group of organisms are blind to the needs of specialists in other groups.

THE INTERPRETATION AND APPLICATION OF THE CODE

The International Code, like any code of laws, presents the rules starkly and with little explanation. Even the experienced specialist is sometimes in doubt about how to interpret the rules, and the beginner is often bewildered. The current edition of the Code, however, is a great advance over previous editions because of its more logical organization, the more explicit wording of many provisions, the more appropriate designation of many articles, the better choice of examples, and the addition of an excellent glossary. Many nomenclatural practices implicit in or tolerated by the older versions of the Code are now carefully articulated. What the Code usually fails to state is the rationale behind the various rules. What are the basic principles underlying the specific provisions? Knowledge of these principles often makes it much easier to apply the rules specified in the articles.

We provide in this chapter a series of comments that may help the practicing taxonomist. When a worker is uncertain which article of the Code to consult, the answer can often be found in its excellent index.

Stability

The statement in the Preamble of the Code that "the object of the Code is to promote stability" of nomenclature is perhaps the most important provision in the entire rules of nomenclature. Specific stability-promoting provisions in the Code can be found in Articles 10*g* (preservation of availability), 23*b* (suppression of priority), 40 (family names), 41 (misidentified genera), 70*b* (misidentified type species), and 79 (plenary powers for any action on behalf of stability).

Not all changes of names result from the application of the rules of nomenclature. Sometimes such changes are necessitated by zoological discoveries, for instance, when

1 Several species (e.g., sibling species) had been referred to by a single name prior to a more penetrating analysis
2 Several infraspecific phena were named in the belief that they represented different species
3 Several authors (usually in different countries) had named the same taxon unbeknown to each other
4 Two or more authors use the identical name for different genera, often in unrelated groups
5 Wrongly placed species must be moved to another genus

Even though total stability of names is the ideal, if zoological nomenclature is to serve as a system of communication, taxonomic uncertainties prevent the achievement of this goal at the present time.

Freedom of Taxonomic Thought

It is axiomatic that the freedom to make taxonomic decisions should not be restricted by rules of nomenclature. This is now expressly guaranteed in the Preamble, which states that none of the provisions of the Code "restricts the freedom of taxonomic thought or action." This basic principle is applied to specific situations throughout the Code. In Article 11*e* it acknowledges that the subjective action of an author in placing a name in synonymy should not determine the nomenclature to be used by others; in Article 10*g* it permits a zoologist to continue using a name considered by another author to be a junior homonym; and in Article 64 it authorizes the zoölogist to choose (as the type of a new family) that genus which the zoologist considers the most typical or characteristic genus of that family but "not necessarily that having the oldest name."

The same is true for type designation on the generic level (Articles 67 through 69).

Conversely, taxonomic mistakes should not affect nomenclature except where this is expressly authorized by the Code. For instance, the erroneous assignment of a species to a genus does not constitute the proposal of a new specific name in that genus (Article 49), the naming of individual variants that do not form a taxon has no validity in nomenclature (Article 45*e*), and the erroneous placing of a name in synonymy does not invalidate it (Article 10*g*).

Priority

In order to become "available," a name has to be published in a manner that conforms to the specifications of Articles 10 through 20. The name of a taxon is "valid" if it "is the oldest available name applied to it" (Article 23). The date of publication of a name (Article 21) thus is of crucial importance. In zoological nomenclature the principle of priority applies only to the categorical levels of species (and subspecies), genus, and family. It does not apply to the higher categories.

Priority thus means priority of publication, not priority of usage. In groups of animals—mammals, birds, butterflies, and mollusks, for example—with an extensive eighteenth-century literature, thousands of names were discovered after the first adoption of the Code that were older than the universally used names for certain taxa. The adoption of a few commonsense principles, such as a rule of limitation, could have prevented the wholesale changing of names that became necessary, but no such measures were adopted. After more than 50 years of turmoil, considerable stability has now been achieved in the nomenclature of most groups with large numbers of previously "forgotten" names. The period of transition, however, was very painful. Even Darwin complained with regard to cirriped nomenclature: "I believe if I were to follow the strict rule of priority more harm would be done than good. . . ." For a history of the controversy about priority, see Mayr (1969:337–339).

First-Reviser Principle

A basic Linnaean principle widely followed by his disciples is that the action of the first reviser should be adopted in all equivocal nomenclatural situations. This principle, although sanctioned by usage, was almost entirely ignored in the drafting of the 1901 Code. It has been incorporated, however, in numerous provisions of the current Code.

There are various ways by which a first reviser can help stabilize nomenclature. In the case of simultaneously published names (Article 24*a* and Recommendation 24) the reviser can select the name that is better known rather than the one that has line or page precedence (which is not priority). If a new name is spelled in more than one way in the original publication, Article 32*b* permits the first reviser to accept the spelling which is most commonly used. If no type species is fixed in the publication in which a new genus was described prior to 1931, the first reviser can select the species whose designation is in the best interest of stability (Article 69). In the case of a composite name based on a series of syntypes consisting of several species, the first reviser can help establish stability by making either of two choices (Article 74*a*). The reviser can designate as *lectotype* a specimen of the species to which the name has always been applied in the past, or if it is a name that is best suppressed for the sake of stability, he or she can designate a lectotype belonging to a species for which a senior synonym is available.

Range of Authority of the Code

The Code applies to both living and extinct animals (Article 1). There are, however, separate codes for botanical and bacteriological nomenclature, and Article 2 of the Zoological Code spells out the relation between the codes and the status of names transferred between kingdoms. Familiarity with the botanical rules is important for authors who work with protists.

Application of Names

Names are given only to taxa, and all taxa are populations or groups of populations. Consequently, only populations are named. Names are given to individuals only as representatives of populations. Names given to individuals as such or to phena within populations have no official status. "Names given to...infrasubspecific forms as such...are excluded" from zoological nomenclature [Article 1*b*(5); see also Article 45*c*]. Also unavailable are all names given to hypothetical concepts, names given to teratological specimens as such, and names proposed for other than taxonomic use (Article 1).

The absence of a convenient non-Linnaean nomenclature for single specimens is a considerable handicap in paleontology, particularly in the study of fossil humans. To have a reference name, many anthropologists created a generic and specific name for every new specimen they found. Simpson (1963) exposed the fallacy of this custom. When dealing with individuals, one must adopt some sort of vernacular nomenclature. Let us use "Olduwai L1" or "Trinil D7" rather than a set of scientific names that implies a nonexistent zoological status. The same is true for infrasubspecific names.

Names of Fossil and Living Animals If a still-living species was first named on the basis of fossil material, that is also the name for the living species. If the same name was independently proposed for a living species and a fossil species, the names are *homonyms*. If a generic name has been used for a fossil animal, it cannot be used again for a different genus of living animals and vice versa. Zoological nomenclators therefore contain the names given to both living and extinct animals so that the possible occurrence of homonymy can be determined.

For some special provisions that apply only to the names of fossil taxa, see Articles 10*d*, 11*f*(4), 13*b*, 20, 23*g*, 55*a*, 56*a*, 56*c*, 57*a*, 67*m*, and 72*c*. Most of these provisions concern ichnotaxa, that is, taxa based on the fossilized work of an animal, including fossilized trails (including dinosaur tracks) and burrows.

Starting Date of Zoological Nomenclature No name is available that was published prior to January 1, 1758. Linnaeus's *Systema Naturae*, 10th edition, and Clerck's *Aranei Svecici* are deemed to have been published on that date. The rules relating to the date of publication of later names are stated in Article 21.

The Binominal Nomenclature of Linnaeus Linnaeus introduced the consistent use of binominalism in 1758. The scientific name of every species should be composed of a generic name and a specific name, the generic name as an aid to the memory and the specific name to express distinctness and uniqueness.

The disadvantage of this system is its instability: The name of a species changes every time a species is transferred to a different genus. In a group of bees of the tribe Paracolletini, a single revision resulted in a change of 288 of a total of 332 binomina and, even worse, in the need for renaming 16 species as a result of secondary homonymy (Michener 1964). However, no one has been able to suggest a superior system.

Concept of Availability

In order to be available, a name must have been published in a way that satisfies the requirements detailed in Articles 8 through 20. The term *available*, as used in the Code, means "legitimate" or, as botanists put it, "effectively published." A name may be available in this technical sense even though, as a junior primary homonym, it is automatically invalid at the moment of publication (Articles 53 and 59*a*). It is important to understand clearly the difference between *available* and *valid* names. To have this distinction made clear is one of the major advances of the new Code. To employ the name of a taxon one must make sure that it is available (properly published) as well as valid (not preoccupied by a senior synonym or homonym).

Articles 11 to 19 list the prerequisites for making a specific name available. Particularly important is Article 13*a*(i), according to which names published after 1930 must be "accompanied by a description that states in words characters that are purported to differentiate the taxon." It is good taxonomic practice to include in the original description not only such diagnostic information but also a differential diagnosis that consists of an actual comparison with other closely related species.

A name published without satisfying the conditions of availability is generally called a *nomen nudum,* particularly if it fails to satisfy the conditions of Articles 12 through 16. A *nomen nudum* has no standing in zoological nomenclature and is best never recorded, even in synonymy. There is always the danger that such a listing will provide an "indication" in the sense of Article 12 and thus inadvertently make the manuscript name available.

If editors of scientific journals were more familiar with the provisions of the Code, they would not permit the publication of so many names that are not available because they violate one or another of Articles 8 through 20.

Differential Diagnosis

To be available, every new scientific name published after 1930 must be accompanied by a description containing characters that purport to differentiate the taxon (Article 13*a*).

Respectable taxonomists go well beyond this minimal requirement by comparing the newly proposed taxon with its closest relative(s) and describing the diagnostic differences carefully. Even though such a *differential diagnosis* is not a mandatory provision of the Code, it will be supplied by every conscientious worker.

Names Given to Hybrids

Hybrids are normally individuals, not populations, and hence are not taxa. Article 1*b*(3) states clearly that a name given to an animal that is later found to be a hybrid remains available only for purposes of homonymy, not for purposes of synonymy. A name given to a species hybrid cannot be applied to either of the parental species.

Validity

Validity is a term that refers to the rights of names in relation to homonyms and synonyms. *Synonyms* are different names for the same thing. The earliest published synonym is referred to as the *senior synonym*; all later synonyms are *junior synonyms*.

Two kinds of synonyms can be distinguished. One consists of names that objectively refer to the same thing, such as a new name for a supposedly preoccupied name, or names based on the same specimen or illustration. These names are called *objective synonyms*. The other kind consists of names based on different type material; taxonomists may nevertheless decide that they belong to the same taxon. These are called *subjective synonyms*. A generic lumper, for instance, may consider as synonyms certain generic names that a splitter would consider valid.

At a given time only one name can be the valid name of a taxon. This is normally the oldest available name that is not preoccupied by a senior synonym or homonym and has not been suppressed by the Commission for the sake of nomenclatural stability. The provisions are clearly stated in Article 23. Suppression by the Commission converts a previously valid name into an invalid name.

Protection of Well-Established Names

Demands have been made for more than 100 years for the incorporation in the rules of an automatic provision for the protection of well-established names. Such a *statute of limitation* was indeed adopted by the Copenhagen and London congresses but was eliminated in 1963 by the Washington congress. It was incorporated in the 1961 Code as Article 23*b* and was valid from November 6, 1961, to January 1, 1973. It permitted one to ignore a senior name (*nomen oblitum*) that had not been used during the immediately preceding 50 years while the competing junior name had been universally used during that period. To make rare names ineligible, it was later specified that the junior name had to be used by at least five different authors and in at least 10 publications during this 50-year period.

The great advantage of this statute of limitations was that it provided badly needed relief to the Secretariat of the Commission, which is overburdened with applications. It did not preclude applications to the Commission for the resurrection of senior names, but it placed the burden of work on the taxonomist who wanted to upset established nomenclature rather than on the one who wanted to preserve stability.

As it now stands, according to Article 79*c*, an application for the protection of a universally used name must be made to the Commission. It is our opinion that stability and simplicity in nomenclature would be greatly advanced if the statute of limitation were, after careful rewording, restored to the Code (see also Mayr 1969:350–352).

Page and Line Precedence

If two names for the same taxon are published in the same publication, they are considered to be "published simultaneously" (Article 24*a*). One of these names may have line or page precedence, but their priority is determined by the action of the first reviser. The first reviser must give serious consideration to Recommendation 24A and "select the name that will best ensure stability and universality of nomenclature." Almost invariably one of the two competing names is better known, is based on a better description, is based on better type material, is based on an adult phenotype rather than an immature stage, or has some other nomenclatural advantage. This is the name the first reviser should select. Chronological priority is not involved in the case of simultaneous publication; it is replaced by *designated priority*. The same is true for simultaneous publication in different works (see also Article 24*a*).

Work of an Animal

Prior to 1931 the description of the work of an animal, for instance, its gall, was considered a valid indication [Articles 23*f*(iii) and 72*c*(i)]. After December 31, 1930, a name not based on the animal itself was no longer an available name.

Formation of Names

Mandatory provisions governing the formation of zoological names are given in Articles 25 through 31. Since such names must be either Latin or latinized or at least so constructed that they can be treated as a Latin word (Article 11*b*), it is important for the zoologist to be familiar with the rules that govern the correct transliteration and latinization of words. Three appendixes (B, C, D) to the Code are specifically devoted to these matters. They are not mandatory provisions but have the same status as recommendations. See also Brown (1954).

Appendix B (Code, pp. 183–189) deals with the transliteration and latinization of Greek words. Appendix C (Code, pp. 189–191) deals with the latinization of geographic and proper names, while Appendix D (Code, pp. 193–229) deals quite generally with recommendations on the formation of names. Every zoologist who plans to propose a new name for a taxon should read these recommendations carefully.

A few points need to be stressed. Names that are complex, long, and difficult to pronounce should be avoided. For obvious reasons, a new generic name should differ from any existing generic name in more than a single letter or its termination. In describing a new species, it is very unwise to choose a specific name that is already in use in a related genus. Such a name would become a homonym if a subsequent author combined these genera. For instance, names like *africanus, robustus,* and *capensis* have been given repeatedly to closely related hominids which subsequent authors may well place in the genus *Homo.* Names for protists should also be valid under the botanical rules.

A noun in apposition cannot be treated as an adjective. Words like *longicauda, melanogaster,* and *albipectus,* in which the ending is a noun, cannot be changed when placed in a genus with a different gender.

Compound names are to be united into a single word without a hyphen (Article 31*d*), except that a hyphen may be used when a single letter is used as a character, as in *c-album* [Article 31*d*(iii).]. Diacritical marks are never used (Article 27).

Gender of Generic Names

If the specific name is an adjective in the nominative singular, it must agree in gender with the generic name. To apply this rule correctly the zoologist must understand the rules governing the gender of Latin nouns, whether of Latin, Greek, or nonclassical origin. These rules are given in Article 30. Various handbooks on scientific names are helpful, and standard Greek and Latin dictionaries are indispensable (see also Appendixes B, C, and D).

Latin and Greek grammar are full of pitfalls. A Latin noun ending in -*us* is by no means necessarily a noun of the second declension and thus automatically masculine: It may be feminine like *domus* or neuter like *pectus*; moreover, the Latin suffix -*cola* is masculine. A latinized Greek word takes the gender appropriate to its Latin termination [see the examples in Article 30*a*(i)(3)].

A zoologist who proposes a new name in the genus group should give the etymology and the gender of the new name (Appendix E, 16).

Family Names

The names of taxa above genus rank are always uninominal and given in the plural. Provisions relating to names in the family group are found in Articles 11*e,* 29, 35 through 41, and 62 through 65. To avoid grammatical mistakes, remember that these names are in the plural. One cannot say,

"The family Fringillidae are the largest family of songbirds" but must say, "The Fringillidae are the largest. . . ." The same is true for the names of orders, classes, and other higher taxa: They all are in the plural.

An author proposing a new taxon of family rank has the taxonomic freedom to select the genus which he or she considers central (most typical) for the new taxon. Seniority of the name of the type genus is irrelevant. Provisions for the formation of family names are given in Article 29.

The 1961 Code introduced limited priority for family names as a result of a decision made by the Copenhagen congress. However, since family names are widely used in zoology, even by nonspecialists, stability is even more important on the family level than it is on those of genus and species. The Code therefore provides for two important stabilizing devices. Article 4 decrees that "if stability and continuity in the meaning of a family-group name are threatened. . . the case is to be referred to the Commission for a ruling" (Articles 65*b* and 70*b*).

The other provision favoring stability relates to name changes in the type genus. The name of a taxon in the family group must be formed from the stem of the name of the type genus (Article 29). The type of a family taxon is the zoological object identified by the name of the type genus. If a nominal type genus is rejected as a junior synonym, a family group name based on it is not to be changed and continues to be the valid name of the family group taxon (Article 40).

When a zoologist finds that the name of the type genus of a family is threatened because of the availability of a senior synonym, that zoologist may want to bring this situation to the attention of the Commission. It is sometimes advisable to suppress the senior name in order to retain the convenient tautonymous relation between the name of the family and that of its type genus.

Names for Higher Taxa (above the Family Group Rank)

The 1985 Code does not contain rules dealing with names of taxa in categories above the family group. There are a few general principles concerning the names of higher taxa above the family rank. Such names are single words in the plural. Although they are not based on a type genus, there is generally a consensus as to the central (most typical) component of a higher taxon. Zoologists maintain well-established names of higher taxa even though they may find that some other word is "more appropriate." When a composite higher taxon is divided, the original name should be retained for the more typical or larger group and a new name should be applied to the newly recognized group.

Genus Group Names

Provisions dealing with generic names are found in Articles 11*g*, 13*b*, and 42 through 44. A genus group name must be a noun in the nominative singular or be treated as such. A noun that was in the plural when first published takes its date and authorship from its first publication in the singular. An author proposing a new generic name should make certain that the proposal does not omit any of the following five essential points:

1 That it is clearly indicated as a new genus: "*X-us*, new genus."

2 That the proposed generic name does not violate the rules and recommendations concerning the formation of names (consult Appendix D).

3 That the newly proposed name is neither a homonym nor a synonym.

4 That the description contains a clear statement of the characters in which the new genus differs from previously described genera (Article 13*a*). It is desirable that a differential diagnosis be added in which a direct comparison is made with that genus or those genera which are believed to be most closely related to the newly described one.

5 That the type species is clearly and definitely stated (Article 13*b*). The generic limits may be interpreted differently by succeeding authors, but the type species will forever anchor the concept of the stated genus to a clearly defined zoological object.

Names of Taxa in the Species Group

The rules concerning these names are given in Articles 5, 11*h*, 31, 34*b*, and 45 through 49. There are only two categories in the species group: the species and the subspecies. The *specific name* is the second word in the *binomen* (Article 5), and the *subspecific name*, when employed, is the third word in the trinomen. Since a scientific name in the modern interpretation is considered a recognition symbol formed by a sequence of letters, it is not advisable (and indeed leads to confusion with generic names) to capitalize specific names at the beginning of a sentence.

Infrasubspecific Names

Linnaeus did not recognize the subspecies category. The term *variety* used by him and his followers referred to a medley of deviations from the type of a species (Mayr 1963:334–346). No distinction was made by Linnaeus between different kinds of infraspecific variants. The subspecies designates a genuine taxonomic category based on populations; variety names are names for phena and aberrant individuals and have no standing in nomenclature [Articles 1 and 45*e*(ii)].

Any name proposed after 1960 is unavailable if not clearly given to a subspecies taxon. In the case of infrasubspecific names given prior to 1961, corroborating evidence is to be used, such as the statement that a taxon is "characteristic of a particular geographical area or geological horizon" (Article 45*f*). A name first established with infrasubspecific rank becomes an available name (Article 10*c*) when it is used for a taxon of the species group and takes the date and authorship from the time of such a change in usage. It is taxonomic practice to give the benefit of the doubt to authors who introduced "varieties" prior to 1961 (Article 45*g*).

Authorship

The meaning of authorship and the rules concerning the citation of names are given in Articles 50 and 51.

The author of a name is the person who alone is responsible both for the name and for the conditions that make it available (Article 50), that is, for the diagnostic description. Consequently, if someone other than the author of the paper is responsible for a name and its availability, this other contributor shall be considered the author of the name. It is sometimes difficult to determine "responsibility" when the unfinished manuscript of a zoologist after his or her death is published by another zoologist. If the manuscript contains only the names and the editor supplied the diagnostic descriptions that make the names available, the editor is to be regarded as the author of these names (Article 50).

Homonymy

Homonyms are identical names for two or more different taxa. The earliest of such names is the *senior homonym*; later ones are *junior homonyms*. Articles 52 through 60 deal with the validity of homonyms and with replacement names for junior homonyms. This area is one of the most difficult in zoological nomenclature, and future congresses may modify some of the provisions, particularly where they seem to be in conflict with other provisions or with taxonomic freedom or stability.

A junior homonym in the genus group is always invalid (Articles 53 and 56). However, even a difference involving a single letter prevents homonymy of generic names.

There are some real conflicts with respect to the application of the law of homonymy to species group names. Homonymous specific names originally published in the same genus (primary homonyms) necessitate the renaming of the junior homonym. However, when two identical specific names are secondarily brought together under one generic name, such homonymy may be a matter of purely subjective taxonomic judgment. For an author who rejects the lumping of the two genera into one, no homonymy exists and there is no need for that author to reject the junior name. It remains an available name (Article 10*g*). Articles 57*c*(i) and 59*b* of the current Code state that subjective junior secondary homonyms, or at least those replaced before 1961 by some author, are permanently invalid. These provisions not only are a violation of the Preamble, which guarantees freedom of taxonomic thought or action, but also are contrary to universal practice in many branches of zoology, for instance, ornithology (Mayr 1987*c*). Until these erroneous articles are modified, some relief is given in Article 59*d*, at least for junior secondary homonyms rejected after 1960.

Replacement of Rejected Homonyms

Provisions governing the publication of replacement names are given in Article 60. Before proposing a new name as a replacement name (*nomen novum*) for one which is preoccupied, an author must be certain that it meets the following conditions:

1 That there is no other name available for the species (or genus). In the past a few nomenclaturists published replacement names for all junior homonyms whenever a catalog or nomenclator was published. Since in most of these cases replacement names had already been provided by specialists in the respective groups, such wholesale renaming resulted in nothing but an added burden to synonymies and a loss of respect for the authors of these unnecessary names.

2 That the original author of the preoccupied name is no longer alive. According to paragraph 3 of the Code of Ethics (Appendix A, p. 93), a zoologist should communicate with the author of a junior homonym, if that author is still alive, and give that author a reasonable opportunity to publish a replacement name.

3 That the new name is proposed in the form recommended in the Code. As stated in Article 13*a*(ii) and recommended by several zoological congresses, there must be "a full bibliographic reference" (not merely "Smith 1907") to the original citation of the preoccupied name and, in the case of preoccupied generic names, to the name of the type species (Appendix E, 15, p. 145).

4 That it is desirable to propose a replacement name. For instance, there is no excuse for renaming preoccupied names that are invalid synonyms. More important, a replacement name automatically takes the type and type locality (in case of a species) of the preoccupied name. If the type of the preoccupied name is no longer in existence or if there is the slightest doubt about the identity of the species with the preoccupied name, it is sometimes preferable simply to redescribe the taxon in question as if it were entirely new and thus to provide it with an unambiguous type and type locality.

A genus (or species) divided into subtaxa takes the name of the next oldest available subtaxon if the name of the nominate subtaxon is found to be a junior homonym (Articles 44*b* and 47*b*).

The Type Method

In taxonomic practice doubt often arises as to the identity of taxa to which names are attached. For species, the descriptions are frequently insufficient to establish identity, particularly the rather short descriptions of earlier authors. Sometimes a description applies equally well to several later discovered species because the species-specific diagnostic characters were not mentioned in the early description. In the case of higher taxa, the identity can become unclear as the contents of these taxa change and additional species are discovered. When such a higher taxon is split, the question arises, Which of the components should retain the name? It is obvious that a secure standard of reference is needed to tie taxonomic names unequivocally to definite, objectively recognizable taxa. These standards are the types, and the method of using types to tie names to taxa is called the *type method*. The principles and procedures of the type method have changed drastically since the days of Linnaeus. For a short history of these changes, see Mayr (1969:368–369; 1989).

Species consist of variable populations, and no single specimen can represent this variability. No single specimen can be typical in an essentialistic sense. As Simpson (1961) stated perceptively, the only function of the type specimen is to be a "name bearer." The younger generation of taxonomists understands clearly that a type is nothing more and nothing less than a specimen (or taxon) which tells us to which taxon a given name should be attached.

A type is always a zoological object, never a name. The type of a genus is a species, and the type of a family is a genus (Article 61). This is important in the case of "misidentification of the type," which occurs when an author designating the type of a new taxon refers to it under a wrong name. Once designated, the type cannot be changed even by the author of the taxon, except—by exercise of the plenary powers of the Commission (Article 79)—through the designation of a *neotype* (Article 75).

Description of a new species is based on the entire material available to the zoologist, including the type specimen. It is *not* the function of the type to serve as the exclusive or primary basis of the description.

Simpson (1961:183–186) discussed this aspect fully. He also introduced the term *hypodigm* for the entire sample of specimens personally known to a given taxonomist at a given time and considered by that taxonomist to consist of unequivocal members of the taxon. Disagreements among taxonomists are often caused by the fact that they have studied different hypodigms.

Correcting the Misidentification of Types

It is unfortunately true that the type species of new genera and the type genera of new taxa of the family group have sometimes been misidentified (misnamed) by the original authors of the new taxa. Articles 41, 49, 65*b,* 67*i,* and 70*b* contain provisions for correcting the error of the original authors. The principle on which such corrections are based is that the type of a taxon is not a name but a zoological object. The type (species or genus) is then the zoological object which the original author had before him or her (when making the type designation), not the name which this author might have erroneously attached to this object.

However, it would be an intolerable burden if the working taxonomist had to make sure in each case that the original type was correctly identified and named. It is therefore provided in the Code that a zoologist must assume that the author correctly identified this type (Articles 65*a* and 70*a*).

If there is strong or clear evidence that a misidentification is involved, the case is to be referred to the Commission (Article 41). In such cases the Commission tends to make a ruling that will maintain stability and continuity.

If reexamination of the type of a well-known species proves that the type actually belongs to a different species, the Commission can by its plenary powers (Article 79) suppress the original type and designate a neotype which conforms to the accepted concept of the species.

Wisely administered, the various provisions on misidentification can do much to preserve stability and continuity in nomenclature.

Kinds of Type Specimens

Since the type specimen is the official standard of reference for the name of a species, it can have full authority only if it is unique. When there are two or more type specimens, it has all too often been found that they actually belong to different species. Which of these species shall have the name? A taxon of the species group (species or subspecies) can have only a single type, either the specimen designated or indicated as the type by the original author at the time of publication of the original description (*holotype*) or one designated from the type series (*lectotype*), or the *neotype*.

Specimens studied by the original describer that are neither the holotype nor the lectotype (of a subsequent author) are conventionally referred to as *paratypes*. Paratypes have no special standing under the Code and do not qualify as types.

Type Designation

Since the information concerning the type specimen is nomenclaturally the most important information given upon the naming of a new taxon, it has become customary in taxonomic literature to record this information immediately following the new name. The information concerning the type takes the same place in the sequence as does the synonymy in a redescription.

The definite fixation of a type species has been, since 1930, a prerequisite for making a name in the genus group available (Article 13*b*). Designation of a type specimen is not a prerequisite for making a name in the species group available. In certain groups, for instance, certain genera of protozoans, it is exceedingly difficult to preserve specimens in such a way that they can serve as a permanent standard of reference. However, even in protozoology the designation of type specimens has increasingly become standard practice (Corliss 1963). For rules concerning types in the species group, see Articles 71 through 75.

Even though the publication of information on the type specimen of a new species name is not mandatory, it is nevertheless expected that the zoologist will clearly designate a single specimen as the holotype of the new species and supply in the original description the following information, in addition to measurements and other descriptive data, characteristic of the type specimen:

1 Precise collecting locality and other relevant data on the labels of the specimen

2 Sex

3 Developmental stage or form (if significant) to which the type is referable

4 In the case of parasites, name of the host species

5 Name of the collector

6 Collection in which the holotype is deposited and, when specimens are numbered, the number assigned to it

7 Altitude of the type locality or depth in meters below sea level at which the holotype was taken

8 In the case of fossil species, geological horizon

If a large hypodigm is available, great care should be exercised by the author to select as the type a specimen which—owing to its state of preservation, sex, age, or locality data—is most suitable as a name bearer.

The diagnostic characters relevant in a given genus are sometimes better accessible in some specimens than they are in others. The type should have maximum usefulness for taxonomic discrimination.

In the case of fossil material, if the hypodigm consists of many individual pieces (e.g., bones), it is advisable to designate the most diagnostic of them as the type, particularly if there is the slightest doubt whether the pieces actually belong to a single individual. Many "types" of formerly described fossil species have on reexamination turned out to be composites of several different species.

The following additional advice is offered with regard to types:

1 Type designation or fixation should always be completed before publication.

2 Type designation should be clear and unambiguous; the deposition (and museum number) of the type should always be recorded.

3 Types of undescribed species should not be distributed prior to publication.

4 Type labels should never be changed or removed.

5 Type fixation for species of older authors should be attempted only by a specialist during revisionary work.

Lectotypes

If the name of a species was based on a series of *syntypes,* formerly called *cotypes,* any zoologist may designate one of these syntypes as the lectotype (Article 74 and Recommendations 74A, 74B, 74C, 74D, and 74E). Selection of a lectotype should be undertaken only by a specialist during revisionary work and ordinarily only if it contributes to the unambiguous affixation of a name to a given taxon. It should never be done merely in order to add a type specimen to the collection. If the description of a species is clearly based on a particular specimen, that specimen should be made the lectotype. If one of the syntypes was illustrated, it should be selected as the lectotype, other things being equal.

Syntypes are often widely scattered as a result of exchanges. This requires special considerations in lectotype selection. When possible, a lectotype should be chosen from syntypes in the collection of a public institution, preferably the institution containing the largest number of syntypes of the species, the collection on which the author of the nominal species worked, or the majority of that author's other types.

Many classical authors clearly designated one specimen as the type in their collections without specifically citing such a specimen as "the type" in the published description. The current wording of Article 73*b* implies that such a specimen does not qualify as the holotype. Such specimens have traditionally been accepted as holotypes, in contrast to cases where an author labeled numerous syntypes as the type.

In cases where the syntypes are from several localities and a previous reviser has already restricted the type locality of the species, a responsible zoologist will give due consideration to this fact in the selection of the lectotype.

Neotypes

In Article 75 the Code regulates the designation of neotypes if through loss or destruction no holotype, lectotype, or syntype exists. This provision specifically forbids the manufacturing of neotypes as a matter of curatorial routine or for the sake of having a type for every species. Most of the older species of birds, for instance, have no types but are such clearly understood taxa that no type is needed. Neotypes should be designated "only in connection with revisory work, and then only in exceptional circumstances, when a neotype is necessary in the interest of stability of nomenclature" (Article 75*b*). Even when the original type has been damaged or lost, it is in most cases unnecessary to designate a neotype.

It is least desirable for a neotype to be designated for a species whose name is not in general use either as a valid name or as a synonym or for any name, not in use, that is a *nomen dubium*. Authors should not designate neotypes until they have carefully checked that their actions conform entirely with all the provisions of Article 75.

The Commission has the power (Article 79) to suppress an existing type in the interest of stability of nomenclature and to designate a neotype to conform with the traditional usage of a name. For instance, some years ago it was discovered that the type of the Hottentot teal (*Anas punctata*) in the Oxford University Museum actually was a Maccoa duck (*Oxyura maccoa*) and that a most confusing switch of names would be inevitable if this type specimen were not suppressed. The Commission designated a neotype for the species, and the stability of names that had existed for the previous 125 years was preserved. If a zoologist finds that the type of a well-known species has been misidentified, that zoologist can, if stability is seriously threatened, apply to the Commission for the suppression of that type and the designation of a neotype conforming to the traditional usage of the name. Such action should be confined to exceptional cases.

Types and First Revisers

There are a number of situations in which a zoologist may have to undertake an action in order to clarify the status of a type.

1 Type of a species

- **a** If a series of syntypes belongs to two or more species, a zoologist must determine whether a previous first reviser restricted the name to one of the components. If not, the zoologist has to select a lectotype from among the syntypes in such a way that it best serves stability of nomenclature and is in accordance with the provisions of Article 74.
- **b** If the type is found not to belong to the species to which the name has been traditionally applied, two avenues are open. In inactive groups the zoologist may prefer simply to shift the name to the correct species. However, if such action would cause a serious disturbance of stability, particularly in the case of names in active and universal use for more than 50 years, the zoologist may ask the International Commission to use its plenary powers to suppress the original type and designate a neotype that conforms to existing usage.

2 Type of a genus

- **a** If no species is included in the original naming of a genus published before 1931, the reviser should follow the provisions of Article 69*a*.
- **b** If none of several included species has been previously designated as type species, the first reviser should again act in accordance with the provisions of Article 69.

3 If the evidence indicates that the name used for the previously designated type species resulted from misidentification, the provisions of Article 70 are to be followed. In such cases the Commission is charged "to designate as the type-species whichever species will in its judgment best serve stability and uniformity of nomenclature."

Type Localities

The type locality is the place where the population from which the type specimen was taken occurs (Article 72*h*). Specimens collected at the type locality are called *topotypes,* and the population that occurs at the type locality is called the *topotypical population*. Recommendation 72H contains advice on their designation or restriction. The designation or restriction of type localities should not be done routinely but only by a specialist in connection with revisionary work (Mayr 1969:375–377).

International Commission on Zoological Nomenclature

The duties, powers, organization, and operation of the Commission are regulated by Articles 76 through 82 of the Code and by the Constitution of the Commission (Code, pp. 237–249).

Since this is often misunderstood, it must be pointed out that the rules of nomenclature are phrased and adopted by the zoological congresses. These are the legislative bodies responsible for the rules. The Commission is authorized by these international congresses, as stated in Articles 76 through 79, to interpret or suspend provisions of the Code in individual cases and to submit to the congresses recommendations for the clarification or modification of the Code.

In particular, the Commission is empowered to suspend the application of any provision of the Code "if such application to a particular case would in its judgment disturb stability or universality or cause confusion" (Preamble and Article 79). Under these plenary powers it may "annul or invalidate any name, type-designation, or other published nomenclatural act, or any publication, and validate or establish replacements" by a two-thirds majority.

The role of zoologists in the updating of the Code has been made quite difficult since the demise of the international zoological congresses (the last took place at Monaco in 1972). Although the functions of the congresses were nominally transferred to the Section on Zoological Nomenclature of the International Union of Biological Sciences (IUBS), the assemblies of IUBS were administrative affairs virtually without participation by zoologists. It has therefore been suggested that a Section on Zoological Nomenclature be established within the International Congress of Systematic and Evolutionary Biology (ICSEB), a body which has become the virtual successor of the defunct International Congress of Zoology (Ride and Younes 1986).

Applications to the Commission

Any zoologist may submit cases involving nomenclatural problems to the Commission (Article 78). It must be remembered, however, that the Commission is a judicial, not a fact-finding, board. It is under no obligation to supplement or verify information contained in applications. A zoologist planning to submit an application should study the style of such applications in recent volumes of the *Bulletin of Zoological Nomenclature* and circulate the application first informally among cospecialists.

Cases under Consideration

For the sake of stability in nomenclature it is of extreme importance that zoologists obey the provisions of Article 80, which states that while an application to the Commission is pending, "existing usage is to be maintained, until the decision of the Commission is published."

Appendixes to the Code

Attached to the Code are five appendixes that serve as a guide to good usage in nomenclature. They are not mandatory, as are Articles 1 through 88 of the Code proper, but rather have the same status as recommendations in the Code.

Appendix A (Code, p. 93) is a code of ethics. Every zoologist should carefully study the eight paragraphs of this appendix to avoid violating well-established conventions and risk losing the respect of colleagues. For instance, no zoologist should publish a replacement name for a junior homonym during the lifetime of its author without following the procedure of paragraph 3 of the code of ethics. A new name proposed in violation of these provisions is, however, available if it otherwise satisfies the provisions of the Code.

Appendix B deals with the transliteration and latinization of Greek words (Code, pp. 183–189). Appendix C deals with the latinization of geographic and proper names (Code, pp. 189–191). Appendix D contains extraordinarily detailed and helpful recommendations on the formation of names (Code, pp. 193–229). Since every zoologist must have a copy of the Code handy at all times, we have not included a corresponding section in this text. Appendix F (Code, pp. 237–249) contains the constitution of the Commission.

General Recommendations

Appendix E (Code, pp. 231–235) contains a useful summary of recommendations to the working taxonomist. Beginners in particular should carefully study the 24 recommendations, because this will surely help them avoid making mistakes or at least will improve the quality of their publications. Recommendations 15, 16, 22, and 24 are of special importance.

Glossary and Index

Zoologists who are in doubt as to the meaning of a word used in the Code should consult the glossary on Code pp. 250–272 and 273–298 (French).

Even the experienced worker sometimes has difficulty finding the exact article containing provisions concerning a given case. Even though there are only 88 articles, there are over 600 individual provisions in the Code and the appendixes (exclusive of the tables). The index of the Code is an invaluable guide to these provisions. Glossary items are included in the index.

GLOSSARY

Accessory sexual characters The structures and organs (except the gonads) of which the genital tract is composed, including accessory glands and external genitalia (cf. **Secondary sexual characters**).

Adaptation The condition of showing fitness for a particular environment as applied to the characteristics of a structure, function, or entire organism; in physiology, the process by which such fitness can be acquired by an individual (cf. **Environment**).

Adaptedness Properties of the phenotype or the parts of the phenotype that give it selective advantage over other members of the population.

Adaptive radiation Evolutionary divergence of members of a single phyletic line into a series of different niches or adaptive zones.

Affinity Relationship. Sometimes misleadingly employed as a synonym for phenetic similarity.

Agamic A species or generation that does not reproduce sexually.

Agamospecies A species without sexual reproduction; an asexual species.

Albinism In zoology, the absence of pigmentation, particularly of melanins, in an animal (cf. **Melanism**).

Algorithm A step-by-step procedure for solving a problem, for instance, in numerical taxonomy.

Allele Any of the alternative expressions (states) of a gene (locus).

Note: Refer to the index for reference to the page where the particular subject is treated comprehensively.

For specialized terms pertaining to the fields of biochemistry, cytology, entomology, genetics, and geology, see relevant glossaries.

Allochronic species Species that do not occur in the same time dimension (cf. **Synchronic species**).

Allometric growth Growth in which the growth rate of one part of an organism is different from that of another part or of the body as a whole.

Allopatric Of populations or species that occupy mutually exclusive geographic areas.

Allopatric hybridization The hybridization of two previously isolated allopatric populations or species in a zone of contact.

Allopatric speciation Species formation during geographic isolation and geographic speciation; occurs in two forms, dichopatric and peripatric speciation (cf. **Sympatric speciation**).

Allospecies A component species of a superspecies (*see* **Semispecies**). A term proposed by Amadon (1966:246) to replace the semispecies concept of Mayr (1942:165) but not the semispecies concept of Mayr (1963:501–502). A component species of a superspecies (*see* **Semispecies**).

Allotetraploid An individual or species with the doubled chromosome number of a (sterile) species hybrid.

Allotype A paratype of the opposite sex to that of the holotype (cf. **Paratype**).

Allozyme An enzyme that is the gene product of one of a series of alleles at a given gene locus.

Alpha taxonomy The level of taxonomy concerned with the characterization and naming of species (see Mayr 1969:15).

Alternation of generations The alternation of a bisexual generation with a unisexual (parthenogenetic) generation.

Amphiploid A polyploid produced by the chromosome doubling of a species hybrid, that is, an individual with two rather different chromosome sets; allopolyploid.

Anagenesis Divergent, or upward, evolution.

Anagenetic change The accumulation of changes in ancestor-to-descendant lineages.

Analogy Phenotypic similarity that is due not to common descent but to similarity of function. Often incorrectly cited as the antonym of homology.

Anatomy The science of internal morphology as revealed by dissection.

Antibody A serum globulin that is produced in the blood of an immunized animal in response to the introduction of a foreign antigen (cf. **Antigen, Antiserum, Serology**).

Antigen A substance capable of inducing the formation of antibodies when introduced into the bloodstream of an animal (cf. **Antibody, Precipitin reaction, Serology**).

Antiserum Blood serum containing specific antibodies (cf. **Antibody, Precipitin reaction**).

Apomixis In plants, reproduction without meiosis and fertilization by male gametes, i.e., parthenogenesis.

Apomorphic A more derived state in an evolutionary sequence of homologous characters.

A posteriori weighting The empirical weighting of taxonomic characters on the basis of their proved contribution to the establishment of sound classifications, i.e., monophyletic taxa.

A priori weighting The weighting of taxonomic characters on the basis of preconceived criteria, e.g., their physiological importance.

Archetype A hypothetical ancestral type constructed by means of the elimination of specialized characters (cf. **Phylogeny**).

Artenkreis (Rensch) *See* **Superspecies.**

Artificial classification Classification based on convenient and conspicuous diagnostic characters, without attention to characters indicating relationship; often a classification based on a single arbitrarily chosen character instead of an evaluation of the totality of characters (cf. **Classification, Phylogeny**).

Asexual reproduction Propagation that does not involve zygote formation after the fusion of the nuclei of different gametes.

Atlas In taxonomy, a method of presenting taxonomic materials primarily by means of comparative illustrations rather than comparative descriptions (cf. **Monograph**).

Autapomorphic Pertaining to apomorphic characters found in only one of two sister groups.

Authority citation The custom of citing the name of the author of a scientific name or name combination [e.g., *X-us* Jones, *X-us albus* Jones, *Y-us albus* (Jones)].

Autopolyploid A polyploid originating through the doubling of a diploid chromosome set.

Autosome Any chromosome that is not a sex chromosome.

Available name A name published in a manner that satisfies the requirements specified in Articles 8 through 20 of the Code (cf. **Valid name,** Chapter 14).

Baculum An ossification (bone) in the phallus of some mammals.

Barnacle A group of sessile crustaceans studied by Darwin, forming the subclass Cirripedia.

Batesian mimicry The mimicking (similarity) of a species that is distasteful or dangerous to a predator by an unrelated edible species.

Beta taxonomy The level of taxonomy concerned with the arrangement of species into a natural system of lower and higher taxa (cf. **Alpha taxonomy, Gamma taxonomy**; see Mayr 1969:15).

Bibliographic reference For nomenclatural purposes, the citation of the name of the author and date of publication of a scientific name; a full bibliographic reference also includes the citation of the exact place of publication of a scientific name (i.e., title of book or journal, volume, page, etc.).

Binary Refers to designations consisting of two kinds of names (see Chapter 14; cf. **Binominal nomenclature**).

Binomen The scientific designation of a species; consists of a generic name and a specific name.

Binominal nomenclature The system of nomenclature, adopted by the International Congress of Zoology, by which the scientific name of an animal is designated by both a generic name and a specific name (cf. **Binary**).

Biogeography The study of the geographic distribution of living organisms and the ecological and historical reasons for such distribution.

Biological classification The proposal of an ordered grouping of organisms in accordance with their similarities and consistent with their inferred descent (Chapter 6).

Biological races Noninterbreeding sympatric populations which differ in biology but not, or scarcely, in morphology; supposedly prevented from interbreeding by preference for different food plants or other hosts (mostly actually sibling species).

Biological species concept A concept of the species category based on the reproductive isolation of the constituent populations from those of other species.

Biota The flora and fauna of a region (cf. **Fauna, Flora**).

Biotype A group of genetically identical individuals.

Bisexual Of a population composed of functional males and females; sometimes also applied to an individual possessing functional male and female reproductive organs (hermaphrodite).

Blending inheritance The refuted theory of the complete fusion of the genetic factors of the father and mother in their offspring.

Catalog An index to taxonomic literature arranged by taxa so as to provide information about the most important taxonomic and nomenclatural references to each taxon covered (cf. **Checklist**).

Category *See* **Taxonomic category.**

Character *See* **Taxonomic character.**

Character displacement A divergence of equivalent characters in sympatric species resulting from the selective effects of competition.

Character divergence The name given by Darwin to the differences that develop in two or more related species in their area of sympatry as a result of the selective effects of competition.

Character gradient *See* **Cline.**

Character index A numerical value, compounded from the ratings of several characters, that indicates a degree of difference among related taxa; a rating of an individual, particularly a hybrid, in comparison with its most nearly related species (Chapter 5).

Character matrix A table of taxa and characters in which characters are coded to indicate the sequence from ancestral to derived (Chapter 11).

Character state One of at least two specific variations or characters that constitute a signifer, such as blue or brown in eyes (*see* **Signifer, Taxonomic character**).

Character transformation series A series of homologous characters which, when placed into a primitive to derived sequence, becomes an evolutionary transformation series (*see* **Homologous, Polarity**).

Checklist Usually a skeleton classification of a group listed by taxa for quick reference and as an aid in the arrangement of collections (cf. **Catalog**) (Chapter 12).

Cheironym *See* **Manuscript name.**

Chi-square test A statistical test measuring the significance of a proportion or percentage.

Chromatophore A pigment-bearing intracellular body.

Chromosomal inversion Reversal of the linear order of the genes in a segment of a chromosome.
Chromosome A deeply staining DNA-containing body in the nucleus of the cell, best seen during cell division (Mayr 1970); carriers of (nuclear) genes.
Chronocline A character gradient in the time dimension.
Chronospecies A species delimited in the time-dimension division.
Circular overlap The phenomenon in which a chain of contiguous and intergrading populations curves back until the terminal links overlap with each other geographically and behave like good (noninterbreeding) species.
Clade The species of a phyletic lineage that is derived from a single stem species; a branch of a cladogram.
Cladistic analysis Analysis of the characters of organisms to infer the evolutionary branching sequence of a group's phylogeny (phylogenetic analysis of Hennig).
Cladistic classification Classification in which only holophyletic taxa are permitted and categorical rank is determined by the branching pattern of the cladogram (*see* **Holophyletic**).
Cladistics (cladism) A taxonomic theory by which organisms are ordered and ranked exclusively on the basis of joint descent from a single ancestral species (i.e., on the basis of the most recent branching point of the inferred phylogeny) and in which taxa are delimited by holophyly.
Cladogenesis Branching evolution.
Cladogram A dendrogram based on the principles of cladism; a strictly genealogical dendrogram which features the branching points of phyletic lineages but in which rates of evolutionary divergence are not considered (Chapter 9).
Classification The grouping of objects into classes owing to their joint possession of attributes. In biology, the delimitation, ordering, and ranking of taxa (cf. **Taxonomy, Systematics, Horizontal classification, Vertical classification, Artificial classification, Biological classification**).
Clinal Varying gradually; said of characters.
Cline A gradual geographic change of a character in a series of contiguous populations; a character gradient (cf. **Subspecies**).
Clique A method of compatibility analysis (Chapter 11).
Clone All the individuals derived by asexual reproduction from a single sexually produced individual.
Cluster A group of phenetically similar species.
Cluster analysis Grouping of the phenetically most similar taxa on the basis of a similarity matrix.
Clustering methods Methods of grouping related or similar species into species groups or higher taxa.
Coccids Scale insects.
Code *See* **International Code of Zoological Nomenclature.**
Code of ethics A set of recommendations on the propriety of taxonomic actions to guide the taxonomist. Formulated in Appendix A of the International Code of Zoological Nomenclature.
Coefficient of difference Difference of means divided by the sum of the standard deviations (Mayr 1969:189):

$$CD = \frac{Mb - Ma}{SDa + SDb}$$

Coefficient of variation The standard deviation as a percentage of the mean (Chapter 4):

$$CV = SD \times \frac{100}{M}$$

Cohort Designation of a taxonomic category, sometimes used by paleontologists.

Collective group An aggregate of related species whose generic position is uncertain; used principally in paleontology and parasitology.

Commission The International Commission on Zoological Nomenclature (Chapter 14).

Common ancestor The taxon from which two phyletic lineages or descendant taxa are derived.

Common name Colloquial name. *See* **Vernacular name.**

Compatibility A method in which that tree is selected which is compatible with the largest number of characters and is thus believed to have the fewest homoplasies.

Competition The simultaneous seeking of an essential resource of the environment that is in limited supply.

Competitive exclusion The principle that no two species can coexist at the same locality if they have identical ecological requirements (cf. **Exclusion principle**).

Complex A neutral term for a number of related taxonomic units, most commonly involving units in which the taxonomy is difficult or confusing (cf. **Group, Neutral term**).

Congeneric A term applied to species of the same genus (cf. **Genus**).

Conspecific A term applied to individuals or populations of the same species (cf. **Species**).

Continuity In nomenclature, the principle that continuity of usage should take precedence over priority of publication in determining which of two or more competing scientific names should be adopted for a particular taxon (cf. **Priority**).

Continuous variation Variation in which individuals differ from each other by infinitely small steps, such as variation in the quality of expression of a character or group of characters (cf. **Discontinuous variation**).

Convergence The acquisition of a similar character by two taxa whose common ancestor lacked that character.

Convergent polyphyly Derivation of a group or taxon from only distantly related ancestors that also gave rise to other taxa.

Convex group A group of species, whether ancestors or descendants, is convex if for every two of its species, all the species on the path of ancestor-descendant pairs connecting those two species are also in that group (Estabrook 1986).

Cope's rule The generalization that there is a steady increase in size in phyletic series.

Correlated characters Characters that are associated as manifestations of a well-integrated ancestral gene complex (phyletic correlation) or because they are functionally correlated.

Correlation coefficient See Chapter 11.

Cotype *See* **Syntype.**

Crown groups Late or terminal groups in a phyletic lineage (Jefferies 1979).

Cryptic species *See* **Sibling species.** A species whose diagnostic features are not easily perceived.

Cyclomorphosis A cyclic, seasonal change of form in a series of genetically identical populations, such as cladocerans and rotifers (Chapter 4).

Cytogenetics The comparative study of chromosomal mechanisms and behavior in populations and taxa and their effect on inheritance and evolution.

Cytology The study of the structure and physiology of the cell and its parts.

Darwin's principle The more useful a character is in adapting an organism to a specific habitat or niche, the less valuable it is in classification (Darwin 1859:414).

Data matrix A tabulation of differences between species (or other taxa) in rows [characters (signifers)] and columns (taxa).

Delimitation In taxonomy, a formal statement of the characters of a taxon which set its limits (cf. **Description, Diagnosis, Differential diagnosis**).

Deme A local population of a species; the community of potentially interbreeding individuals at a given locality.

Dendrogram A diagrammatic drawing in the form of a branching tree designed to indicate degrees of relationship (cf. **Phylogenetic tree**).

Derived character A character that differs materially from the ancestral condition.

Description In taxonomy, a more or less complete formal statement of the characters of a taxon without special emphasis on those which set limits to the taxon or distinguish it from coordinate taxa (cf. **Delimitation, Diagnosis, Differential diagnosis**).

Designated priority In cases of simultaneous publication of several names (more than one name for a taxonomic entity), the priority established by the first reviser.

Diagnosis In taxonomy, a formal statement of the characters (or most important characters) that distinguish a taxon from other similar or closely related coordinate taxa (cf. **Differential diagnosis, Description**).

Diapause A temporary interruption of growth in the embryos or larvae of insects, usually during hibernation or estivation.

Dichopatric speciation (Cracraft) Division of a parental species by a geographic, vegetational, or other extrinsic barrier, with the isolated portions eventually reaching species status (cf. **Allopatric speciation, Peripatric speciation**).

Dichotomous Divided or dividing into two parts (cf. **Polytomous**).

Differentia Linnaeus's polynomial species diagnosis.

Differential diagnosis A formal statement of the characters that distinguish a given taxon from other specifically mentioned equivalent taxa.

Dimorphism Occurrence of two distinct morphological types (morphs, phena) in a single population (cf. **Sexual dimorphism, Polymorphism**).

Diploid Having a double set of chromosomes (2n); the normal chromosome number of the cells (except for mature germ cells) of a particular organism derived from a fertilized egg (cf. **Haploid, Polyploidy, Chromosome**).

Discontinuous variation Variation in which the individuals of a sample fall into definite classes that do not grade into each other (cf. **Continuous variation**).

Discriminant function The sum of numerical values of certain diagnostic characters multiplied by calculated constants.

Dollo's rule The principle that evolution is irreversible to the extent that once complex structures or functions are lost, they cannot be restored exactly to their prior condition.

Dominant An allele that determines the phenotype of a heterozygote (cf. **Recessive, Homozygous, Heterozygous**).

Downward classification Classification from the largest class downward, using the principles of logical division.

Eclipse plumage Inconspicuous plumage of birds worn in seasons other than breeding seasons, when a bright nuptial plumage is worn.

Ecological isolation Reproductive isolation dependent on habitat or host selection (cf. **Reproductive isolation, Geographic isolation**).

Ecological race A local race that owes its most conspicuous attributes to the selective effect of a specific environment (*see* **Ecotype**).

Ecology The study of the interactions between organisms and their environment.

Ecophenotype A nongenetic modification of the phenotype in response to a particular environmental condition.

Ecospecies "A group of populations so related that they are able to exchange genes freely without loss of fertility or vigor in the offspring" (Turesson 1922).

Ecotype A descriptive term applied to plant races of varying degrees of distinctness that owe their most conspicuous characters to the selective effects of the local environment (cf. **Subspecies, Ecological race**).

Edaphic race A race that is affected by the properties of the substrate (soil) rather than by other environmental factors.

Eidos Any of the fixed types (ideas) that Plato thought underlie the apparent variability of phenomena.

Electronic data processing (EDP) The sorting and storage of data with the help of computers.

Electrophoresis A process of separating different molecules, particularly polypeptides, by means of their differential rates of migration in an electric field.

Emendation In nomenclature, an intentional modification of the spelling of a previously published scientific name (cf. **Error, *Lapsus calami***).

Environment The totality of physical, chemical, and biotic conditions surrounding an organism.

Epistasis Interaction between nonallelic genes.

Equal weighting The method that treats all taxonomic characters as equally important; a key assumption of phenetics.

Error In nomenclature, an unintentional misspelling of a scientific name, such as a typographical error or an error of transcription (cf. **Emendation, *Lapsus calami***).

Essentialism A school of philosophy, originating with Plato and the Pythagoreans and later maintained by the Thomists and so-called realists among the philosophers, who believed in the reality and constancy of underlying universals or essences; in taxonomy, usually referred to as typology or typological thinking, resulting in a neglect of the role of variation.

Ethological barriers Isolating mechanisms caused by behavioral incompatibilities of potential mates resulting from differences in their visual, tactile, scent, or sound signals.

Ethological (behavioral) isolation Reproductive isolation due to ethological barriers.

Ethology The science of the comparative study of animal behavior.

Euclidean distance A coefficient measuring the distance between two taxa in multidimensional space (Chapter 11).

Eukaryotes Organisms with a well-defined nucleus and meiosis. All higher organisms above the level of **prokaryotes.**

Evolutionary classification Classification following Darwin's recommendation to arrange taxa on the basis of degree of difference but consistent with their inferred genealogy.

Evolutionary novelty A newly acquired structure or other property that permits the performance of a new function.

Evolutionary species A lineage "evolving separately from others and with its own unitary evolutionary role and tendencies" (Simpson 1961).

Evolutionary taxonomy A school of classification, the principles of which are explained in Chapter 10.

Exclusion principle (Gause's principle) The principle stating that two species cannot coexist at the same locality if they have identical ecological requirements.

Ex-group A group descended from a monophyletic group (traditionally defined) which renders that group paraphyletic.

Extrinsic isolation The nongenetic isolation of populations by extrinsic factors, such as geographic isolation and the isolation produced by some host-plant mechanisms.

Eyepiece micrometer A linear scale in the field of vision of the eyepiece (or one of a pair of eyepieces) of a microscope; used as a measuring device.

Family A category for monophyletic taxa composed of one genus or a group of genera or tribes that are separated from related similar units (families) by a decided gap; the size of the gap is in inverse ratio to the size of the family. In the hierarchy of categories the family represents the level between genus and order.

Family name The scientific designation of a taxon of family rank; recognized by the termination *-idae*.

Fauna The animal life of a region (cf. **Flora, Biota**).

Faunal work A publication in which taxa are included on the basis of their occurrence in a specified area rather than on the basis of relationship.

Fecundity Reproductive potential as measured by the quantity of gametes, particularly eggs, produced.

Fertility Reproductive potential as measured by the quantity or percentage of developing eggs or fertile matings.

First reviser The first author to publish a definite choice of one among two or more conflicting names or zoological interpretations which are equally available under the Code; to qualify as first reviser, an author must give evidence of having made a choice between available alternatives.

Flora The plant life of a region (cf. **Fauna, Biota**).

Form A neutral term for a single individual, phenon, or taxon (cf. **Group, Neutral term**).

Formenkreis A collective category of allopatric subspecies or species (Kleinschmidt); in paleontology, a group of related species or variants.

Founder principle The principle that the founders of a new colony (or population) of a species in a previously unoccupied area contain only a small fraction of the total genetic variation of the parental population or species (*see* **Genetic revolution**).

Full bibliographic synonymy A reasonably complete list of references to a given taxon arranged to serve the needs of nomenclature (chronology of names) and zoology (pertinent taxonomic and biological sources) simultaneously (cf. **Synonymy**).

Gamete Male or female germ cell, usually with a haploid chromosome set.

Gametic isolation Male and female gametes not mutually attractive or inviable in the ducts through which they must pass.

Gamma taxonomy The level of taxonomy dealing with various biological aspects of taxa, ranging from the study of intraspecific populations to studies of speciation and evolutionary rates and trends (Mayr 1969:15).

Gause's principle or rule *See* **Exclusion principle.**

Gene A hereditary determiner; the unit of inheritance, carried on a chromosome, transmitted from generation to generation by the gametes and controlling the development (and characteristics) of the individual (cf. **Chromosome**).

Gene flow The exchange of genetic factors between populations as a result of the dispersal of zygotes or gametes, e.g., by pollen.

Gene frequency The percentage of a given gene in a population (cf. **Gene, Local population**).

Gene pool The totality of the genes of a given population existing at a given time.

Genetic characters Characters relating to variation in the genotype. See Chapter 7, under "genetic characters."

Genetic drift Genetic changes in populations caused by stochastic processes rather than by selection, mutation, or immigration (cf. **Local population**).

Genetic homeostasis The property of a population of equilibrating its genetic composition and resisting sudden changes.

Genetic revolution A major reorganization of the genotype that may occur in a small founder population.

Genotype The genetic constitution of an individual or taxon (cf. **Phenotype**). Use of this term in nomenclature for the type species of a genus is confusing and contrary to the terminology of the Code (Recommendation 67A).

Genus (pl. genera) A category for a taxon that includes one species or a group of species, presumably of common phylogenetic origin, which is separated from related similar units (genera) by a decided gap; the size of the gap is in inverse ratio to the size of the unit. The taxonomic category directly above the species in the Linnaean hierarchy.

Geographic barrier Any terrain that prevents gene flow between populations.

Geographic isolate A population or group of populations that is prevented by an extrinsic barrier from free gene exchange with other populations of the species.

Geographic isolation The prevention of gene exchange between a population and other populations by geographic barriers.

Geographic race *See* **Subspecies.** A geographically delimited race, usually a subspecies.

Geographic speciation The acquisition of isolating mechanisms by a population during a period of geographic (allopatric) isolation.

Geographic variation The differences between spatially segregated populations of a species; population differences in the space dimension.

Gloger's rule "Races in warm and humid areas are more heavily pigmented than those in cool and dry areas" (Mayr 1942).

Grade A group of animals similar in level of organization; a level of anagenetic advance.

Ground-plan divergence method A method of phylogenetic analysis developed by the botanist Wagner and described in Chapter 11.

Group A neutral term for a number of related taxa, especially an assemblage of closely related species within a genus (cf. **Complex, Neutral term, Section**).

Gynandromorph An individual in which one part of the body is masculine and the other is feminine; most frequent are bilateral gynandromorphs, in which the left and right halves are of different sexes.

Habitat selection The capacity of a dispersing individual to select an appropriate (species-specific) habitat.

Handbook In taxonomy, a publication designed primarily to aid in field and laboratory identification rather than in the presentation of new taxonomic conclusions (cf. **Monograph**).

Haploid Having only a single set of chromosomes; gametes are usually haploid.

Hardy-Weinberg formula The statement in mathematical terms that the frequency of genes in a population remains constant in the absence of selection, nonrandom mating, immigration, and accidents of sampling.

Heritability The genetic component of phenotypic variability.

Hermaphrodite An individual having both male and female reproductive organs (cf. **Intersex**).

Heterozygote An individual with different genetic factors (alleles) at the homologous (corresponding) loci of the two chromosomes of a diploid chromosome set.

Heterozygous Having different alleles at homologous loci of the two parental chromosomes (cf. **Allele, Locus, Homozygous**).

Hierarchical classification The system of ranks that indicates the categorical level of each taxon (i.e., kingdom to species) (cf. **Taxonomic category, Linnaean hierarchy**).

Higher category A taxonomic category of rank higher than the species: genus, family, order, and so forth (i.e., from subgenus to kingdom) (cf. **Supraspecific**). A class into which all taxa are placed that are ranked at the same level in a hierarchic classification.

Higher taxon A taxon ranked in one of the higher categories; a monophyletic group of species (or a single species) separated from other taxa of the same rank by a gap greater than any found within the taxon.

Histogram A set of rectangles in which the midpoints of class intervals are plotted on the abscissa and the frequencies (usually number of specimens) are plotted on the ordinate.

Holistic Considering a whole as being more than the sum of its parts.

Hollow curve A curve demonstrating an excess over expectancy of very small (e.g., monotypic) and very species-rich higher taxa (Chapter 10).

Holometabolic Of an insect undergoing a complete metamorphosis between the larval and adult stages.

Holophyletic Pertaining to a group that consists of all the descendants of its most recent common ancestor (monophyly of Hennig).

Holosteans A group of fishes ancestral to the teleost fishes.

Holotype The single specimen designated or indicated as the type by the original author at the time of publication of the original description of a species.

Homologous Referring to a structure, behavior, or other character of two taxa that is derived from the same or equivalent feature of the nearest common ancestor (cf. **Analogy**).

Homonym In nomenclature, one of two or more independently proposed names with identical spelling for the same or different taxa (cf. **Senior homonym, Junior homonym, Primary homonym, Secondary homonym**) (Chapter 14).

Homonymy The sameness of serial structures, as in metamerism. Formerly but misleadingly called serial homology.

Homoplasy Possession by two or more taxa of a character derived not from the nearest common ancestor but through convergence, parallelism, or reversal.

Homosequential Of chromosomes with the same banding pattern.

Homozygous Having identical alleles at the two homologous loci of a diploid chromosome set (cf. **Allele, Locus, Heterozygous**).

Horizontal classification Classification which stresses the grouping together of species in a similar stage of evolution rather than species located on the same phyletic line (cf. **Vertical classification**).

Host races Different genetic races of the same species in oligophagous food specialists or parasites that occur on different hosts.

Hybrid belt A zone of interbreeding between two species, subspecies, or other unlike populations; zone of secondary intergradation.
Hybrid breakdown Occurs when viable adults are produced by the interbreeding of unlike parents but the F_2 or backcross generations show reduced fertility or viability.
Hybrid index *See* **Character index** (Chapter 7).
Hybrid inviability Reduced viability in the offspring of unlike parents, particularly parents of different species.
Hybridization The crossing of individuals belonging to two unlike natural populations, principally different species.
Hybrid sterility Production of mature but sterile offspring by unlike parents, particularly individuals of different species.
Hypodigm The entire material of a species available to a taxonomist.
Identification The determination of the taxonomic identity of an individual.
Incipient species Populations or groups of populations, usually geographically isolated, that are in the process of becoming a separate species but have not acquired all attributes of a species.
Indication In nomenclature, the publication of certain types of evidence or cross references that establish the typification of a name and thus make it available (Code, Article 16).
Individual variation Variation within a population.
Industrial melanism The increase in the frequency of melanistic (blackish) individuals (morphs) in populations of lepidopterans in sooty areas.
Infraorder An optional category below the suborder.
Infraspecific Within the species; usually applied to taxa (subspecies) and phena (varieties) (cf. **Subspecies, Variety, Infrasubspecific form**).
Infrasubspecific form An individual or seasonal variant within a single interbreeding population; a phenon.
Infrasubspecific name A name given to an infrasubspecific form.
Intergradation Merging gradually through a continuous series of intermediate forms or populations.
Intergrading Said of populations that are intermediate in characters between populations adjacent on either side.
International Code of Zoological Nomenclature The official set of regulations dealing with zoological nomenclature (Chapter 14).
Intersex An individual more or less intermediate in phenotype between male and female (cf. **Hermaphrodite**).
Intrinsic isolation Reproductive isolation (cf. **Extrinsic isolation**).
Introgressive hybridization The spread of one or more genes of one species into the germ plasm of another species as a result of hybridization (cf. **Hybridization**).
Irreversibility The inability of an evolving group of organisms or a structure of an organism to return to an ancestral condition; the theory of irreversibility is that a given structure or adaptation that has been lost in evolution cannot be restored exactly to its prior condition.
Irreversibility rule *See* **Dollo's rule.**

Isolate A population or group of populations that is separated from other populations.

Isolating mechanism Properties of individuals that tend to prevent the interbreeding of different populations that are actually or potentially sympatric.

Isophene A line connecting points of equal expression of a geographically variable character; a line on a map at right angles to a cline (cf. **Cline**).

Junior homonym The more recently published of two or more identical names for the same taxon or different taxa (cf. **Homonym, Senior homonym**).

Junior synonym The more recently published of two or more available names for the same taxon (cf. **Synonym, Senior synonym**).

Karyological character A character involving chromosome structure or number.

Karyotype The chromosome complement.

Key A tabulation of diagnostic characters of species or other taxa in dichotomous couplets that facilitate rapid identification.

Key character In taxonomy, a diagnostic character of special utility in a key.

Lapsus calami In nomenclature, a slip of the pen, especially an error in spelling (cf. **Error, Emendation**).

Lectotype One of a series of syntypes which, subsequent to the publication of the original description, is selected and designated through publication to serve as the type.

Line precedence Occurrence of a name on an earlier line of the same page than another name for the same taxon.

Linnaean hierarchy The hierarchical arrangement of categorical ranks for taxa in which each category except the lowest includes one or more subordinate categories.

Local population The individuals of a given locality which potentially form a single interbreeding community (cf. **Deme**).

Locus The position of a gene on a chromosome (cf. **Gene, Chromosome**).

Lumper A taxonomist who emphasizes relationship in the delimitation of taxa and tends to recognize large taxa (cf. **Splitter**).

Macrogenesis Evolution by the sudden origin of new types through saltation.

Macrotaxonomy The classification of higher taxa.

Manhattan distance A similarity coefficient consisting of the absolute differences between the characters of each signifer for every possible pair of taxa.

Manuscript name In nomenclature, an unpublished scientific name.

Matching coefficient A coefficient of association of the characters of two taxa (Chapter 11).

Material In taxonomy, the sample available for taxonomic study (cf. **Series, Hypodigm**).

Mechanical isolation Reproductive isolation resulting from mechanical incompatibility of male and female genitalic structures.

Meiofauna The fauna of the interstitial spaces in the substrate of the ocean floor and of beaches.

Meiosis The two successive divisions of the nucleus preceding the formation of gametes.

Melanism An unusual darkening of color caused by increased amounts of black pigment; sometimes a racial character, sometimes restricted to a certain percentage of individuals within a population, giving rise to polymorphism (cf. **Industrial melanism, Albinism**).

Mendelian population A population with unrestricted interbreeding of individuals and free reassortment of genes.

Meristic variation Numerical variation in characters that can be counted, such as vertebrae, scales, and fin rays.

Metamorphosis A drastic change of form during development, as when a tadpole changes into a frog or an insect larva changes into an imago.

Metric system A decimal system of measures (with the meter as the base) and weights (with the gram as the base); the universal system in science for reporting measurements and weights.

Microgeographic race A local race restricted to a very small area.

Microtaxonomy The discrimination of species and their subdivisions; taxonomy at the species level.

Mimetic polymorphism Polymorphism (best known in Lepidoptera) in which the various morphs resemble other species distasteful or poisonous to a predator; often restricted to females.

Mimicry Resemblance in color or structure to other species that are distasteful or poisonous to a predator.

Minimal spanning tree A tree in which each taxon is connected by a line to its most similar neighbor; the length of the lines represents the taxonomic distance between them (Chapter 11).

Molecular clock The hypothesis that the rate of evolutionary change in DNA and other molecules is essentially constant over long periods of geological time and can be calibrated with the help of the fossil record or biogeographic evidence.

Monogenic Determined by a single gene (cf. **Polygenic**).

Monograph In taxonomy, an exhaustive treatment of a higher taxon in terms of all available information pertinent to taxonomic interpretation; usually involves full systematic treatment of the comparative anatomy, biology, ecology, and detailed distributional analyses of all included taxa (cf. **Revision, Synopsis**).

Monophyletic Pertaining to a group whose members are all descended from the nearest common ancestor.

Monophyly (Simpson's definition) The derivation of a taxon through one or more lineages from one immediately ancestral taxon of the same or lower rank.

Monothetic Said of a group defined in terms of a single feature or set of features that is both necessary and sufficient for inclusion in that group (antonym of **Polythetic**).

Monotypic A taxon containing only one immediately subordinate taxon, as a genus containing only one species or a species containing only one (the nominate) subspecies.

Morph Any of the genetic forms (individual variants) that account for polymorphism in a population or taxon.

Morphospecies A typological species recognized merely on the basis of morphological difference (cf. **Phenon**).

Morphotype A phenotype recognizable by morphological characters.
Mosaic evolution Different rates of evolutionary change in the same group of organisms for different structures, organs, or other components of the phenotype.
Muellerian mimicry Similarity (usually consisting of similar warning coloration) of several species that are distasteful, poisonous, or otherwise harmful.
Multivariate analysis The simultaneous analysis of several variable characters.
Mutation A change in the genetic material; most often a change in a single gene (gene mutation), consisting of a replacement, duplication, or deletion of one or several base pairs in the DNA.
Myrmecophily The utilization by other insects, mostly beetles, of ant colonies as a domicile and source of food.
Natural selection The unequal contribution of genotypes to the gene pool of the next generation through differential mortality and differences in reproductive success.
Natural system A system of classification which endeavors to reflect most closely the actual relationship of the included species.
Neo-Darwinism Weismann's theory of evolution, in which any inheritance of acquired characters is rejected; sometimes, any modern evolutionary theory featuring natural selection and a rejection of "soft" inheritance.
Neontology The study or science of recent organisms; antonym of **Paleontology.**
Neoteny Attainment of sexual maturity in an immature or larval stage.
Neotype A specimen selected as type subsequent to the original description in cases where the original types are known to have been destroyed or were suppressed by the Commission.
Neutral term A taxonomic term of convenience, such as *form* or *group,* which can be employed without reference to the formal taxonomic hierarchy of categories and which has no nomenclatural significance.
New name A replacement name for a preoccupied name (cf. **Substitute name**).
New systematics The populational, biological approach to systematics (Chapter 3).
Niche (ecological) The multidimensional resource space of a species; its ecological requirements; its specific way of utilizing the environment.
Nomenclator A work containing a list of scientific names assembled for nomenclatural rather than taxonomic purposes (cf. **Catalog**).
Nomenclature A system of names (Chapter 14).
Nomen conservandum A name preserved by action of the Commission and placed on the appropriate official list.
Nomen dubium The name of a nominal species for which available evidence is insufficient to permit recognition of the zoological species to which it was applied.
Nomen oblitum A name that lost its validity under the statute of limitation of the pre-1973 Code (Article 23*b*).
Nominal taxon (species, genus, etc.) A named taxon, objectively defined by its type.
Nominalism A school of philosophy that denies the existence of universals and emphasizes the importance of artificial names for the grouping of individuals.

Nominate Of a subordinate taxon (subspecies, subgenus, etc.) which contains the type of the subdivided higher taxon and bears the same name; such a taxon is sometimes incorrectly designated as typical.

Nondimensional species The species concept, represented by the noninterbreeding of a species at a given place and time, not involving longitude, latitude, or time.

Numerical phenetics The methodology of assembling individuals into taxa on the basis of an estimate of unweighted overall similarity.

Objective synonym Each of two or more names based on the same type.

Official index A list of names or works suppressed or declared invalid by the Commission.

Official list A list of names or works declared as available by the Commission.

Oligogenic character A character determined by only a few genes.

Oligolectic Said of bees; collecting the pollen of only a few kinds of flowers.

Oligophagous Feeding on few species of food plants.

Onomatophore Name bearer, or type (Simpson 1961).

Ontogeny The developmental history of an individual organism from egg to adult, particularly its embryogenesis.

Open population A population freely exposed to gene flow and subject to a great input of alien genes as a result of immigration.

Operationalism A philosophical methodology, proposed by the physicist Bridgman, according to which the meaning of a concept is defined by a set of operations.

Original description A statement of characters accompanying the proposal of a name for a new taxon in conformance with Articles 12 and 13 of the Code.

Orthogenesis The hypothesis that phyletic lines may follow a predetermined rectilinear pathway whose direction is not determined by natural selection.

Operational taxonomic unit (OTU) A name for individuals, populations, species, or higher taxa classified by numerical methods.

Out-group A taxon, outside a given study group and preferably a sister group, "that is examined in the course of a phylogenetic study to determine which of two homologous characters" found within the study group "may be inferred to be apomorphic" (Wiley 1981:7).

Overall similarity A usually numerical value of similarity calculated by the summation of similarities in numerous individual characters (Chapter 8).

Page precedence Occurrence of a name on an earlier page in the same publication than another name (synonym) for the same taxon or the same name (homonym) for a different taxon.

Paleontology The science that deals with the life of past geological periods (cf. **Neontology**).

Panmictic Of populations that randomly interbreed.

Parallel character A similar character derived independently in each of two or more taxa with a similar genetic background.

Parallelism The independent acquisition of the same or similar characters in closely related evolutionary lines (cf. **Convergence**).

Parallelophyly Multiple independent derivation from the nearest common ancestral taxon.

Parapatric speciation The progressive divergence of two contiguous parapatric populations until they have become two different species.

Parapatry Nonoverlapping geographic contact (contiguity) of populations with or without interbreeding (*see* **Allopatric**).

Paraphyletic Pertaining to a monophyletic group that does not contain all the descendants (derivatives) of that group.

Parasitoid Wasps and flies, the larvae of which parasitize and usually kill individuals of the host species.

Paratype A specimen other than the holotype which was before the author at the time of preparation of the original description and was so designated or indicated by the original author; paratypes have no standing in nomenclature.

Parsimony The principle stating that that tree is best which is "shortest," that is, has the smallest number of character state changes (branching points).

Parthenogenesis The production of offspring from unfertilized eggs.

Particulate inheritance The theory that the genetic factors received from the mother and father do not blend or fuse but retain their integrity from generation to generation.

Patristic character Equals ancestral character.

Patronymic In nomenclature, a dedicatory name, a name based on that of a person or persons.

Peripatric speciation Achievement of species status by the descendants of a founder population established beyond the periphery of the parental species range.

Peripheral isolate A population isolated at or beyond the periphery of the species range.

Phage Bacterial virus.

Phenetic ranking Ranking into categories, based strictly on degree of overall similarity.

Phenetics The delimitation and ranking of taxa on the basis of the overall similarity in unweighted characters of the included species.

Phenocopy A nongenetic modification of the phenotype (owing to special environmental conditions) that resembles a change of the phenotype caused by a mutation.

Phenogram A diagram indicating degree of overall similarity (usually unweighted) among taxa.

Phenon (pl. phena) A sample of phenotypically similar specimens in a population; any different phenotype within a population.

Phenotype The totality of characteristics of an individual (its appearance); results from interaction between genotype and environment.

Philopatry The drive (tendency) of an individual to return to (or stay in) its home area (birthplace or another adopted locality).

Phyletic Pertaining to a line of descent (cf. **Phylogeny**).

Phyletic correlation Correlation of characters that are phenotypic manifestations of a well-integrated ancestral gene complex.

Phyletic speciation Not a multiplication of species; a misnomer for phyletic evolution.

Phyletic weighting An assessment of the taxonomic importance of a character on the basis of its phyletic information content.

Phylogenetic systematics Hennig's proposal (1950) to base classifications entirely on genealogy, that is, on the branching pattern of phylogeny; later renamed cladistics.

Phylogenetic tree A diagrammatic presentation of inferred lines of descent; based on paleontological, morphological, molecular, or other evidence.

Phylogeny The inferred lines of descent of a group of organisms, including a reconstruction of the common ancestor and the amount of divergence (anagenesis) of the various branches; "the science of the changes of form through which the phyla or organic lineages pass through the entire time of their discrete existence" (Haeckel 1866).

Phylogram A dendrogram indicating both cladistic branching and the relative amount of progressive (anagenetic) change that has taken place between internodes.

Pie graph A graphic method most useful for illustrating the percent contribution of polymorphic characters in a population or species (Chapter 4).

Pleiotropy The capacity of a gene to affect several characters, that is, several aspects of the phenotype.

Pleistocene refuges Favorable areas south of the borders of the ice, where species and populations survived periods of glaciation.

Plenary powers Special powers granted to the Commission to set aside provisions of the Code, usually for the sake of stability.

Plesiomorphic (Hennig) Of characters; an ancestral (primitive, patristic) character state.

Plesion Rank assigned to a fossil taxon in a cladistic classification.

Poikilothermal Ectothermal.

Polarity The evolutionary direction of a character transformation series from ancestral toward derived (*see* **Transformation series**).

Polygenes Genes that control a character jointly with other genes.

Polygenic Of a character controlled by several genes.

Polymorphism The simultaneous occurrence of several different alleles or discontinuous phenotypes in a population, with the frequency of even the rarest type higher than can be maintained by recurrent mutation.

Polynominal nomenclature A system of nomenclature in which the specific epithet of a species consists of several words.

Polyphagous Feeding on many different kinds of food, such as species of host plants.

Polyphenism The occurrence of several phenotypes in a population; the differences among them are not the result of genetic differences.

Polyphyly Derivation of a taxon from two or more ancestral sources (cf. **Convergent polyphyly, Parallelophyly**).

Polyploidy A condition in which the number of chromosome sets in the nucleus is a multiple (greater than 2) of the haploid number.

Polythetic Of taxa which are based on the greatest number of shared characters; no single character is either essential or sufficient to make an organism a member of the group, and no member of the taxon necessarily has all the attributes which jointly characterize the taxon.

Polytomous Divided into more than two parts; said of a branching point giving rise to more than two branches.

Polytopic A taxon containing two or more taxa in the immediately subordinate category, such as a genus with several species or a species with several subspecies (cf. **Monotypic**).

Polytypic Occurring in different places, as, for instance, a subspecies composed of widely separated but phenotypically identical populations.

Population *See* **Local population.**

Population thinking The concept according to which classes of biological phenomena are composed of unique individuals and show a characteristic variation which has reality. The calculated mean values of these populations and of samples from them do not reflect an underlying essence (antonym of **Essentialism**).

Preadaptation The possession of properties that permit a shift into a new niche or habitat. A structure is preadapted if it can assume a new function without interference with the original function.

Precipitin reaction The formulation of a visible precipitate at the interface when an antigen and the corresponding antiserum are brought together (cf. **Antigen, Antiserum, Antibody**).

Predictive value The capacity of a classification to make predictions about newly employed characters or newly discovered taxa (Chapter 4).

Pre-Linnaean name A name published prior to January 1, 1758, the starting point of zoological nomenclature.

Prim network Cf. **Minimal spanning tree.**

Primary homonym Each of two or more identical species group names which at the time of original publication were proposed in combination with the same generic name (e.g., *X-us albus* Smith, 1910, and *X-us albus* Jones, 1920).

Primary intergradation A zone of intermediacy between two phenotypically different populations that have developed in situ as a result of selection (cf. **Secondary intergradation**).

Primary zoological literature Literature dealing with animals or zoological phenomena; not merely a listing of names; literature in which valid species are carefully distinguished from synonyms.

Principal component analysis See statistics texts.

Priority The principle that of two competing names for the same taxon (below the rank of infraorder), ordinarily that one is valid which was published first.

Program, open (or closed) A behavior program which when open is able and when closed is unable to incorporate additional experiential information.

Progression rule As a lineage of organisms moves to new areas, evolving new species and characters as it migrates, the path of migration is marked by the presence of sequentially more derived characters.

Prokaryotes Microorganisms (various kinds of bacteria and blue-green algae) that lack well-defined nuclei and meiosis and have their nucleic acid organized into a single string (cf. **Eukaryotes**).

Pseudogamy (Gynogenesis) Parthenogenetic development of the egg cell after the egg membrane has been penetrated by a male gamete, which does not contribute to the genotype of the zygote.

Q technique An analysis of association of pairs of taxa in a data matrix.

Race *See* **Subspecies**; local populations not sufficiently different to be formally designated as subspecies.

Ranking The placement of a taxon in the appropriate category in the hierarchy of categories.

Rassenkreis (Rensch) The German equivalent of polytypic species; not "a circle of races."

Recapitulation The theory that ontogeny repeats the stages of phylogeny (cf. **Ontogeny, Phylogeny**).

Recent Of taxa which still exist; antonym of **fossil**.

Recessive Of a gene that is not expressed in the phenotype of a heterozygote.

Reductionism The belief that complex phenomena can be understood by reducing them to the smallest possible component parts and explaining the components.

Redundant characters Characters so closely correlated with other, already used characters that they do not contribute new information to the analysis.

Regression analysis A form of multivariate analysis.

Regressive character A character that is being reduced or lost in the course of phylogeny, sometimes independently in several related lines, resulting in character reversal.

Relationship For meaning in classification, see Chapter 6.

Reproductive community A community of individuals that are not reproductively isolated.

Reproductive isolation A condition in which interbreeding between two or more populations is prevented by intrinsic factors (cf. **Isolating mechanism**).

Reticulate evolution Evolution dependent on repeated intercrossing between a number of lines, and thus both convergent and divergent at once.

Reticulate speciation Speciation via hybridization which "results in the origin of one new species from two previously existing species" (Wiley 1981:8).

Reversal The reappearance in phylogeny of an ancestral character as a result of the loss of an advanced (apomorphic) character.

Revision In taxonomy, the presentation of new material or new interpretations integrated with previous knowledge through summary and reevaluation (cf. **Synopsis, Monograph**).

Robertsonian fusion Fusion of two acrocentric chromosomes into a single metacentric one.

R technique An analysis of association of characters in a data matrix.

Saltation Discontinuous variation produced in a single step (leap) by a major mutation (cf. **Mutation**).

Sample That portion of a true population which is actually available to the taxonomist.

Scala naturae The belief in a linear progression in kinds of organisms from the simplest to the most perfect.
Scatter diagram A bivariate or multivariate graphic method of population analysis (Chapter 4).
Scientific name The binominal or trinominal designation of an organism; the formal nomenclatural designation of a taxon (cf. **Vernacular name**).
Secondary homonym Each of two or more identical specific names which at the time of original publication were proposed in combination with different generic names but which, through subsequent transference, reclassification, or combination of genera proposed by a taxonomist, have come to bear the same (or an identical) combination of a generic and a specific name.
Secondary intergradation The intergradation or hybridization of two distinct populations or groups of populations along a zone of secondary contact.
Secondary sexual characters Characters that distinguish the two sexes of the same species but do not (like gonads and accessory sexual characters) function directly in reproduction (cf. **Sexual dimorphism**).
Section A neutral term usually employed with reference to a subdivision of a taxon or a series of related elements in one portion of a higher taxon (cf. **Higher category, Neutral term, Group**).
Selection *See* **Natural selection.**
Semigeographic speciation (parapatric speciation) The splitting apart of species along a line of secondary intergradation of two incipient species or along a line of strong ecological contrast (an ecological escarpment).
Semispecies Borderline cases between species and subspecies; populations that have acquired some but not yet all the attributes of species rank.
Senior homonym The earliest published of two or more identical names for the same or different taxa (cf. **Homonym, Junior homonym**).
Senior synonym The earliest published of two or more available synonyms for the same taxon (cf. **Synonym, Junior synonym**).
Sequencing A cladistic method in which sister taxa are given the same categorical rank, the sequence to be determined by the amount of divergence from the ancestral stem species.
Series In taxonomy, the sample that the collector takes in the field or the sample available for taxonomic study (cf. **Material, Hypodigm**).
Serology The study of the nature and interactions of antigens and antibodies (cf. **Antigen, Antibody**).
75 percent rule The rule that one population can be considered subspecifically distinct from another if at least 75 percent of the individuals of the first population are different from ''all'' the individuals of the second (Chapter 4).
Sex chromosome A special chromosome that does not occur in identical number or structure in the two sexes and is concerned with sex determination; usually an X and a Y chromosome (cf. **Chromosome, Autosome**).
Sex-limited character A character occurring in only one sex (cf. **Secondary sexual characters, Sex-linked character**).
Sex-linked character A character controlled by a gene located in the sex chromosome (cf. **Sex chromosome**).

Sexual dimorphism The phenotypic difference between the two sexes of a species.

Sexual reproduction Reproduction resulting in a diploid zygote with maternal and paternal chromosome sets.

Shortest spanning tree *See* **Minimal spanning tree.**

Sibling species Morphologically similar or identical populations that are reproductively isolated; cryptic species (cf. **Species**).

Sickle-cell anemia An anemia due to a mutation in a single base pair of a hemoglobin gene; found mostly in tropical areas and lethal in homozygotes.

Signifer A feature that varies from one organism to another.

Signifer state The particular expression of a signifer in a given organism. In taxonomic analysis, a signifer state is most often the same as a character as traditionally defined (Chapter 7).

Sister groups In a dichotomous cladogram, the two holophyletic groups that descended from their common ancestor.

Sonagram A graphic representation of the vocalization of an animal.

Speciation The process of the multiplication of species; the origin of incompatibility between populations caused by the acquisition of reproductive isolating mechanisms (cf. **Allopatric speciation, Sympatric speciation**).

Species Groups of interbreeding natural populations that are reproductively isolated from other such groups (cf. **Subspecies, Local population, Reproductive isolation**).

Species group A group of closely related species, usually with partially overlapping ranges.

Species recognition The exchange of appropriate (species-specific) stimuli and responses among individuals, particularly during courtship.

Specific name The second component of the binominal name of a species.

Specimen A preserved individual (or part of one) studied and classified in taxonomy.

Splitter In taxonomy, one who divides taxa very finely to express every shade of difference and relationship through the formal recognition of separate taxa and their elaborate categorical ranking.

Spontaneous generation The sudden, spontaneous origin of organisms from inert matter; a discredited concept.

Standard deviation (SD) The square root of the sum Σ of the squared deviations d from the mean, divided by N:

$$\mathrm{SD} = \sqrt{\frac{\Sigma d^2}{N}}$$

Standard error (of the mean) Standard deviation divided by the square root of the sample size N:

$$\mathrm{SE} = \frac{\mathrm{SD}}{\sqrt{N}}$$

Stasipatric speciation Sympatric speciation through chromosomal restructuring.

Statute of limitation A provision in the Code (valid prior to 1973) to protect universally adopted junior names against the revival of forgotten senior synonyms.

Stem groups Basal (early) groups (usually fossil) in a phyletic lineage.

Step cline An abrupt break in the rate of change in a cline over a short geographic distance.

Stochastic processes Processes involving chance (randomness) or probability.

Stratigraphic sequence Occurring in a sequence of geological strata.

Strickland Code A code of nomenclature prepared by a committee of the British Association for the Advancement of Science under the secretaryship of H. E. Strickland; first published in 1842.

Stylops A genus of parasitoids of beetles; belongs to the order of Strepsiptera.

Subfamily A category of the family group subordinate to the family; an individual taxon ranked in the subfamily category.

Subgeneric name The name of an optional category between the genus and the species; enclosed in parentheses when cited in connection with a binominal or trinominal combination and therefore excluded from consideration in determining the number of words of which a specific or subspecific name is composed; e.g., *X-us (Y-us) albus rufus* is a trinominal.

Subjective synonym Each of two or more synonyms based on different types but regarded as referring to the same taxon by zoologists who hold them to be synonyms.

Subspecies An aggregate of local populations of a species inhabiting a geographic subdivision of the range of the species and differing taxonomically from other populations of the species.

Substitute name A name proposed to replace a preoccupied name; automatically takes the same type and type locality (*see* **New name**).

Substrate race A local race selected to agree in its coloration with that of the substrate, for example, a black race on a lava flow.

Superfamily The taxonomic category immediately above the family and below the order; an individual taxon ranked in this category.

Superspecies A monophyletic group of closely related and entirely or largely allopatric species that are too distinct to be included in a single species or that demonstrate their reproductive isolation in a zone of contact (cf. **Allopatric, Semispecies**).

Supraspecific A term applied to a category or evolutionary phenomenon above the species level.

Sympatric hybridization The occasional production of hybrid individuals between two otherwise well-defined sympatric species.

Sympatric speciation Speciation without geographic isolation; the acquisition of isolating mechanisms within a deme.

Sympatry The occurrence of two or more populations in the same area; more precisely, the existence of a population in breeding condition within the cruising range of individuals of another population.

Symplesiomorphy The sharing of ancestral characters by different taxa (Hennig 1950) (Chapter 10).

Synapomorphic Pertaining to a uniquely derived apomorphic character that is found in two or more taxa under consideration.

Synapomorphy A homologous character shared by two or more taxa and inferred to have been present in the nearest common ancestor but not in earlier ancestors nor in taxa outside this group.

Synchronic species Species that occur in the same time dimension (cf. **Allochronic species**).

Synonym In nomenclature, each of two or more different names for the same taxon (cf. **Senior synonym, Junior synonym, Objective synonym, Subjective synonym**).

Synonymy A chronological list of the scientific names given to a taxon, with the dates of their publication and their authors.

Synopsis In taxonomy, a brief summary of current knowledge about a group.

Synthetic theory The currently adopted assemblage of evolutionary theories; evolutionary change is explained as being due to the interaction of genetic variation and selection, and the origin of diversity results from the divergence of populations and speciation.

Syntype Every specimen in a type series in which no holotype has been designated.

Systematics The science dealing with the diversity of organisms; "the scientific study of the kinds and diversity of organisms and of any and all relationships among them" (Simpson 1961).

Taxon (pl. taxa) A monophyletic group of populations or lower taxa that can be recognized by their sharing of a definite set of characters; such a group is sufficiently distinct to be worthy of a name and to be ranked in a definite taxonomic category.

Taxonomic category Rank in a hierarchy of levels to which taxa are assigned, such as subspecies, species, and genus. A class whose members are all taxa assigned a given rank.

Taxonomic character Any attribute of a member of a taxon by which it differs or may differ from a member of a different taxon.

Taxonomy The theory and practice of classifying organisms (cf. **Classification, Systematics**).

Teleonomic Said of a process or behavior that owes its goal-directedness to the operation of a program.

Temporal isolation Isolation effected by differences in the season or time of day when mating takes place.

Temporal subspecies Subspecies in the time dimension.

Teratology The study of structural abnormalities, especially monstrosities and malformations.

Territory An area defended by an animal against other members of its species and occasionally against members of other species.

Thelytoky Parthenogenesis in which only females are produced.

Therapsid reptiles The reptilian group from which the mammals evolved.

Topotype A specimen collected at the type locality.

Transformation series A series of homologous characters some of which were derived from others during evolution (cf. **Polarity**).

Tribe A taxonomic category intermediate between the genus and the subfamily.

Trinominal nomenclature An extension of the binominal system of nomenclature to permit the designation of subspecies by a three-word name.

Triploid A cell or individual with three haploid chromosome sets; one of the forms of polyploidy.

Trivial name An obsolete designation by Linnaeus for the specific name; a synonym for a vernacular name.

Type In nomenclature, a zoological object that serves as the basis for the name of a taxon. In morphology, the term refers to the basic Bauplan of a higher taxon, that is, a generalized, idealized structural pattern from which one can derive all the variation observed within the taxon; an archetype.

Type designation Determination of the type of a genus under Articles 62 through 65 of the Code.

Type locality The locality at which a holotype, lectotype, or neotype was collected (cf. **Topotype**).

Type method The method by which the name for a taxon is unambiguously associated with a definite zoological object belonging to that taxon.

Type selection *See* **Type designation.**

Type species The species that was designated as the type of a nominal genus.

Typological thinking A concept in which variation is neglected and the members of a population or species are considered to be as replicas of the type, or the Platonic *eidos*.

Typologist One who disregards variation and regards the members of a population as replicas of the type.

Uninominal nomenclature The designation of a taxon by a scientific name consisting of a single word; required for taxa above species rank.

Univariate analysis Biometric analysis of a single character.

UPGMA (unweighted pair-group method using arithmetic averages) A clustering method of numerical taxonomy for determining relationship among taxa (Chapter 11).

Upward classification The assembling of populations and phena into species, and of species into the taxa of higher categorical rank.

Valid name An available name that is not preoccupied by a valid senior synonym or homonym.

Variance The square of the standard deviation; a statistic relating to deviations from the mean.

Variation, ecophenotypic Variation caused by nongenetic responses of the phenotype to local conditions of habitat, season, or climate.

Variety An ambiguous term of classical (Linnaean) taxonomy for a heterogeneous group of phenomena including nongenetic variations of the phenotype, phena, domestic breeds, and geographic races.

Vicariance Discontinuous (allopatric) geographic occurrence of the same or related taxa.

Vernacular name The colloquial designation of a taxon (cf. **Scientific name**).

Vertical classification Classification that stresses common descent and tends to unite ancestral and descendant groups of a phyletic line in a single higher taxon, separating them from contemporaneous taxa that have reached a similar grade of evolutionary change (cf. **Horizontal classification**).

Wagner tree A dendrogram based on Wagner's groundplan method (Chapter 11).

Weighting The evaluation of the probable contribution of a character to a sound classification on the basis of its phyletic information content.

Zygote A fertilized egg; the individual that results from the union of two gametes and their nuclei.

BIBLIOGRAPHY

Abbott, L. A., F. A. Bisby, and D. J. Rogers. 1985. *Taxonomic Analysis in Biology*. New York: Columbia University Press.

Adanson, M. 1763. *Familles des Plantes*. Vol. 1. Paris: Vincent.

Agassiz, L., and H. E. Strickland. 1848. *Bibliographia Zoologiae et Geologiae*. Vols. 1–4. London: Ray Society.

Albrecht, F. O. 1962. Physiologie, comportement, et écologie des acridiens, etc. *Colloq. Int. Centre Nat. Rech. Sci.* **114**:283–297.

Alexander, R. D. 1962. The role of behavioral study in cricket classification. *Syst. Zool.* **11**:53–72.

Amadon, D. 1967. The superspecies concept. *Syst. Zool.* **15**:245–249.

Ander, K. 1942. Die Insektenfauna des baltischen Bernsteins nebst damit verknüpften zoogeographischen Problemen. *Fysiogr. Sällsk. Handl.* NS 53 **4**:1–83.

Anderson, E. 1936. Hybridization in American Tradescantias. *Ann. Mo. Bot. Garden* **23**:511–525.

———. 1954. Efficient and inefficient methods of measuring specific differences, pp. 93–106 in O. Kempthorne et al. (eds.), *Statistics and Mathematics in Biology*. Ames: Iowa State College Press.

Anderson, N. M. 1978. Some principles and methods of cladistic analysis with notes on the use of cladistics in classification and biogeography. *Z. Zool. Syst. Evolforsch.* **16**:243–255.

Anderson, R. M. 1965. Methods of collecting and preserving vertebrate animals. 4th ed. Ottawa: *Bull. Natl. Mus. Canada,* Dept. Mines, No. 69, Biol. Ser. 18.

Anonymous. 1936. *Instructions for Collectors*. British Museum (Nat. Hist.).

Appel, T. A. 1987. *The Cuvier-Geoffroy Debate*. New York and Oxford: Oxford University Press.

Arkell, W. J., and J. A. Moy-Thomas. 1940. Paleontology and the taxonomic problem, pp. 395–410 in J. S. Huxley (ed.), *The New Systematics*. Oxford: Clarendon Press.

Ashlock, P. D. 1971. Monophyly and associated terms. *Syst. Zool.* **20:**63–69.

———. 1972. Monophyly again. *Syst. Zool.* **21:**430–438.

———. 1974. The uses of cladistics. *Ann. Rev. Ecol. System.* **5:**81–99.

———. 1979. An evolutionary systematist's view of classification. *Syst. Zool.* **28:**441–450.

———. 1984. Monophyly, its meaning and importance, pp. 39–46 in T. Duncan and T. Stuessy (eds.), *Cladistics: Perspectives on the Reconstruction of Evolutionary History*. New York: Columbia University Press.

———. 1985. A revision of the Bergidea group: A problem in classification and biogeography. *J. Kansas Ent. Soc.* **57:**675–688.

———. 1987. Classification: Philosophies and methods, pp. 42–51 in R. S. Boardman, A. H. Cheetham, and A. J. Rowell (eds.), *Fossil Invertebrates*. Palo Alto, CA: Blackwell Scientific Publications.

Atchley, W. R., and E. H. Bryant (eds.). 1975. *Multivariate Statistical Methods*. Vol. 1: *Among-Groups Covariation*. New York: Academic Press.

Avise, J. C. and R. A. Lansman. 1983. Polymorphism of mitochondrial DNA in populations of higher animals, pp. 147–164 in M. Nei and R. K. Koehn (eds.), *Evolution of Genes and Proteins*. Sunderland, MA: Sinauer.

Ax, P. 1984. *Das Phylogenetische System*. Stuttgart and New York: Gustav Fischer. (English ed. 1987. Chichester: John Wiley).

Baer, J. G. (ed.). 1957. First symposium on host specificity among parasites of vertebrates. Neuchatel: Zoological Institute of the University of Neuchatel.

Baker, A. N., F. W. E. Rowe, and H. E. S. Clark. 1986. A new class of Echinodermata. *Nature* **321:**862–863.

Baker, R. J., M. B. Qumsieh, and C. S. Hood. 1987. Role of chromosomal banding patterns in understanding mammalian evolution. *Curr. Mammal.* **1:**67–96.

Barber, H. S. 1951. North American fireflies of the genus Photuris. *Smithson. Misc. Coll.* **117**(1):1–58.

Barigozzi, C. (ed.) 1982. *Mechanisms of Speciation*. New York: Alan R. Liss.

Baroin, A., R. Perasso, L.-H. Qu, G. Brugerolle, J.-P. Bachellerie, and A. Adoutte. 1988. Partial phylogeny of the unicellular eukaryotes etc. *PNAS* **85:**3474–3478.

Barrowclough, G. F. 1983. Biochemical studies of microevolutionary processes, pp. 223–261 in A. H. Brush and G. A. Clark, Jr. (eds.), *Perspectives in Ornithology*. New York: Cambridge University Press.

———. 1985. Museum collections and molecular systematics, pp. 43–54 in E. H. Miller (ed.), *Museum Collections: Their Role and Future in Biological Research*. Brit. Columbia Prov. Occas. Papers Series, No. 25.

Barton, N. H., and G. M. Hewitt. 1989. Adaptation, speciation, and hybrid zones. *Nature* **341:**497–502.

Beatty, J. 1982. Classes and cladists. *Syst. Zool.* **31:**25–34.

Beckner, M. 1959. *The Biological Way of Thought*. New York: Columbia University Press.

Beer, J. E. de, and E. F. Cook. 1957. A method for collecting ectoparasites from birds. *J. Parasitol.* **43:**445.

Bessey, C. E. 1908. The taxonomic aspect of the species. *Amer. Nat.* **42:**218–224.

Beverley, S. M., and A. C. Wilson. 1985. Ancient origin for Hawaiian Drosophilinae inferred from protein comparisons. *Proc. Nat. Acad. Sci. USA* **82:**4753–4757.

Bianco, S. L. 1899. The methods employed at the Naples zoological station for the preservation of marine animals (trans. from Italian by E. O. Hovey). *U.S. Nat. Mus. Bull.* No. 39, part M:3–42.

Bickham, J. W., and J. L. Carr. 1983. Taxonomy and phylogeny of the higher categories of cryptodiran turtles based on a cladistic analysis of chromosomal data. *Copeia* 1983:918–932.

Bishop, M. J. 1982. Criteria for the determination of the direction of character state changes. *Zool. J. Linn. Soc.* **74:**197–206.

Blackwelder, R. E. 1972. *Guide to the Taxonomic Literature of Vertebrates*. Ames: Iowa State University Press.

Blair, W. F. (ed.). 1961. *Vertebrate Speciation*. Austin: University of Texas Press.

Blaker, A. A. 1977. *Handbook of Scientific Photography*. 2nd ed. San Francisco: W. H. Freeman.

Bock, H. H. (ed.). 1988. *Classification and Related Methods of Data Processing*. Amsterdam: North Holland.

Bock, W. J. 1968. Review of Hennig, "Phylogenetic Systematics," *Evolution* **22:**646–648.

———. 1973. Philosophical foundations of classical evolutionary classification. *Syst. Zool.* **22:**375–392.

———. 1977. Foundations and methods of evolutionary classification, pp. 851–895 in M. Hecht, P. C. Goody, and B. M. Hecht (eds.), *Major Patterns in Vertebrate Evolution*. New York: Plenum Press.

———. 1978. Comments on classifications as historical narratives. *Syst. Zool.* **27:**362–364.

———. 1981. Functional-adaptive analysis in evolutionary classification. *Amer. Zool.* **21:**5–20.

Bocquet, C., J. Génermont, and M. Lamotte (eds.). 1976–1980. Les problèmes de l'espèce dans le règne animal. *Mem. Soc. Zool. France* 38 (1976), 39 (1977), 40 (1980).

Boudreaux, H. B. 1979. *Arthropod Phylogeny with Special Reference to Insects*. New York: John Wiley & Sons.

Boyce, A. J. 1964. The value of some methods of numerical taxonomy with reference to hominid classification. London: The Systematics Association, Publ. No. 6:47–65.

Bradbury, S. 1984. *An Introduction to the Optical Microscope*. New York: Oxford University Press.

Bretsky, S. S. 1975. Allopatry and ancestors: A response to Cracraft. *Syst. Zool.* **24:**113–119.

———. 1979. Recognition of ancestor-descendant relationships in invertebrate paleontology, pp. 113–163 in J. Cracraft and N. Eldredge (eds.). *Phylogenetic Analysis and Paleontology*. New York: Columbia University Press.
Brewer, G. J. 1970. *An Introduction to Isozyme Techniques*. New York: Academic Press.
Bridgman, P. W. 1927. *The Logic of Modern Physics*. New York: Macmillan.
———. 1936. *The Nature of Physical Theory*. Princeton, NJ: Princeton University Press.
———. 1938. Operational analysis. *Philosophy Sci.* **5**:114–131.
Briggs, J. C. 1974. *Marine Zoogeography*. New York: McGraw-Hill.
Brinkmann, R. 1929. Statistisch-biostratigraphische Untersuchungen an mitteljurassischen Ammoniten über Artbegriff und Stammesentwicklung. *Abhandl. Ges. Wiss. Göttingen*, Math. Nat. Kl. (N.F.), **13**:1–249.
British Museum (Nat. Hist.). 1936. *Instructions for Collectors*.
Britten, R. J. 1986. Rates of DNA sequence evolution differ between taxonomic groups. *Science* **231**:1393–1398.
Brooks, D. R., and E. O. Wiley. 1985. Theories and methods in different approaches to phylogenetic systematics. *Cladistics* **1**:1–11.
Brothers, D. J. 1975. Phylogeny and classification of the aculeate Hymenoptera, with special reference to Mutillidae. *U. Kansas Sci. Bull.* **50**:483–648.
Brown, J. H., and A. C. Gibson. 1983. *Biogeography*. St. Louis: C V Mosby.
Brown, R. W. 1954. *Composition of Scientific Words: A Manual of Methods and a Lexicon of Materials for the Practice of Logotechnics*. Publ. by the author, Washington, DC: U.S. Nat. Mus.
Brundin, L. 1966. Transantarctic relationships and their significance as evidenced by the chironomid midges, with a monograph of the subfamilies Podonominae, Aphrotaeninae, and the austral Heptagiae. *Kungl. Svensk. Vetenskap. Handl.* **11**:1–472.
Bryant, E. H., and W. R. Atchley (eds.). 1975. *Multivariate Statistical Methods*. Vol. II: *Within-Groups Covariation*. New York: Academic Press.
Bryson, V., and H. J. Vogel (eds.). 1965. *Evolving Genes and Proteins*. New York and London: Academic Press.
Buchner, P. 1966*a*. *Endosymbiosis of Animals with Plant Microorganisms*. New York: Wiley Interscience.
———. 1966*b*. Die Symbiosen der Palaeococcoidea. *Z. Morph. Ökol. Tiere* **56**:275–362.
Buffon, G. L. 1749. *Histoire naturelle, générale et particulière*. Paris: Imprimerie Royale.
Bullini, L., G. Nascetti, S. Ciaffrè, F. Rumore, and E. Biocca. 1978. Richerche cariologiche ed elettroforetiche su *Parascaris univalens* e *Parascaris equorum*. *Rend. Classe Scienze*. Ser. VIII, **LXV**:151–156.
Burma, B. H. 1948. The species concept: A semantic review. *Evolution* **3**:369–370.
Bush, G. L. 1975. Modes of animal speciation. *Ann. Rev. Ecol. Syst.* **6**:339–364.
Busnel, R. G. 1963. *Acoustic Behavior of Animals*. Amsterdam and New York: Elsevier.

Buth, D. G. 1984. The application of electrophoretic data in systematic studies. *Ann. Rev. Ecol. Syst.* **15:**501–522.

Butlin, R. K. 1985. Speciation by reinforcement, pp. 84–113 in J. Gosálves, C. López-Fernández, and C. García de la Vega (eds.), *Orthoptera*. Vol. 1. Fundacion Ramon Areces, Madrid.

———. 1989. Reinforcement of premating isolation, pp. 158–179 in D. Otte and J. A. Endler (eds.), *Speciation and Its Consequences*. Sunderland, MA: Sinauer.

Cain, A. J. 1956. The genus in evolutionary taxonomy. *Syst. Zool.* **5:**97–109.

———. 1958. Logic and memory in Linnaeus' system of taxonomy. *Proc. Linn. Soc. London* **169:**144–163.

———. 1959*a*. Deductive and inductive methods in post-Linnaean taxonomy. *Proc. Linn. Soc. London* **170:**185–217.

———. 1959*b*. Taxonomic concepts. *Ibis* **101:**302–318.

———. 1982. On homology and convergence, pp. 1–19 in K. A. Joysey and A. E. Friday (eds.), *Problems of Phylogenetic Reconstruction*. London: Academic Press.

———, and G. A. Harrison. 1958. An analysis of the taxonomist's judgment of affinity. *Proc. Zool. Soc. London* **131:**85–98.

———, and ———. 1960. Phyletic weighting. *Proc. Zool. Soc. London* **135:**1–31.

Camin, J. H., and R. R. Sokal. 1965. A method for deducing branching sequences in phylogeny. *Evolution* **19:**311–326.

Camp, W. H., and C. L. Gilly. 1943. The structure and origin of species. *Brittonia* **4:**323–385.

Candolle, A. P. de. 1813. *Theorie elémentaire de la botanique*. Paris: Chez Deterville.

Capanna, E. 1982. Robertsonian numerical variation in animal speciation: *Mus musculus,* an emblematic model, pp. 155–177 in C. Barigozzi (ed.), *Mechanisms of Speciation*. New York: Alan R. Liss.

Carlquist, S. 1974. *Island Biology*. New York: Columbia University Press.

Carpenter, G. D. H. 1949. *Pseudacrea eurytus* (L.). (Lep. Nymphalidae): A study of a polymorphic mimic in various degrees of speciation. *Trans. Roy. Entomol. Soc. London* **100:**71–133.

Carson, H. L., F. E. Clayton, and H. D. Stalker. 1967. Karyotypic stability and speciation in Hawaiian Drosophila. *Proc. Nat. Acad. Sci.* **57:**1280–1285.

Cartmill, M. 1981. Hypothesis testing and phylogenetic reconstruction. *Zeitschr. Zool. Syst. Evolutionsforsch.* **19:**73–96.

Cato, P. S. 1986. *Guidelines for managing bird collections. Museology*. No. 7. Lubbock, TX: Texas Tech University. 78 pp.

Catzeflis, F. M., F. M. Sheldon, J. E. Ahlquist, and C. G. Sibley. 1987. DNA-DNA hybridization evidence of the rapid rate of muroid rodent DNA evolution. *Molec. Biol. Evol.* **4:**242–253.

Cavalier-Smith, T. 1987. Eukaryotes with no mitochondria. *Nature* **326:**332–333.

Champion, A. B., E. M. Prager, D. Wachter, and A. C. Wilson. 1974. Micro-complement fixation, pp. 397–416 in C. A. Wright (ed.), *Biochemical and Immunological Taxonomy of Animals*. London: Academic Press.

Charig, A. J. 1982. Systematics in biology: A fundamental comparison of some major schools of thought, pp. 363–440 in K. A. Joysey and A. E. Friday (eds.), *Problems of Phylogenetic Reconstruction*. London: Academic Press.

Chicago. 1987. *Chicago Guide to Preparing Electronic Manuscripts*. Chicago: Chicago University Press.

Claridge, M. F. 1985. Acoustic signals in the Homoptera: Behavior, taxonomy, and evolution. *Ann. Rev. Ent.* **30**:297–317.

Clay, T. 1949. Some problems in the evolution of a group of ectoparasites. *Evolution* **3**:279–299.

———. 1958. Revisions of Mallophaga genera. *Bull. Brit. Mus. (Nat. Hist.) Ent.* **7**:123–208.

Cleveland, W. S. 1985. *The Elements of Graphing Data*. Monterey, CA: Wadsworth Advanced Books.

Cole, A. J. 1969. *Numerical Taxonomy: Proceedings of the Colloquium in Numerical Taxonomy Held in the University of St. Andrew,* September, 1968. London: Academic Press.

Colless, D. H. 1972. A note on Ashlock's definition of "monophyly." *Syst. Zool.* **21**:126–128.

———. 1981. Predictivity and stability in classifications: Some comments on recent studies. *Syst. Zool.* **30**:325–331.

———. 1983. Wagner trees in theory and practice, pp. 259–278 in J. Felsenstein (ed.), *Numerical Taxonomy*. Berlin: Springer.

Collier, G. E., and S. J. O'Brien. 1985. A molecular phylogeny of the Felidae: Immunological distance. *Evolution* **39**:473–487.

Corliss, J. O. 1959. An illustrated key to the higher groups of the ciliated protozoa, with definition of terms. *J. Protozool.* **6**:265–284.

———. 1963. Establishment of an international type-slide collection for the ciliate protozoa. *J. Protozool.* **10**:247–249.

———. 1979. *The Ciliated Protozoa: Characterization, Classification, and Guide to the Literature*. New York: Pergamon Press.

———. 1981. What are the taxonomic and evolutionary relationships of the Protozoa to the Protista? *BioSystems* **14**:445–459.

———. 1986. The kingdoms of organisms—from a microscopist's point of view. *Trans. Amer. Microsc. Soc.* **105**:1–10.

Council of Biology Editors. 1988. *Illustrating Science*. Bethesda, MD: Council of Biology Editors.

Coyne, J. A., H. A. Orr, and D. J. Futuyma. 1988. Do we need a new definition of species? *Syst. Zool.* **37**:190–200.

Cracraft, J. 1983. The significance of phylogenetic classifications for systematic and evolutionary biology, pp. 1–17 in J. Felsenstein (ed.), *Numerical Taxonomy*. Berlin: Springer.

———, and N. Eldredge (eds.). 1979. *Phylogenetic Analysis and Paleontology*. New York: Columbia University Press.

Cronquist, A. 1968. *The Evolution and Classification of Flowering Plants*. Boston: Houghton Mifflin.

———. 1987. A botanical critique of cladism. *Bot. Rev.* **53**:1–52.

Cutler, E. B., and P. E. Gibbs. 1985. A phylogenetic analysis of higher taxa in the phylum Sipuncula. *Syst. Zool.* **34**:162–173.

Dall, W. H. 1898. Contributions to the tertiary fauna of Florida. *Trans. Wagner Free Inst. Sci. Phila.* **3**:675–676.

Daly, H. V., and S. S. Balling. 1978. Identification of Africanized honeybees in the Western Hemisphere by discriminant analysis. *J. Kansas Entomol. Soc.* **51**:857–868.

Darlington, P. J. 1957. *Zoogeography: The Geographical Distribution of Animals.* New York: John Wiley & Sons.

———. 1965. *Biogeography of the Southern End of the World.* Cambridge, MA: Harvard University Press.

Darwin, C. 1859. *On the Origin of Species by Means of Natural Selection, or the Preservation of Favoured Races in the Struggle for Life.* London: John Murray. [Facsimile ed. Mayr, E. (ed.). 1964. Cambridge, MA: Harvard University Press.]

Davis, D. D. 1964. The giant panda: A morphological study of evolutionary mechanisms. *Fieldiana, Zool. Mem.* 3. Chicago: Chicago Nat. Hist. Museum.

Dayhoff, M. O. (ed.). 1973, 1976, 1979. *Atlas of Protein Sequence and Structure.* Vol. 5, Supplements 1, 2, and 3. Washington, DC: National Biomedical Research Foundation.

Desmond, A. J. 1982. *Archetypes and Ancestors: Paleontology in Victorian London, 1850–1875.* London: Blond and Briggs.

Dessauer, H. C., and M. S. Hafner. 1984. *Collections of Frozen Tissues.* Washington, DC: Association of Systematics Collections.

Dethier, V. G. 1947. *Chemical Insect Attractants and Repellents.* New York: McGraw-Hill.

Diamond, J. M. 1965. Zoological classification system of a primitive people. *Science* **151**:1102–1104.

Dice, L. R., and H. J. Leraas. 1936. A graphic method for comparing several sets of measurements. *Contrib. Lab. Vert. Genet. Univ. Michigan* **3**:1–3.

Dowler, R. C., and H. H. Genoways. 1976. Supplies and suppliers for vertebrate collections. *Museology* No. 4. Lubbock, TX: Texas Tech University. 83 pp.

Downey, J. C. 1962. Host-plant relations as data for butterfly classification. *Syst. Zool.* **11**:150–159.

Dubois, A. 1982. Les notions de genre, sous-genre, et groupe d'espèces en zoologie à la lumière de la systèmatique èvolutive. *Monitore Zool. Ital.* (NS) **16**:9–65.

———. 1983. Hybridation interspecifique, similarité genetique, parenté phylogenetique et classification supraspecifique en zoologie. *Ann. Biol.* **22**:37–68.

———, and R. Günther. 1982. Klepton and Synklepton: Two new evolutionary systematics categories in zoology. *Zool. Jb. Syst.* **109**:290–305.

Duncan, T., and B. R. Baum. 1981. Numerical phenetics: Its uses in botanical systematics. *Ann. Rev. Ecol. Syst.* **12**:387–404.

Duncan, T., R. B. Phillips, and W. H. Wagner. 1980. A comparison of branching diagrams derived by various phenetic and cladistic methods. *Syst. Bot.* **5**:264–293.

Duncan, T., and T. F. Stuessy (eds.). 1984. *Cladistics: Perspectives on the Reconstruction of Evolutionary History*. New York: Columbia University Press.
———, and ——— (eds.). 1985. *Cladistic Theory and Methodology*. New York: Van Nostrand Reinhold.
Dunn, G., and B. S. Everitt. 1982. *An Introduction to Mathematical Taxonomy*. Cambridge: Cambridge University Press.
Dupuis, C. 1979. Permanence et actualité de la systematique: La systematique phylogenetique de W. Hennig. *Cahiers des Naturalistes* **34**(1978):1–69.
———. 1984. Willi Hennig's impact on taxonomic thought. *Ann. Rev. Ecol. Syst.* **15:**1–24.
Eberhard, W. G. 1982. Behavioral characters for the higher classification of orb weaving spiders. *Evolution* **36:**1067–1095.
———. 1985. *Sexual Selection and Animal Genitalia*. Cambridge, MA: Harvard University Press.
Edmondson, W. T. 1949. A formula key to the rotatorian genus Ptygura. *Trans. Amer. Microsc. Soc.* **68:**127–135.
Edwards, A. W. F., and L. L. Cavalli-Sforza. 1964. Reconstruction of evolutionary trees, pp. 67–76 in V. H. Heywood and J. McNeill (eds.), *Phenetic and Phylogenetic Classification*. London: The Systematics Association.
Ehlers, U. 1985. *Das Phylogenetische System der Platyhelminthes*. Stuttgart: G. Fischer.
Ehrlich, P. R. 1964. Some axioms of taxonomy. *Syst. Zool.* **13:**109–123.
Eickwort, G. C., and S. F. Sakagami. 1979. A classification of nest architecture of bees in the tribe Angochlorini (Halictidae). *Biotropica* **11:**28–37.
Eldredge, N., and J. Cracraft. 1980. *Phylogenetic Patterns and the Evolutionary Process*. New York: Columbia University Press.
———, and S. J. Gould. 1972. Punctuated equilibria: An alternative to phyletic gradualism, pp. 82–115 in T. J. M. Schopf (ed.), *Models in Paleobiology*. San Francisco: Freeman, Cooper.
Elton, C. 1947. *Animal Ecology*. London: Sidgwick & Jackson.
Emden, F. I. van. 1957. The taxonomic significance of the characters of immature insects. *Ann. Rev. Entomol.* **2:**91–106.
Endler, J. A. 1977. *Geographic Variation, Speciation, and Clines*. Princeton, NJ: Princeton University Press.
Engelmann, W. 1846. *Bibliotheca Historico-Naturalis. Verzeichnis der Bücher über Naturgeschichte 1700–1846*. Leipzig: W. Engelmann.
Erwin, T. L. 1983. Tropical forest canopies: The last biotic frontier. *Bull. Ent. Soc. Amer.* Spring 1983:14–19.
Estabrook, G. F. 1972. Cladistic methodology: A discussion of the theoretical basis for the induction of evolutionary history. *Annu. Rev. Ecol. System.* **3:**427–456.
———. 1978. Some concepts for the estimation of evolutionary relationships in systematic botany. *Syst. Bot.* **3:**146–150.
———. 1986. Evolutionary classification using convex phenetics. *Syst. Zool.* **35:**560–570.
———, C. S. Johnson, and F. R. McMorris. 1975*a*. An idealized concept of the true cladistic character. *Math. Biosci.* **23:**263–272.

———, ———, and ———. 1975*b*. An algebraic analysis of cladistic characters. *Discrete Math.* **16:**141–147.

———, and L. R. Landrum. 1975. A simple test for the possible simultaneous evolutionary divergence of two amino acid positions. *Taxon* **24:**609–613.

Evans, H. E. 1964. The classification and evolution of digger wasps as suggested by larval characters. *Ent. News* 75:225–237.

———. 1966. *The Comparative Ethology and Evolution of the Sand Wasps.* Cambridge, MA: Harvard University Press.

Everitt, B. S. 1980. *Cluster Analysis.* 2nd ed. London: Heinemann.

Faith, D. P. 1985. Distance methods and the approximation of most-parsimonious trees. *Syst. Zool.* **34:**312–325.

Farris, J. S. 1969. A successive approximatious approach to character weighting. *Syst. Zool.* **18:**374–385.

———. 1970. Methods for computing Wagner trees. *Syst. Zool.* **19:**83–92.

———. 1972. Estimating phylogenetic trees from distance matrices. *Amer. Nat.* **106:**645–668.

———. 1974. Formal definitions of paraphyly and polyphyly. *Syst. Zool.* **23:**548–554.

———. 1977. On the phenetic approach to vertebrate classification, pp. 823–850 in M. K. Hecht, P. C. Goody, and B. M. Hecht (eds.), *Major Patterns in Vertebrate Evolution.* New York: Plenum Press.

———. 1979*a*. On the naturalness of phylogenetic classification. *Syst. Zool.* **28:**200–214.

———. 1979*b*. The information content of the phylogenetic system. *Syst. Zool.* **28:**483–519.

———. 1981. Distance data in phylogenetic analysis, pp. 3–22 in V. A. Funk and D. R. Brooks (eds.), *Advances in Cladistics: Proceedings of the First Meeting of the Willi Hennig Society.* New York: New York Botanical Garden.

———. 1983. The logical basis of phylogenetic analysis. *Adv. Cladistics* **2:**7–36.

———. 1985. Distance data revisited. *Cladistics* **1:**67–85.

———. 1986. On the boundaries of phylogenetic systematics. *Cladistics* **2:**14–27.

———, A. G. Kluge, and M. J. Eckardt. 1970. A numerical approach to phylogenetic systematics. *Syst. Zool.* **19:**172–189.

Felsenstein, J. 1978. Cases in which parsimony and compatibility methods will be positively misleading. *Syst. Zool.* **27:**401–410.

———. 1979. Alternative methods of phylogenetic inference and their interrelationships. *Syst. Zool.* **28:**49–62.

———. 1981. A likelihood approach to character weighting and what it tells us about parsimony and compatibility. *Biol. J. Linn. Soc.* **16:**183–196.

———. 1982. Numerical methods for inferring evolutionary trees. *Quart. Rev. Biol.* **57:**379–404.

———. 1983*a*. *Numerical Taxonomy.* NATO Advanced Study Institute Series. Berlin: Springer.

———. 1983*b*. Methods for inferring phylogenies: A statistical view, pp. 315–334 in J. Felsenstein, *Numerical Taxonomy.* Berlin: Springer.

———. 1983*c*. Parsimony in systematics: Biological and statistical issues. *Ann. Rev. Ecol. Syst.* **14:**313–333.

———. 1984. Distance methods for inferring phylogenies: A justification. *Evolution* **38:**16–24.

———. 1988*a*. Perils of molecular introspection. *Nature* **335:**118.

———. 1988*b*. Phylogenies and quantitative characters. *Annu. Rev. Evol. Syst.* **19:**445–471.

———, and E. Sober. 1986. Parsimony and likelihood: An exchange. *Syst. Zool.* **35:**617–626.

Ferris, G. F. 1928. *The Principles of Systematic Entomology*. Stanford, CA: Stanford University Press.

Field, K. G., G. J. Olsen, D. L. Lane, S. J. Giovannoni, M. T. Ghiselin, E. C. Raff, N. R. Pace, and R. A. Raff. 1988. Molecular phylogeny of the animal kingdom. *Science* **239:**748–753.

Fingerman, M. 1963. *The Control of Chromatophores*. New York: Pergamon Press.

Fisher, D. R., and F. J. Rohlf. 1969. Robustness of numerical methods and errors in homology. *Syst. Zool.* **18:**33–36.

Fisher, R. A. 1936. The use of multiple measurements in taxonomic problems. *Ann. Eugenics* **7:**179–188.

———. 1938. The statistical use of multiple measurements. *Ann. Eugenics* **8:**376–386.

Fitch, W. M. 1971. Toward defining the course of evolution: Minimum change for a specific tree morphology. *Syst. Zool.* **20:**406–416.

———. 1976. Molecular evolutionary clocks, pp. 160–178 in F. J. Ayala (ed.), *Molecular Clocks*. Sunderland, MA: Sinauer.

———. 1977. On the problem of discovering the most parsimonious tree. *Amer. Nat.* **111:**223–257.

———. 1981. A nonsequential method for constructing trees and hierarchical classifications. *J. Mol. Evol.* **18:**30–37.

———. 1984. Cladistic and other methods: Problems, pitfalls, and potentials, pp. 221–252 in T. Duncan and T. F. Stuessy (eds.), *Cladistics: Perspectives on the Reconstruction of Evolutionary History*. New York: Columbia University Press.

———, and E. Margoliash. 1967. The construction of phylogenetic trees. *Science* **155:**279–284.

Flegelman, A., and J. J. Hewes. 1983. *Writing in the Computer Age*. Garden City, NY: Anchor Press, Doubleday.

Foltz, D. W., H. Ochman, J. S. Jones, S. M. Evangelisti, and R. K. Selander. 1982. Genetic population structure and breeding systems in arionid slugs (Mollusca: Pulmonata). *Biol. J. Linn. Soc.* **17:**225–241.

Ford, E. B. 1945. Polymorphism. *Biol. Rev.* **20:**73–88.

———. 1965. *Genetic Polymorphism*. Cambridge, MA: MIT Press.

Friday, A. E. 1982. Parsimony, simplicity, and what actually happened. *Zool. J. Linn. Soc.* **74:**329–375.

Friedmann, H. (ed.). 1950. *The Birds of North and Middle America*. Part XI. Washington, DC: Smithsonian Institution Bulletin 50.

Frost, D. R., and J. W. Wright. 1988. The taxonomy of uniparental species with special reference to parthenogenetic Cnemidophorus (Squamata). *Syst. Zool.* **37**:200–209.

Funk, V. A., and D. R. Brooks (eds.). 1981. *Advances in Cladistics*. New York: New York Botanical Garden.

———, and Q. D. Wheeler. 1986. Symposium: Character weighting, cladistics, and classification. *Syst. Zool.* **35**:100–134.

Futuyma, D. J. 1986. *Evolutionary Biology,* 2nd ed. Sunderland, MA: Sinauer.

———, and G. C. Mayer. 1980. Non-allopatric speciation in animals. *Syst. Zool.* **29**(3):254–271.

———, and M. Slatkin (eds.). 1983. *Coevolution*. Sunderland, MA: Sinauer.

Galigher, A. E., and E. N. Kozloff. 1964. *Essentials of Practical Microtechnique*. Philadelphia: Lea & Febiger.

Gans, C., and F. H. Pough. 1982. *Biology of the Reptilia*. New York: Academic Press.

Gardiner, B. G. 1982. Tetrapod classification. *Zool. J. Linn. Soc.* **74**:207–232.

Gauthier, J. A., A. G. Kluge, and T. Rowe. 1988. Amniote phylogeny and the importance of fossils. *Cladistics* **4**:105–209.

Gersch, M. 1964. *Vergleichende Endokrinologie der wirbellosen Tiere*. Leipzig: Akad. Verlagsges., Geest & Portig.

Ghiselin, M. T. 1966. On psychologism in the logic of taxonomic controversies. *Syst. Zool.* **15**:207–215.

———. 1974. A radical solution to the species problem. *Syst. Zool.* **23**:536–544.

———. 1980. Natural kinds and literary accomplishments. *Mich. Quart. Rev.* **19**:73–88.

———. 1981. Categories, life, and thinking. *Behav. Brain Sci.* **4**:269–283.

———. 1984. Narrow approaches to phylogeny: A review of nine books of cladism. *Oxford Surv. Evol. Biol.* **1**:209–222.

———. 1987. Species concepts, individuality, and objectivity. *Biol. Philosophy* **2**:127–143.

Gilbert, P., and C. J. Hamilton. 1983. *Entomology: A Guide to Information Sources*. London: Mansell Publishing.

Gillespie, J. H. 1984. The status of the neutral theory. *Science* **224**:732–733.

———. 1986. Rates of molecular evolution. *Annu. Rev. Ecol. Syst.* **17**:637–665.

Gilmour, J. S. L. 1940. Taxonomy and philosophy, pp. 461–474 in J. S. Huxley (ed.), *The New Systematics*. Oxford: Clarendon Press.

———. 1961. Taxonomy, pp. 27–45 in A. M. MacLeod and L. S. Cobley, *Contemporary Botanical Thought*. Edinburgh: Oliver & Boyd.

Gingerich, P. D. 1979. Stratophenetic approach to phylogeny reconstruction in vertebrate paleontology, pp. 41–79 in J. Cracraft and N. Eldredge (eds.), *Phylogenetic Analysis and Paleontology*. New York: Columbia University Press.

Glaessner, M. F. 1984. *The Dawn of Animal Life: A Biohistorical Study*. Cambridge: Cambridge University Press.

Goldschmidt, R. 1933. Lymantria. *Bibliog. Genet.* **11**:1–185.

———. 1945. Mimetic polymorphism, a controversial chapter of Darwinism. *Quart. Rev. Biol.* **20**:147–164, 205–230.

———. 1952. Evolution, as viewed by one geneticist. *Amer. Sci.* **40:**84–98.
Good, D. A. 1987. An allozyme analysis of anguid subfamilial relationships. *Copeia* 1987:696–701.
Goodman, M. (ed.), 1982. *Macromolecular Sequences in Systematic and Evolutionary Biology*. New York: Plenum Press.
Gordl, G., and P. DeBach. 1978. Courtship behavior in the *Aphytis linguanensis* group, its potential usefulness in taxonomy, and a review of sexual behavior in the parasitic Hymenoptera. *Hilgardia* **46**(2):37–75.
Gosliner, T. M., and M. T. Ghiselin. 1984. Parallel evolution in opistobranch gastropods and its implications for phylogenetic methodology. *Syst. Zool.* **33:**255–274.
Gould, S. J. 1966. Allometry and size in ontogeny and phylogeny. *Biol. Rev. (Cambridge Phil. Soc.)* **41:**587–640.
———, and N. Eldredge. 1977. Punctuated equilibria: The tempo and mode of evolution reconsidered. *Paleobiology* **3:**115–151.
Grant, P. R. 1986. *Ecology and Evolution of Darwin's Finches*. Princeton, NJ: Princeton University Press.
Grant, V. 1957. The plant species in theory and practice, pp. 39–80 in E. Mayr (ed.), The species problem, *Amer. Assoc. Adv. Sci. Publ.*, No. 50. Washington, DC.
———. 1981. *Plant Speciation*. 2nd ed. New York: Columbia University Press.
Gray, P. (ed.). 1973. *The Encyclopedia of Microscopy and Microtechnique*. New York: Van Nostrand Reinhold.
Guide to Museum Pest Control. 1987. Washington, DC: Assoc. Syst. Collections.
Haeckel, E. 1866. *Generelle Morphologie der Organismen,* II. Berlin: Georg Reiner.
Hall, A. V. 1988. A joint phenetic and cladistic approach for systematics. *Biol. J. Linn. Soc.* **33:**367–382.
Hallam, A. 1988. The contribution of paleontology to systematics and evolution, pp. 128–147 in D. L. Hawksworth (ed.), *Prospects in Systematics*. Oxford: Clarendon Press.
Halstead, L. B. 1982. Evolutionary trends and the phylogeny of the Agnatha, pp. 159–196 in K. A. Joysey and A. E. Friday (eds.), *Problems of Phylogenetic Reconstruction*. London: Academic Press.
Hand, D. J. 1981. *Discrimination and Classification*. London: John Wiley & Sons.
Handler, P. (ed.). 1964. Biochemistry symposium: Biochemical evolution. *Fed. Proc.* **23:**1229–1266.
Harper, C. W. 1976. Phylogenetic inference in paleontology. *J. Paleont.* **50:**180–193.
———, and N. I. Platnick. 1978. Phylogenetic and cladistic hypotheses: A debate. *Syst. Zool.* **27:**354–362.
Harris, H., and D. A. Hopkinson. 1976. *Handbook of Enzyme Electrophoresis in Human Genetics*. Amsterdam: North-Holland.
Harrison, G. A. 1959. Environmental determination of the phenotype. *Syst. Assoc.* **3:**81–86.
Hartigan, J. A. 1973. *Clustering Algorithms*. New York: John Wiley & Sons.

Haszprunar, G. 1986. Die klado-evolutionäre Klassifikation—Versuch einer Synthese. *Z. Systematik Evolutionsforsch.* **24:**89–109.

Hatheway, W. H. 1962. A weighted hybrid index. *Evolution* **16:**1–10.

Hecht, M. K., and J. L. Edwards. 1977. The methodology of phylogenetic inference above the species level, pp. 3–51 in M. K. Hecht, P. C. Goody, and B. M. Hecht (eds.), *Major Patterns in Vertebrate Evolution.* New York: Plenum Press.

Hempel, C. G. 1965. *Aspects of Scientific Explanation.* New York: Free Press.

Hendy, M. D., and D. Penny. 1982. Branch and bound algorithms to determine minimal evolutionary trees. *Math. Biosci.* **59:**277–290.

Hennig, W. 1950. *Grundzüge einer Theorie der Phylogenetischen Systematik.* Berlin: Deutscher Zentralverlag.

———. 1965. Phylogenetic systematics. *Ann. Rev. Entom.* **10:**97–116.

———. 1966*a*. *Phylogenetic Systematics* (transl. D. D. Davis and R. Zangerl). Urbana: University of Illinois Press.

———. 1966*b*. The Diptera fauna of New Zealand as a problem in systematics and zoogeography. *Pac. Inst. Monogr.* **9:**1–81.

———. 1969. *Die Stammesgeschichte der Insekten.* Frankfurt am Main: Waldemar Kramer.

———. 1981. *Insect Phylogeny.* New York: John Wiley & Sons.

———. 1984. *Aufgaben und Probleme stammesgeschichtlicher Forschung* (ed. W. Hennig). Berlin and Hamburg: Parey.

Hewitt, G. M. 1989. The subdivision of species by hybrid zones, pp. 85–110 in D. Otte and J. A. Endler (eds.). *Speciation and Its Consequences.* Sunderland, MA: Sinauer.

Heywood, V. H., and J. McNeill (eds.). 1964. *Phenetic and Phylogenetic Classification.* London: The Systematics Association, Publ. No. 6. 164 pp.

Higgins, R. P., and H. Thiel (eds.). 1988. *Introduction to the Study of Meiofauna.* Washington, DC: Smithsonian Institution Press.

Highton, R. 1962. Revision of North American salamanders of the genus *Plethodon. Bull. Fla. St. Mus.* (*Biol. Sci.*) **6:**235–367.

Hillis, D. M. 1985. Evolutionary genetics of the Andean lizard genus *Pholidobolus* (Sauria: Gymnophthalmidae): Phylogeny, biogeography, and a comparison of tree construction techniques. *Syst. Zool.* **34:**109–126.

———. 1987. Molecular versus morphological approaches to systematics. *Annu. Rev. Ecol. Syst.* **18:**23–42.

———, and C. Moritz (eds.). 1990. *Molecular Systematics.* Sunderland, MA: Sinauer

———, and S. K. Davis. 1986. Evolution of ribosomal DNA: Fifty million years of recorded history in the frog genus *Rana. Evolution* **40:**1275–1288.

———, J. S. Frost, and D. A. Wright. 1983. Phylogeny and biogeography of the *Rana pipiens* complex: A biochemical evaluation. *Syst. Zool.* **32:**132–143.

Holland, G. P. 1964. Evolution, classification, and host relationships of Siphonaptera. *Ann. Rev. Entomol.* **9:**123–146.

Holmes, E. B. 1980. Reconsideration of some systematic concepts and terms. *Evol. Theory* **5:**35–87.

———. 1985. Lungfishes, a sister group of tetrapods? *Biol. J. Linn. Soc.* **25**:379–397.
Holmquist, R., M. M. Miyamoto, and M. Goodman. 1988. Analysis of higher primate phylogeny from transversion differences in nuclear and mitochondrial DNA by Lake's methods of evolutionary parsimony and operator metrics. *Mol. Biol. Evol.* **5**:217–236.
Hopkins, G. H. E. 1949. The host-association of the lice on mammals. *Proc. Zool. Soc. London* **119**:387–604.
Hopson, E. B., and H. R. Barghusen. 1986. An analysis of therapsid relationships, pp. 83–106 in N. Hotton III, P. D. MacLean, J. J. Roth, and E. C. Roth (eds.), *The Ecology and Biology of Mammal-like Reptiles*. Washington, DC: Smithsonian Institution Press.
Hull, D. L. 1964. Consistency and monophyly. *Syst. Zool.* **13**:1–11.
———. 1967. Certainty and circularity in evolutionary taxonomy. *Evolution* **21**:174–189.
———. 1968. The operational imperative—sense and nonsense in operationalism. *Syst. Zool.* **17**:438–457.
———. 1970. Contemporary systematic philosophies. *Annu. Rev. Ecol. Syst.* **1**:19–54.
———. 1979. The limits of cladism. *Syst. Zool.* **28**:416–440.
———. 1983. Karl Popper and Plato's metaphor. *Adv. Cladistics* **2**:177–189.
———. 1988. *Science as a Process*. Chicago: Chicago University Press.
Humason, G. L. 1979. *Animal Tissue Techniques*. San Francisco: W. H. Freeman.
Hungay, G., and M. Dingley. 1985. *Biological Museum Methods*. Vol. 1: *Vertebrates*. Vol. 2: *Plants, Invertebrates, and Techniques*. New York: Academic Press.
Hurt, P. 1949. *Bibliography and Footnotes: A Style Manual for College and University Students*. Rev. Ed. Berkeley: University of California Press.
Huxley, J. S. 1939. Clines: An auxiliary method in taxonomy. *Bijdr. Dierk.* **27**:491–520.
———. 1940. *The New Systematics*. Oxford: Clarendon Press.
———. 1958. Evolutionary processes and taxonomy, with special reference to grades. *Uppsala University Arsskr.* **1958**:6, 21–39.
Imbrie, J. 1957. The species problem with fossil animals, pp. 125–153 in E. Mayr (ed.), The species problem. *Amer. Assoc. Adv. Sci. Publ.*, No. 50. Washington, D.C.
Inger, R. F. 1958. Comments on the definition of genera. *Evolution* **12**:370–384.
———. 1961. Problems in the application of the subspecies concept in vertebrate taxonomy, pp. 262–285 in W. F. Blair (ed.), *Vertebrate Speciation*. Austin: University of Texas Press.
———. 1967. The development of a phylogeny of frogs. *Evolution* **21**:369–384.
Irwin, M. R. 1947. Immunogenetics. *Adv. in Genetics* **1**:133–159.
Iwasuki, K., P. H. Raven, and W. J. Bock (eds.). 1986. *Modern Aspects of Species*. Tokyo: University of Tokyo Press.
Jardine, N., and R. Sibson. 1971. *Mathematical Taxonomy*. London: John Wiley & Sons.
Jefferies, R. P. S. 1979. The origin of chordates—a methodological essay, pp.

443–477 in M. R. House (ed.), *The Origin of Major Invertebrate Groups*. London: Academic Press.

———. 1986. *The Ancestry of the Vertebrates*. London: British Museum (Nat. Hist.).

John, B. (ed.). 1974 et seq. *Animal Cytogenetics*. Berlin: Gebrüder Borntraeger.

Johnson, L. A. S. 1968. Rainbow's end: The quest for an optimal taxonomy. *Proc. Linn. Soc. New South Wales* **93:**1–45.

Johnson, N. K., R. M. Zink, G. F. Barrowclough, and J. A. Marten. 1984. Suggested techniques for modern avian systematics. *Wilson Bull.* **96:**543–560.

Jordan, K. 1905. Der Gegensatz zwischen geographischer und nichtgeographischer Variation. *Zeitschr. Wissensch. Zool.* **83:**151–210.

Joysey, K. A., and A. E. Friday. 1982. *Problems of Phylogenetic Reconstruction*. Syst. Assoc. Special Vol. 21. London: Academic Press.

Keast, A. 1961. Bird speciation on the Australian continent. *Bull. Mus. Comp. Zool.* **123:**305–495.

Kemp, T. S. 1988. Haemothermia or Archosauria: The interrelationship of mammals, birds, and crocodiles. *Zool. J. Linn. Soc.* **92:**67–104.

Kennedy, J. S. (ed.). 1961. Insect polymorphism. *Symp. Roy. Entomol. Soc. London* **1:**1–115.

Kerfoot, W. C. 1980. Perspectives on cyclomorphosis: Separation of phenotypes and genotypes, pp. 470–496 in W. C. Kerfoot (ed.), *Evolution and Ecology of Zooplankton Communities*. Hanover, NH, and London: University Press of New England.

Key, K. 1981. Species, parapatry, and the morabine grasshoppers. *Syst. Zool.* **30:**425–458.

Kim, K. C., B. W. Brown, and E. F. Cook. 1966. A quantitative taxonomic study of the *Hoplopleura hesperomydis* complex (Anoplura, Hoplopleuridae), with notes on *a posteriori* taxonomic characters. *Syst. Zool.* **15:**24–25.

———, and M. A. Burgman. 1988. Accuracy of phylogenetic estimation methods under unequal evolutionary rates. *Evolution* **42:**596–602.

———, and H. W. Ludwig. 1982. Parallel evolution, cladistics, and classification of parasitic Psocodea. *Ann. Ent. Soc. Amer.* **75:**537–548.

Kimura, M. 1983. *The Neutral Theory of Molecular Evolution*. Cambridge: Cambridge University Press.

King, M.-C., and A. C. Wilson. 1975. Evolution at two levels in humans and chimpanzees. *Science* **188:**107–116.

Kinsey, A. C. 1930. The gallwasp genus *Cynips*. *Indiana Univ. Studies* **16:**1–577.

Kirby, H. 1950*a*. *Materials and Methods in the Study of Protozoa*. Berkeley: University of California Press.

———. 1950*b*. Systematic differentiation and evolution of flagellates in termites. *Rev. Soc. Mex. Hist. Nat.* **10:**57–79.

Kitching, I. J. 1985. Early stages and the classification of the milkweed butterflies (Lepidoptera: Danainae). *Zool. J. Linn. Soc.* **85:**1–97.

Kitts, D. B. 1977. Karl Popper, verifiability, and systematic zoology. *Syst. Zool.* **26:**185–194.

Kluge, A. G. 1971. Concepts and principles of morphologic and functional stud-

ies, pp. 3–41 in A. J. Waterman (ed.), *Chordate Structure and Function*. New York: Macmillan.
———. 1984. The relevance of parsimony to phylogenetic inference, pp. 24–38 in T. Duncan and T. F. Stuessy (eds.), *Cladistics: Perspectives on the Reconstruction of Evolutionary History*. New York: Columbia University Press.
———. 1985. Ontogeny and phylogenetic systematics. *Cladistics* **1**:13–28.
———, and J. S. Farris. 1969. Quantitative phyletics and the evolution of anurans. *Syst. Zool.* **18**:1–32.
———, and R. E. Strauss. 1985. Ontogeny and systematics. *Annu. Rev. Ecol. Syst.* **16**:247–268.
Knudsen, J. W. 1966. *Biological Techniques: Collecting, Preserving, and Illustrating Plants and Animals*. New York: Harper & Row.
Knutson, L. (ed.). 1978. *Biosystematics in Agriculture. Beltsville Symposia in Agricultural Research*. Montclair, NJ: Allanheld, Osmun & Co. (John Wiley & Sons, distributor).
———, and W. L. Murphy. 1988. *Systematics: Relevance, Resources, Services and Management: A Bibliography*. Washington, DC: Assoc. Syst. Collections.
Kohn, A. J. 1959. The ecology of *Conus* in Hawaii. *Ecol. Monographs* **29**:47–90.
———, and G. H. Orians. 1962. Ecological data in the classification of closely related species. *Syst. Zool.* **11**:119–127.
Kraus, F. 1988. An empirical evaluation of the use of the ontogeny polarization criterion in phylogenetic inference. *Syst. Zool.* **37**:106–141.
Kristensen, N. P. 1981. Phylogeny of insect orders. *Annu. Rev. Entomol.* **26**:135–157.
Kummel, B., and D. Raup. 1965. *Handbook of Paleontological Techniques*. San Francisco: W. H. Freeman.
Lack, D. 1947. *Darwin's Finches*. London: Cambridge University Press.
Lake, J. A. 1987. A rate-independent technique for analysis of nucleic acid sequences: Evolutionary parsimony. *Mol. Biol. Evol.* **4**:167–191.
Lee, W. L., B. M. Bell, and J. F. Sutton. 1982. *Guidelines to Acquisition and Management of Biological Specimens*. Washington, DC: Ass. Syst. Coll.
Leone, C. A. (ed.). 1964. *Taxonomic Biochemistry and Serology*. New York: Ronald Press.
LeQuesne, W. J. 1969. A method of selection of characters in numerical taxonomy. *Syst. Zool.* **18**:201–205.
———. 1972. Further studies based on the uniquely derived character concept. *Syst. Zool.* **21**:281–288.
———. 1974. The uniquely evolved character concept and its cladistic application. *Syst. Zool.* **23**:512–517.
Levi, C. 1956. Étude des Halisarca de Roscoff. *Arch. Zool. Exptl. Gén.* **93**:1–181.
Levi, H. W. 1966. The care of alcoholic collections of small invertebrates. *Syst. Zool.* **15**:183–188.
Lewontin, R. C. 1966. On the measurement of relative variability. *Syst. Zool.* **15**:141–142.
Li, W. H. 1981. Simple method for constructing phylogenetic trees from distance matrices. *Proc. Natl. Acad. Sci. USA* **78**:1085–1089.

Lidicker, W. Z., Jr. 1962. The nature of subspecific boundaries in a desert rodent and its implications for subspecific taxonomy. *Syst. Zool.* **11:**160–171.

Linnaeus, C. 1758. *Systema naturae per regna tria naturae, secundum classes, ordines, genera, species cum characteribus, differentiis, synonymis, locis.* Editio decima, reformata, Tom. I. Laurentii Salvii, Holmiae.

Linsley, E. G. 1937. The effect of stylopization on *Andrena porterae. Pan-Pacific Entomol.* **13:**157.

Lipscomb, D. L., and J. O. Corliss. 1982. Stephanopogon, a phylogenetically important "ciliate," shown by ultrastructural studies to be a flagellate. *Science* **215:**303–304.

Lloyd, J. E. 1983. Bioluminescence and communication in insects. *Annu. Rev. Entomol.* **28:**131–160.

———. 1985. Firefly communication and deception: "Oh, what a tangled web," pp. 113–128 in R. W. Mitchell and N. S. Thompson (eds.), *Deception: Perspectives on Human and Non-human Deceit.* Albany, NY: SUNY Press.

Luckow, M., and R. A. Pimentel. 1985. An empirical comparison of numerical Wagner computer programs. *Cladistics* **1**(1):47–66.

Lyal, C. H. C. 1985. Phylogeny and classification of the Psocodea, with particular reference to the lice (Psocodea: Phthiraptera). *Syst. Ent.* **10:**145–165.

Lynes, H. 1930. Review of the genus Cisticola. *Ibis.* (suppl.).

MacFayden, A. 1955. A comparison of methods for extracting soil arthropods. *Soil Zool.* 1955:315–322.

Maddison, W. P., M. J. Donoghue, and D. R. Maddison. 1984. Outgroup analysis and parsimony. *Syst. Zool.* **33:**83–103.

Maerz, A., and M. R. Paul. 1950. *A Dictionary of Color.* 2nd ed. New York: McGraw-Hill.

Manwell, R. D., H. W. Stunkard, M. B. Chitwood, and G. W. Wharton. 1957. Intraspecific variation in parasitic animals. *Syst. Zool.* **6:**2–28.

Martin, J. E. H. (compiler). 1977. *The Insects and Arachnids of Canada.* Part 1: *Collecting, Preparing, and Preserving Insects, Mites, and Spiders.* Ottawa: Biosystematics Research Institute, Publ. 1643.

Maslin, T. P. 1952. Morphological criteria of phylogenetic relationship. *Syst. Zool.* **1:**49–70.

Matsuda, R. 1976. *Morphology and Evolution of the Insect Abdomen: With Special Reference to Developmental Patterns and their Bearings upon Systematics.* Oxford: Pergamon Press.

Matthew, W. D. 1915. Climate and evolution. *Ann. N.Y. Acad. Sci.* **24:**171–318.

Mayr, Ernst. 1931. Birds collected during the Whitney South Sea expedition: 12. Notes on *Halycon chloris* and some of its subspecies. *Amer. Mus. Novitates* **469:**1–10.

———. 1942. *Systematics and the Origin of Species.* New York: Columbia University Press.

———. 1943. p. 102 in J. A. Oliver, The status of *Uta ornata lateralis* Boulenger. *Copeia* **2:**97–107.

——— (ed.). 1957. The species problem. *Amer. Assoc. Adv. Sci. Publ.,* No. 50. Washington, D.C. 395 pp.

———. 1958. Behavior and systematics, pp. 349–380 in A. Roe and G. G. Simpson (eds.), *Behavior and Evolution*. New Haven: Yale University Press.
———. 1963. *Animal Species and Evolution*. Cambridge, MA: Harvard University Press.
———. 1964. The new systematics, pp. 13–32 in C. A. Leone (ed.), *Taxonomic Biochemistry and Serology*. New York: Ronald Press.
———. 1965*a*. Numerical phenetics and taxonomic theory. *Syst. Zool.* **14**:73–97.
———. 1965*b*. Classification, identification and sequence of genera and species. *L'Oiseau* 35 Special No., pp. 90–95.
———. 1968*a*. The role of systematics in biology. *Science* **159**:595–599.
———. 1968*b*. Theory of biological classification. *Nature* **220**:545–548.
———. 1969. *Principles of Systematic Zoology*. New York: McGraw-Hill.
———. 1970. Diverse approaches to systematics. *Evol. Biol.* **4**:1–38.
———. 1971. Methods and strategies in taxonomic research, *Syst. Zool.* **20**:426–433.
———. 1974*a*. Cladistic analysis or cladistic classification? *Z. Zool. Syst. Evolforsch.* **12**:94–128.
———. 1974*b*. The challenge of diversity. *Taxon* **23**:3–9.
———. 1976. *Evolution and the Diversity of Life*. Cambridge, MA: Harvard University Press.
———. 1981. Biological classification: Toward a synthesis of opposing methodologies. *Science* **214**:510–516.
———. 1982*a*. *The Growth of Biological Thought*. Cambridge, MA: Harvard University Press.
———. 1982*b*. Of what use are subspecies? *Auk* **99**(3):593–595.
———. 1982*c*. Processes of speciation in animals, pp. 1–19 in C. Barigozzi (ed.), *Mechanisms of Speciation*. New York: Alan R. Liss.
———. 1985. Darwin and the definition of phylogeny. *Syst. Zool.* **34**:97–98.
———. 1986. Uncertainty in science: Is the giant panda a bear or a raccoon? *Nature* **323**:769–771.
———. 1987*a*. The species as category, taxon, and population, pp. 303–320 in J. Roger and J. L. Fischer (eds.), *Histoire du Concept d'Espèce dans les Sciences de la Vie*. Paris: Fondation Singer-Polignac.
———. 1987*b*. The ontological status of species: Scientific progress and philosophical terminology. *Biol. Philosophy* **2**:145–166, 212–225. Reprinted in Mayr 1988*a*, pp. 335–358.
———. 1987*c*. The status of subjective junior homonyms. *Syst. Zool.* **36**:85–86.
———. 1988*a*. *Toward a New Philosophy of Biology*. Cambridge, MA: Harvard University Press.
———. 1988*b*. The why and how of species. *Biology and Philosophy* **3**:431–441.
———. 1989. Attaching names to objects, pp. 235–243 in M. Ruse (ed.), *What the Philosophy of Biology Is: Essays for David Hull*. Dordrecht: Kluwer Academic.
———, and R. Goodwin. 1956. *Biological Materials:* Part I: *Preserved Materials and Museum Collections*. Washington, DC: Nat. Acad. Sci., Natl. Res. Counc., Publ. No. 399.

———, E. G. Linsley, and R. L. Usinger. 1953. *Methods and Principles of Systematic Zoology*. New York: McGraw-Hill.

———, and R. J. O'Hara. 1986. The biogeographic evidence supporting the Pleistocene forest refuge hypothesis. *Evolution*. **40:**55–67.

———, and W. Provine (eds.). 1980. *The Evolutionary Synthesis*. Cambridge, MA: Harvard University Press.

———, and L. L. Short, Jr. 1970. *Species Taxa of North American Birds: A Contribution to Comparative Systematics*. Cambridge, MA: Nuttall Orn. Club, Publ. No. 9.

McKenna, M. C. 1975. Toward a phylogenetic classification of the Mammalia, pp. 21–46 in W. P. Luckett and F. S. Szalay (eds.), *Phylogeny of the Primates*. New York: Plenum Press.

McKinney, F. 1965. The comfort movements of Anatidae. *Behaviour* **25:**120–220.

McMorris, F. R. 1977. On the compatibility of binary qualitative taxonomic characters. *Bull. Math. Biol.* **39:**133–138.

McNeill, J. 1979. Purposeful phenetics. *Syst. Zool.* **28:**465–482.

———. 1982. Phylogenetic reconstruction and phenetic taxonomy. *Zool. J. Linn. Soc.* **74:**337–344.

Meacham, C. A. 1981. A probability measure for character compatibility. *Math. Biosci.* **57:**1–18.

———, and T. Duncan. 1987. The necessity of convex groups in biological classification. *Syst. Bot.* **12:**78–90.

———, and G. F. Estabrook. 1985. Character compatibility analysis. *Annu. Rev. Syst. Ecol.* **16:**431–446.

Meise, W. 1936. Zur Systematik und Verbreitungsgeschichte der Haus- und Weidensperlinge, *Passer domesticus* (L.) und *hispaniolensis* (T.) *J. Ornithol.* **84:**634–672.

Michener, C. D. 1949. Parallelism in the evolution of saturnid moths. *Evolution* **3:**129–141.

———. 1957. Some bases for higher categories in classification. *Syst. Zool.* **6:**160–173.

———. 1963. Some future developments in taxonomy. *Syst. Zool.* **12:**151–172.

———. 1964. The possible use of uninominal nomenclature to increase the stability of names in biology. *Syst. Zool.* **13:**182–190.

———. 1970. Diverse approaches to systematics. *Evol. Biol.* **4:**1–38. New York: Appleton-Century-Crofts.

———. 1977. Discordant evolution and the classification of allodapine bees. *Syst. Zool.* **26:**32–56.

———. 1978. Dr. Nelson on taxonomic methods. *Syst. Zool.* **27:**112–128.

———, and R. R. Sokal. 1957. A quantitative approach to a problem in classification. *Evolution* **11:**130–162.

Miller, E. H. (ed.). 1985. *Museum Collections: Their Role and Future in Biological Research*. Vancouver, BC: Brit. Columbia Prov. Museum Occas. Papers Series, No. 25.

Minkoff, E. C. 1965. The effects on classification of slight alterations in numerical technique. *Syst. Zool.* **14:**196–213.

Moritz, C., T. E. Dowling, and W. M. Brown. 1987. Evolution of animal mitochondrial DNA: Relevance for population biology and systematics. *Annu. Rev. Ecol. Syst.* **18:**269–292 (see also pp. 489–522).
Morse, L. E. 1974. Computer programs for specimen identification, key construction and description printing using taxonomic data matrices. *Publ. Museum Biol. Ser.* **5**(1):1–128. Michigan State University.
Moynihan, M. 1959. A revision of the family Laridae (Aves). *Amer. Mus. Novit.* No. 1928, pp. 1–42.
Munroe, E. 1960. An assessment of the contribution of experimental taxonomy to the classification of insects. *Rev. Canad. Biol.* **19:**293–319.
Murphy, D. D., and P. R. Ehrlich. 1983. Butterfly nomenclature, stability, and the rule of obligatory categories. *Syst. Zool.* **32**(4):451–453.
———. 1984. On butterfly taxonomy. *J. Res. Lepidoptera* **23:**19–34.
Murphy, W. L. 1984. *Functions of Taxonomic Research, Resources, and Services—A Bibliography.* Document 1006M: Insect Identification and Beneficial Insect Introduction Institute. Beltsville, MD: U.S. Dept. of Agriculture.
Nagel, E. 1961. *The Structure of Science.* New York: Harcourt, Brace, & World.
Nei, M., and R. K. Koehn (eds.). 1983. *Evolution of Genes and Proteins.* Sunderland, MA: Sinauer.
Nelson, G. 1972. Comments on Hennig's "Phylogenetic Systematics." *Syst. Zool.* **21:**364–374.
———. 1974*a*. Cladistic analysis and synthesis: principles and definitions. *Syst. Zool.* **22:**1–21.
———. 1974*b*. Classification as an expression of phylogenetic relationships. *Syst. Zool.* **22:**344–359.
———. 1978. Ontogeny, phylogeny, paleontology, and the biogenetic law. *Syst. Zool.* **27:**324–345.
———, and N. Platnick. 1981. *Systematics and Biogeography: Cladistics and Vicariance.* New York: Columbia University Press.
Nevo, E. 1982. Speciation in subterranean mammals, pp. 191–218 in C. Barigozzi (ed.), *Mechanisms of Speciation.* New York: Alan R. Liss.
———. 1983. Adaptive significance of protein variation, pp. 239–282 in G. S. Oxford and D. Rollinson (eds.), *Protein Polymorphism: Adaptive and Taxonomic Significance.* New York: Academic Press.
Newell, N. D. 1947. Infraspecific categories in invertebrate paleontology, *Evolution* **1:**163–171.
———. 1956. Fossil populations, pp. 63–82 in Sylvester-Bradley, P. C. (ed.), *The Species Concept in Paleontology.* London: *Syst. Assoc.,* Publ. No. 2.
Nitecki, M. H. (ed.). 1982. *Coevolution.* Chicago: University of Chicago Press.
O'Hara, R. J. 1989. Diagrammatic classification of birds, 1819–1901: Views of the natural system in nineteenth century British ornithology. *Acta 19th International Ornithological Congress* **II:**2746–2759.
Oldroyd, H. 1958. *Collecting, Preserving, and Studying Insects.* New York: Macmillan.
Oliver, J. A. 1943. The status of *Uta ornata lateralis* Boulenger. *Copeia* **2:**97–107.

Orton, G. L. 1957. The bearing of larval evolution on some problems in frog classification. *Syst. Zool.* **6**:79–86.

Osborn, H. F. 1936. *Proboscidea.* New York: American Museum of Natural History.

Osche, G. 1982. Rekapitulationsentwicklung und ihre Bedeutung für die Phylogenetik. *Verh. Naturwiss. Ver. Hamburg (NF)* **25**:5–31.

——— (ed.). 1984. Symposium über Artbegriff und Artbildung. *Zeitschr. Zool. Syst. Evolutionsforsch.* **22**(3):161–288.

Otte, D., and J. A. Endler. 1989. *Speciation and Its Consequences.* Sunderland, MA: Sinauer.

Owen, R. 1866. On the Anatomy of Vertebrates. 3 vols. London: Longman.

Panchen, A. L. 1982. The use of parsimony in testing phylogenetic hypotheses. *Zool. J. Linn. Soc.* **74**:305–328.

Pankhurst, R. J. 1978. *Biological Identification: The Principles and Practices of Identification Methods in Biology.* London: Edward Arnold.

Papp, C. S. 1976. *Scientific Illustration: Theory and Practice.* Dubuque, IA: William C. Brown.

Parker, S. P. (ed.). 1982. *Synopsis and Classification of Living Organisms.* Vols. 1 and 2. New York: McGraw-Hill.

Paterson, H. E. 1982. Perspective on speciation by reinforcement. *S. Afri. J. Sci.* **78**:53–57.

———. 1985. The recognition concept of species, pp. 21–29 in E. S. Vrba (ed.), *Species and Speciation.* Pretoria: Transvaal Museum Monograph No. 4.

Patterson, C. 1980. Cladistics. *Biologist* **27**:234–240.

———. 1981. Significance of fossils in determining evolutionary relationships. *Annu. Rev. Ecol. Syst.* **12**:195–223.

——— (ed.). 1982. Symposium on methods of phylogenetic reconstruction. *Zool. J. Linn. Soc.* **74**:277–292.

———. 1988. Homology in classical and molecular biology. *Mol. Biol. Evol.* **5**:603–625.

Patton, J. L., and S. Y. Yang. 1977. Genetic variation in Thomomys bottae pocket gophers: Macrogeographic patterns. *Evolution* **31**:697–720.

Payne, R. B. 1986. Bird songs and avian systematics. *Curr. Ornithol.* **3**:87–126.

Pearse, A. G. E. 1980. *Histochemistry: Theoretical and Applied.* 4th ed. Edinburgh and New York: Churchill Livingstone.

Pellegrin, P. 1986. *Aristotle's Classification of Animals.* Berkeley: University of California Press.

Pemberton, C. E. 1941. Contributions of the entomologist to Hawaii's welfare. *Hawaiian Planter's Record* **45**:107–119.

Penny, D. 1982. Graph theory, evolutionary trees, and classification. *Zool. J. Linn. Soc.* **74**:277–292.

Perrier, E. 1893–1932. *Traité de zoologie.* Vols. 1–10. Paris: Masson.

Pielou, E. C. 1979. *Biogeography.* New York: John Wiley & Sons.

Pimentel, R. A., and R. Riggins. 1987. The nature of cladistic data. *Cladistics* **3**:201–209.

Platnick, N. I. 1976. Are monotypic genera possible? *Syst. Zool.* **25**:198–199.

———. 1977*a*. Parallelism in phylogeny reconstruction. *Syst. Zool.* **26:**93–96.

———. 1977*b*. Paraphyletic and polyphyletic groups. *Syst. Zool.* **26:**195–200.

———. 1977*c*. Cladograms, phylogenetic trees, and hypothesis testing. *Syst. Zool.* **26:**438–442.

———. 1979. Philosophy and the transformation of cladistics. *Syst. Zool.* **28:**537–546.

———. 1985. Philosophy and the transformation of cladistics revisited. *Cladistics* **1:**87–94.

———. 1987. An empirical comparison of microcomputer parsimony programs. *Cladistics* **3:**121–144.

———, and V. A. Funk (eds.). 1983. *Advances in Cladistics*. Vol. 2. New York: Columbia University Press.

Postke, M. T., K. S. Howard, A. H. Johnson, and K. L. McMichael. 1980. *Scanning Electron Microscopy—A Student's Handbook*. Burlington, VT: Ladd Research Industries.

Presch, W. 1979. Phenetic analysis of a single data set. *Syst. Zool.* **28:**366–371.

Prosser, C. L., and F. A. Brown. 1973. *Comparative Animal Physiology*. 2nd ed. Philadelphia: W B Saunders.

Prothero, D. R., and D. B. Lazarus. 1980. Planktonic microfossils and the recognition of ancestors. *Syst. Zool.* **29:**119–129.

Prum, R. O. 1988. Phylogenetic interrelationships of the barbets (Aves: Capitonidae) and toucans (Aves: Ramphastidae) based on morphology with comparisons to DNA-DNA hybridization. *Zool. J. Linn. Soc.* **92:**313–343.

Queiroz, K. de. 1985. The ontogenetic method for determining character polarity and its relevance to phylogenetic systematics. *Syst. Zool.* **34:**280–299.

Raff, R. A., J. A. Anstrom, C. J. Huffman, D. S. Leaf, J.-H. Loo, R. M. Showman, and D. E. Wells. 1984. Origin of a gene regulatory mechanism in the evolution of echinoderms. *Nature* **310:**312–314.

Raup, D. M. 1962. Crystallographic data in echinoderm classification. *Syst. Zool.* **11:**97–108.

———, and R. E. Crick. 1981. Evolution of single characters in the Jurassic ammonite *Kosmosceras*. *Paleobiology* **7:**200–215 (see also *Paleobiology* 1982, **8:**90–100).

Raven, P. H., and D. I. Axelrod. 1974. Angiosperm biogeography and past continental movements. *Ann. Missouri Bot. Garden* **61:**539–673.

Reeck, G. R., et al. 1987. Homology in proteins and nucleic acids: A terminology muddle and a way out of it. *Cell* **50:**667.

Reif, W.-E. 1984. Artabgrenzung und das Konzept der evolutionären Art in der Paläontologie. *Z. Zool. Syst. Evolforsch.* **22:**263–286.

Remane, A. 1952. *Die Grundlagen des natürlichen Systems, der vergleichenden Anatomie und der Phylogenetik*. Leipzig: Akad. Verlagsges.

Rensch, B. 1929. *Das Prinzip geographischer Rassenkreise und das Problem der Artbildung*. Berlin: Borntraeger.

———. 1934. *Kurze Anweisung für zoologisch-systematische Studien*. Leipzig: Akad. Verlagsges.

———. 1960. *Evolution above the Species Level.* New York: Columbia University Press.

Ride, W. D. L., and T. Younes (eds.). 1986. *Biological Nomenclature Today.* IUBS Monograph Series No. 2. Oxford: IRL Press.

Ridgway, R. 1912. *Color Standards and Color Nomenclature.* Washington, DC: A. Hoen.

Riech, E. 1937. Systematische, anatomische, ökologische und tiergeographische Untersuchungen über die Süsswassermollusken Papuasiens und Melanesiens. *Archiv Naturgesch.* (*N.S.*) **6**:37–153.

Roger, J., and J. L. Fischer (eds.). 1987. *Histoire du Concept d'Espèce dans les Sciences de la Vie.* Paris: Fondation Singer-Polignac.

Rogers, J. S. 1986. Deriving phylogenetic trees from allele frequencies: A comparison of nine genetic distances. *Syst. Zool.* **35**:297–310.

Rohlf, F. J. 1963. Classification of *Aedes* by numerical taxonomic methods (Diptera: Culicidae). *Ann. Entomol. Soc. Amer.* **56**:798–804.

———. 1964. Congruence of larval and adult classifications in *Aedes* (Diptera: Culicidae). *Syst. Zool.* **12**:97–117.

———, D. H. Colless, and G. Hart. 1983. Taxonomic congruence reexamined. *Syst. Zool.* **32**:144–158.

———, and R. H. Sokal. 1981. Comparing numerical taxonomic studies. *Syst. Zool.* **30**:459–490.

Rosen, D. E., P. L. Forey, B. G. Gardiner, and C. Patterson. 1981. Lungfishes, tetrapods, paleontology, and plesiomorphy. *Bull. Amer. Mus. Nat. Hist.* **167**:159–276.

Ruse, M. 1979. Falsifiability, consilience, and systematics. *Syst. Zool.* **28**:530–536.

Russell, H. 1963. *Notes on Methods for the Narcotization, Killing, Fixation and Preservation of Marine Organisms.* Woods Hole, MA: Systematics-Ecology Program, Marine Biological Library.

Saitou, N., and M. Nei, 1987. The neighboring-joining method: A new method for reconstructing phylogenetic trees. *Mol. Biol. Evol.* **4**:406–425.

Salt, G. 1927. The effects of stylopization on aculeate Hymenoptera. *J. Exp. Zool.* **48**:223–231.

Salvini-Plawen, L. V., and E. Mayr. 1977. On the evolution of photo-receptors and eyes. *Evol. Biol.* **10**:207–263.

Sankoff, D., and D. Rousseau. 1975. Locating the vertices of a Steiner tree in arbitrary space. *Math. Progr.* **9**:240–246.

Sarasan, L., and A. M. Neuner. 1983. *Museum Collections and Computers* Washington, DC: Ass. Syst. Coll.

Sattath, S., and A. Tversky. 1977. Additive similarity trees. *Psychometrika* **42**:319–345.

Schindewolf, O. H. 1969. Über den 'Typus' in der morphologischen und phylogenetischen Biologie. *Abhandl. Akad. Wiss. Lit. Mainz, Math-Naturw. Kl.* **4**:58–131.

Schmidt-Nielsen, K. 1979. *Animal Physiology: Adaptation and Environment.* 2nd ed. Cambridge: Cambridge University Press.

Schnitter, H. 1922. Die Najaden der Schweiz. *Rev. Hydrol.* (suppl.) **2**:1–200.

Schoch, R. M. 1986. *Phylogeny Reconstruction in Paleontology*. New York: Van Nostrand Reinhold.

Scudder, G. G. E. 1971. Comparative morphology of insect genitalia. *Annu. Rev. Entomol.* **16:**379–406.

Sebeok, T. A. 1977. *How Animals Communicate*. Bloomington, IN: Indiana University Press.

Selander, R. K., R. F. Johnston, and T. H. Hamilton. 1965. Colorimetric methods in ornithology. *Condor* **66:**491–495.

———, and R. F. Johnston. 1967. Evolution in the house sparrow: I. Intrapopulation variation in N. America. *Condor* **69:**217–258.

———, and T. S. Whittam. 1983. Protein polymorphism and the genetic structure of populations, pp. 89–114 in M. Nei and R. K. Koehn (eds.), *Evolution of Genes and Proteins*. Sunderland, MA: Sinauer.

Sheppard, P. M., J. R. G. Turner, K. S. Brown, W. W. Benson, and M. C. Singer. 1985. Genetics and the evolution of Muellerian mimicry in *Heliconius* butterflies. *Phil. Trans. R. Soc.* [B]**308:**433–613.

Sherborn, C. D. 1902–1933. *Index Animalium*. London: British Museum (Natural History).

Short, L. L., Jr. 1965. Hybridization in the flickers (*Colaptes*) of North America. *Bull. Amer. Mus. Nat. Hist.* **129:**307–428.

———. 1969. Taxonomic aspects of avian hybridization. *Auk* **86:**84–105.

Sibley, C. G. 1954. Hybridization in the red-eyed towhees of Mexico. *Evolution* **8:**252–290.

———, and J. E. Ahlquist. 1981. Instructions for specimen preservation for DNA extraction. *Assoc. Syst. Coll. Newsl.* **9:**44–55.

———, and ———. 1983. The phylogeny and classification of birds based on the data of DNA-DNA hybridization. *Current Orn.* **1:**245–292.

———, and ———. 1985. The phylogeny and classification of the passerine birds, based on comparisons of the genetic material, DNA, pp. 83–121 in V. D. Ilyichev (ed.), *Proceedings of the 18th International Ornithological Congress*. Moscow: Nauka Publications.

———, ———, and B. L. Monroe. 1988. A classification of the living birds of the world based on DNA-DNA hybridization studies. *Auk* **105:**409–423.

Simpson, G. G. 1945. The principles of classification and a classification of mammals. *Bull. Amer. Mus. Nat. Hist.* **85:**1–350.

———. 1959*a*. Anatomy and morphology: Classification and evolution: 1859 and 1959. *Proc. Amer. Phil. Soc.* **103**(2):286–306.

———. 1959*b*. The nature and origin of supraspecific taxa. *Cold Spring Harbor Symp. Quant. Biol.* **24:**255–271.

———. 1961. *Principles of Animal Taxonomy*. New York: Columbia University Press.

———. 1962. Primate taxonomy and recent studies of nonhuman primates. *Ann. N.Y. Acad. Sci.* **102:**497–514.

———. 1963. The meaning of taxonomic statements, pp. 1–31 in S. L. Washburn (ed.), *Classification and Human Evolution*. Viking Fund Publ. in *Anthropology,* No. 37.

———. 1964. Numerical taxonomy and biological classification. *Science* **144:**712–713.
———. 1965. *The Geography of Evolution*. Philadelphia: Chilton.
———. 1971. Mesozoic mammals revisited, pp. 181–198 in D. M. Kermack and K. A. Kermack (eds.), *Early Mammals,* Suppl. 1, *Zool. J. Linn. Soc.* **50.**
———, A. Roe, and R. C. Lewontin. 1960. *Quantitative Zoology*. rev. ed. New York: Harcourt, Brace & World.
Sims, R. W. 1980. *Animal Identification: A Reference Guide*. Vol. 1: *Marine and Brackish Water Animals*. Vol. 2: *Land and Freshwater Animals*. London: British Museum (Nat. Hist.).
Skarbilovich, T. S. 1985. *Bibliography of Russian Literature on Nematodes: Published between 1874 and 1975*. New Delhi: Oxonian Press.
Sloan, P. R. 1987. From logical universals to historical individuals: Buffon's idea of biological species, pp. 101–140 in J. Roger and J. L. Fischer (eds.), *Histoire du Concept d'Espèce les Sciences de la Vie*. Paris: Fondation Singer-Polignac.
Sluys, R. 1989. Rampant parallelism: An appraisal of the use of nonuniversal derived character states in phylogenetic reconstruction. *Syst. Zool.* **38:**350–370.
Smith, H. M. 1965. More evolutionary terms. *Syst. Zool.* **14:**57–58.
Smith, R. C., and W. M. Reid. 1972. *Guide to the Literature of the Life Sciences*. 8th ed. Minneapolis, MN: Burgess.
Smith, R. F., T. E. Mittler, and C. N. Smith (eds.). 1973. History of Entomology. *Annu. Rev. Entomol.* (Suppl). 517 pp.
Sneath, P. H. A. 1957. The application of computers to taxonomy. *J. Gen. Microbiol.* **17:**201–226.
———. 1962. The construction of taxonomic groups, pp. 289–332 in G. C. Ainsworth and P. H. A. Sneath (eds.), *Microbial Classification*. Cambridge: Cambridge University Press.
———. 1989. Analysis and interpretation of sequence data for bacterial systematics: The view of a numerical taxonomist. *System Appl. Microbiol.* **12:**15–31.
———, and R. R. Sokal. 1973. *Numerical Taxonomy*. San Francisco: W. H. Freeman.
Sober, E. 1983. Parsimony in systematics: Philosophical issues. *Annu. Rev. Ecol. Syst.* **14:**335–357.
———. 1988. *Reconstructing the Past: Parsimony, Evolution, and Inference*. Cambridge, MA: MIT Press.
———. 1989. Systematics and circularity, pp. 263–273 in M. Ruse (ed.), *What the Philosophy of Biology Is*. Dordrecht, Netherlands: Kluwer.
Sokal, R. R. 1975. Mayr on cladism—and his critics. *Syst. Zool.* **24:**257–262.
———. 1983. A phylogenetic analysis of the Caminalcules I–IV. *Syst. Zool.* **32:**159–201.
———. 1985*a*. The principles of numerical taxonomy: Twenty-five years later, pp. 1–20 in *Computer Assisted Bacterial Systematics*. Society for General Microbiology.
———. 1985*b*. The continuing search for order. *Amer. Nat.* **126:**729–749.
———. 1986. Phenetic taxonomy: Theory and methods. *Annu. Rev. Ecol. Syst.* **17:**423–442.

———. 1988. Unsolved problems in numerical taxonomy, pp. 45–56 in H. H. Bock (ed.), *Classification and Related Methods of Data Processing*. Amsterdam: North Holland.

———, and C. A. Braumann. 1980. Significance tests for coefficients of variation and variability profiles. *Syst. Zool.* **29**:50–66.

———, and C. D. Michener. 1958. A statistical method for evaluating systematic relationships. *Univ. Kansas Sci. Bull.* **38**:1409–1438.

———, and F. J. Rohlf. 1962. The comparison of dendrograms by objective methods. *Taxon* **11**:33–40.

———, and ———. 1981. *Biometry*. 2nd ed. San Francisco: W. H. Freeman.

———, and P. H. A. Sneath. 1963. *Principles of Numerical Taxonomy*. San Francisco: W. H. Freeman.

Sotavalta, O. 1964. Studies on the variation of the wing venation of certain tiger moths. *Ann. Acad. Sci. Fenn. A IV (Biol.)* **74**:1–41.

Sourdis, J., and C. Krimbas. 1987. Accuracy of phylogenetic trees estimated from DNA sequence data. *Mol. Biol. Evol.* **4**:159–166.

———, and M. Nei. 1988. Relative efficiencies of the maximum parsimony and distance-matrix method. *Mol. Biol. Evol.* **5**:298–311.

Spencer, M. 1982. *Fundamentals of Light Microscopy*. New York: Cambridge University Press.

Spieth, H. T. 1952. Mating behavior within the genus Drosophila (Diptera). *Bull. Amer. Mus. Nat. Hist.* **99**:399–474.

Stanley, S. M. 1979. *Macroevolution: Pattern and Process*. San Francisco: W. H. Freeman.

Steel, M. A., M. D. Hendy, and D. Penny. 1988. Loss of information in genetic distances. *Nature* **336**:118.

Stevens, P. F. 1980. Evolutionary polarity of character states. *Annu. Rev. Ecol. Syst.* **11**:333–358.

———. 1984. Metaphors and typology in the development of botanical systematics 1690–1960. *Taxon* **33**:169–211.

———. 1986. Evolutionary classification in botany, 1960–1985. *J. Arnold Arboretum* **67**:313–339.

Steyskal, G. C., W. L. Murphy, and E. M. Hoover (eds.). 1987. *Insects and Mites: Techniques for Collection and Preservation*. Washington, DC: USDA/ARS Misc. Publ. No. 1443.

Stiassny, M. L. J. 1986. The limits and relationships of the acanthomorph teleosts. *J. Zool. London (B)* **1**:411–460.

———, and J. S. Jensen. 1987. Labroid intrarelationships revisited: Morphological complexity, key innovations, and the study of comparative diversity. *Bull. Mus. Comp. Zool.* **151**:269–319.

Storey, M., and N. J. Wilmowsky. 1955. Curatorial practices in zoological research collections: I. Preliminary report on containers and closures for storing specimens preserved in liquid. *Circ. Nat. Hist. Mus. Stanford Univ.* **3**:1–22.

Straney, D. O. 1978. Variance partitioning and non-geographic variation. *J. Mammal.* **59**:1–11.

Stunkard, H. W. 1957. Intraspecific variation in parasitic flatworms. *Syst. Zool.* **6:**7–18.

Svenson, H. K. 1945. On the descriptive method of Linnaeus. *Rhodora* **47:**273–302, 363–388.

Swofford, D. L. 1981. On the utility of the distance Wagner procedure, pp. 25–43 in V. A. Funk and D. R. Brooks (eds.), *Advances in Cladistics: Proceedings of the First Meeting of the Willi Hennig Society*. New York: New York Botanical Garden.

———. 1984. Phylogenetic analysis using parsimony. Users Manual: Version 2.3. Urbana: Illinois Natural History Survey.

———, and S. H. Berlocher. 1987. Inferring evolutionary trees from gene frequency data under the principle of maximum parsimony. *Syst. Zool.* **36:**293–325.

———, and W. P. Maddison. 1987. Reconstructing ancestral character states under Wagner parsimony. *Math. Biosci.* **87:**199–229.

Sylvester-Bradley, P. C. 1951. The subspecies in paleontology. *Geol. Mag.* **88:**88–102.

———. 1956. The species concept in paleontology. *Syst. Assoc. Publ. London* **2:**145.

———. 1958. Description of fossil populations. *J. Paleontol.* **32:**214–235.

Sytsma, K. J., and L. D. Gottlieb. 1986. Chloroplast DNA evidence for the origin of the genus *Heterogaura* from a species of *Clarkia*. *PNAS* **83:**5554–5557.

Tassy, P. (ed.). 1986. *L'Ordre et la Diversité du Vivant*. Paris: Fayard-Fondation Diderot.

Tateno, Y., M. Nei, and F. Tajima. 1982. Accuracy of estimated phylogenetic trees from molecular data: I. Distantly related species. *J. Mol. Evol.* **18:**387–404.

Tavolga, W. N., and W. N. Lanyon (eds.). 1960. *Animal Sounds and Communication*. Washington, DC: Amer. Inst. Biol. Sci., Publ. No. 7. 443 pp.

Thorpe, J. P. 1982. The molecular clock hypothesis: Biochemical evolution, genetic differentiation and systematics. *Annu. Rev. Ecol. Syst.* **13:**139–168.

Thorpe, R. S. 1980. Microevolution and taxonomy of European reptiles with particular reference to the grass snake *Natrix natrix* and the wall lizards *Podarcis sicula* and *P. melisellensis*. *Biol. J. Linn. Soc.* **14:**215–233.

Throckmorton, L. H. 1965. Similarity versus relationship in Drosophila. *Syst. Zool.* **14:**221–236.

———. 1969. Concordance and discordance of taxonomic characters in *Drosophila* classification. *Syst. Zool.* **17:**355–387.

Tinbergen, N. 1959. Comparative studies of the behavior of gulls (Laridae): A progress report. *Behaviour* **15:**1–70.

Trelease, S. F. 1951. *The Scientific Paper. How to Prepare It. How to Write It.* Baltimore: Williams & Wilkins.

Tuomikoski, R. 1967. Notes on some principles of phylogenetic analysis. *Ann. Entomol. Fenn.* **33:**137–147.

Turesson, G. 1922. The genotypic response to the plant species to the habitat. *Hereditas* **3**:211–350.

Turner, J. R. G. 1981. Adaptation and evolution in *Heliconius:* A defense of NeoDarwinism. *Annu. Rev. Ecol. Syst.* **12**:99–121.

Tuxen, S. L. 1958. Relationships of protura. *Proc. 10th Int. Congress Ent. (1956)* **1**:493–497.

Udvardy, M. D. F. 1969. *Dynamic Zoogeography.* New York: Van Nostrand Reinhold.

Ueshima, N., and P. D. Ashlock. 1980. Cytotaxonomy of the Lygaeidae. *Univ. Kansas Sci. Bull.* **51**:717–801.

Underwood, G. 1982. Parallel evolution in the context of character analysis. *Zool. J. Linn. Soc.* **74**:245–266.

Usinger, R. L. 1966. Monograph of Cimicidae. Vol. 7. Washington, DC: The Thomas Say Foundation (Ent. Soc. America).

Vachon, M. 1952. *Études sur les scorpions.* Algiers: Institut Pasteur d'Algérie.

Van Tyne, J. 1952. Principles and practices in collecting and taxonomic work. *Auk* **69**:27–33.

Van Valen, L. 1976. Ecological species, multispecies, and oaks. *Taxon* **25**:233–239.

———. 1978. Why not be a cladist? *Evol. Theory* **3**:285–299.

Vaupel Klein, J. C. V. 1984. A primer of a phylogenetic approach to the taxonomy of the genus *Euchirella. Crustaceana* (suppl.) **9**:1–194.

Vaurie, C. 1949. A revision of the bird family Dicruridac. *Bull. Amer. Mus. Nat. Hist.* **93**:205–342.

———. 1955. Pseudo-subspecies. Basel: *Acta XI Congr. Int. Orn., 1954.*

Verheyen, R. 1958. Contribution à la systématique des Alciformes. *Inst. Roy. Sci. Nat. Belgique* **34**:1–15.

Villalobos-Dominguez, C., and J. Villalobos. 1947. *Atlas de los Colores.* Buenos Aires: El Ateneo.

Voss, E. G. 1952. The history of keys and phylogenetic trees in systematic biology. *J. Sci. Labs. Denison Univ.* **43**:1–25.

Wagner, R. P. 1944. *Nutritional differences in the Mulleri-group [of Drosophila].* Austin: University of Texas Publ., No. 4920, pp. 39–41.

Wagner, W. H. 1961. Problems in the classification of ferns, pp. 841–844 in *Recent Advances in Botany,* Vol. 1 (Lectures and Symposia from IX Internt. Bot. Congr., Montreal, 1957). Toronto: University of Toronto Press.

———. 1969. The construction of a classification, pp. 67–90 in C. G. Sibley (ed.), *Systematic Biology.* Washington, DC: National Academy of Science Publication 1692.

———. 1980. Origin and philosophy of the groundplan-divergence method of cladistics. *Syst. Bot.* **5**(2):173–193.

Wagstaffe, R., and J. H. Fidler. 1955. *The Preservation of Natural History Specimens.* Vol. I: *Invertebrates.* London: H. F. and G. Witherby.

Walker, T. J. 1964. Cryptic species among sound-producing ensiferan Orthoptera (Gryllidae and Tettigoniidae). *Quart. Rev. Biol.* **39**:345–355.

Walsh, B. D. 1864. On phytophagic varieties and phytophagic species, with remarks on the unity of coloration in insects. *Proc. Ent. Soc. Philadelphia* **5:**194–215.

Warburton, F. E. 1967. The purposes of classifications. *Syst. Zool.* **16:**241–245.

Waring, H. 1963. *Color Change Mechanisms of Cold-Blooded Vertebrates.* New York: Academic Press.

Watrous, L. E., and Q. D. Wheeler. 1981. The outgroup comparison method of character analysis. *Syst. Zool.* **30:**1–11.

Wheeler, Q. D. 1986. Character weighting and cladistic analysis. *Syst. Zool.* **35:**102–109.

Whewell, W. 1840. *The Philosophy of the Inductive Sciences.* London: Parker.

Whiffin, T., and M. W. Bierner. 1972. A quick method for computing Wagner trees. *Taxon* **21:**83–90.

White, M. J. D. 1973. *Animal Cytology and Evolution.* 3rd ed. Cambridge: Cambridge University Press.

———. 1978. *Modes of Speciation.* San Francisco: W. H. Freeman.

Whitehead, A. N. 1948. *Science and the Modern World.* New York: Macmillan.

Whittington, H. B. 1985. *The Burgess Shale.* New Haven, CT: Yale University Press.

Wickler, W. 1967. Vergleichende Verhaltensforschung und Phylogenetik, pp. 420–508 in G. Heberer (ed.), *Evolution der Organismen.* 3rd ed. Stuttgart: G. Fischer.

Wiens, J. A. 1982. Forum: Avian subspecies in the 1980s. *Auk* **99:**593–615.

Wiley, E. O. 1979. An annotated Linnaean hierarchy, with comments on natural taxa and competing systems. *Syst. Zool.* **28:**308–337.

———. 1980. Is the evolutionary species fiction?—A consideration of class, individuals, and historical entities. *Syst. Zool.* **29:**76–80.

———. 1981. *Phylogenetics: The Theory and Practice of Phylogenetic Systematics.* New York: John Wiley & Sons.

Williams, S. L., R. Laubach, and C. M. Laubach. 1979. *A Guide to the Literature Concerning the Management of Recent Mammal Collections.* Lubbock: Texas Tech.

Willmann, R. 1983. Widersprüchliche Rekonstruktion der Phylogenese am Beispiel der Ordnung Mecoptera. *Paläont. Z.* **57:**285–308.

———. 1985. *Die Art in Raum und Zeit.* Berlin and Hamburg: Parey.

Wilson, E. O. 1965. A consistency test for phylogenies based on contemporaneous species. *Syst. Zool.* **14:**214–222.

———. 1985. Time to revive systematics. *Science* **230:**1227 [see also *Science* **231**(1986):1057].

——— (ed.). 1988. *Biodiversity.* Washington, DC: National Academy Press.

———, and W. L. Brown. 1953. The subspecies concept and its taxonomic application. *Syst. Zool.* **2:**97–111.

Woese, C. R. 1987. Bacterial evolution. *Microbiol. Rev.* **51:**221–271.

Wright, C. A. (ed.). 1974. *Biochemical and Immunological Taxonomy of Animals.* London: Academic Press.

Zimmerman, E. 1948. *Insects of Hawaii.* Vol. I. Honolulu: University of Hawaii Press.

Zuckerkandl, E., and L. Pauling. 1962. Molecular diseases, evolution, and genic heterogeneity, in M. Kasha and B. Pullman (eds.), *Horizons in Biochemistry*. Chicago: Academic Press.

Zweifel, F. W. 1988. *A Handbook of Biological Illustration*. 2nd ed. Chicago: Phoenix Books, University of Chicago Press.

Zycherman, L. A. (ed.). 1987. *Guide to Museum Pest Control*. Washington, DC: Assoc. Syst. Collections.

———, and J. R. Schrock. 1988. *A Guide to Museum Pest Control*. Washington, DC: Assoc. Syst. Collections.

INDEX